Aerosol and Atmospheric Correction

Aerosol and Atmospheric Correction

Editors

Xingfa Gu
Jing Wei
Shuaiyi Shi

MDPI

Basel • Beijing • Wuhan • Barcelona • Belgrade • Novi Sad • Cluj • Manchester

Editors

Xingfa Gu	Jing Wei	Shuaiyi Shi
Aerospace Information Research Institute	Earth System Science Interdisciplinary Center	Aerospace Information Research Institute
Chinese Academy of Sciences	University of Maryland	Chinese Academy of Sciences
Beijing	Baltimore	Beijing
China	United States	China

Editorial Office
MDPI AG
Grosspeteranlage 5
4052 Basel, Switzerland

This is a reprint of articles from the Special Issue published online in the open access journal *Remote Sensing* (ISSN 2072-4292) (available at: https://www.mdpi.com/journal/remotesensing/special_issues/Aerosol_Atmospheric_Correction).

For citation purposes, cite each article independently as indicated on the article page online and as indicated below:

Lastname, A.A.; Lastname, B.B. Article Title. *Journal Name* **Year**, *Volume Number*, Page Range.

ISBN 978-3-7258-2487-8 (Hbk)
ISBN 978-3-7258-2488-5 (PDF)
doi.org/10.3390/books978-3-7258-2488-5

Cover image courtesy of Shuaiyi Shi

Contents

remote sensing

Editorial

An Editorial for the Special Issue "Aerosol and Atmospheric Correction"

Shuaiyi Shi [1,*], Xingfa Gu [2] and Jing Wei [3]

1 State Key Laboratory of Remote Sensing Science, Aerospace Information Research Institute, Chinese Academy of Sciences, Beijing 100094, China
2 School of Geography and Remote Sensing, Guangzhou University, Guangzhou 510006, China; guxf@gzhu.edu.cn
3 Earth System Science Interdisciplinary Center, University of Maryland, College Park, MD 20740, USA; weijing@umd.edu
* Correspondence: shisy01@radi.ac.cn

Aerosol is an important atmospheric component that severely influences the global climate and air quality of our planet [1–4]. In quantitative remote sensing, aerosol is also a key factor in atmospheric correction of remote sensing data to obtain accurate surface information [5–7]. The radiation signal received by the sensor is surface–atmosphere coupled, including the signal of path radiance, surface reflection, and surface–atmosphere interaction, a phenomenon which impedes quantitative information acquisition from both surface and atmospheric aspects. Accurate aerosol estimation and atmospheric correction are needed to solve this problem.

In this Special Issue, the studies cover several important topics, mainly involving aerosol retrieval, aerosol emission and regional transfer, and atmospheric correction. The goal of this Special Issue is to discuss the accurate retrieval and estimation of aerosols to help with precise atmospheric correction and facilitate various corresponding scientific studies focusing on the development of new technologies, instruments, and methods.

Aerosol product quality limits their data applications. Some advancements are made in this Special Issue that improve aerosol detection and retrieval accuracy. Focusing on the characteristics of Coherent Doppler Wind Lidar (CDWL), a novel method for the calibration and quantitative assessment of aerosol properties is proposed [8]. The result is verified through comparison with synchronous Rayleigh–Mie–Raman Lidar (RMRL) data, resulting in good agreement, proving the ability of CDWL to retrieve aerosol properties accurately. Meanwhile, exploring aerosol retrieval of single-angle and multi-band polarization instruments containing short-wave infrared bands, surface and atmosphere decoupling without prior information about the surface is conducted based on optimal estimation theory [9]. The method can avoid the inversion error caused by the untimely updating of the surface reflectance database and the error in spatiotemporal matching. After being applied to the Particulate Observing Scanning Polarimeter (POSP) and validated by AErosol RObotic NETwork (AERONET) measurements, the effectiveness of the proposed algorithm under different geographical regions and pollution conditions is verified. Another independent article thoroughly examines MODIS aerosol retrieval accuracies under different land cover types, aerosol types, and observation geometries based on AERONET measurements involving three different algorithms, namely Dark Target (DT), Deep Blue (DB), and Multi-Angle Implementation of Atmospheric Correction (MAIAC), each with unique characteristics [10]. This Special Issue also contains studies aimed toward the identification of specific aerosol types. A novel MERSI haze mask (MHAM) algorithm to directly categorize haze pixels in addition to cloudy and clear ones has been designed based on the Medium Resolution Imaging Spectrometer II (MERSI-II) on board the FY-3D satellite [11]. The algorithm can illustrate the boundary of the haze region with high reliability, remaining consistent with the true color image. Determining the threshold value for background aerosol optical depth

Citation: Shi, S.; Gu, X.; Wei, J. An Editorial for the Special Issue "Aerosol and Atmospheric Correction". *Remote Sens.* **2024**, *16*, 3745. https://doi.org/10.3390/rs16193745

Received: 30 September 2024
Accepted: 1 October 2024
Published: 9 October 2024

(BAOD) is crucial for identifying aerosol types. A statistical method to select the best BAOD threshold value using VIIRS DB AOD products is proposed in this Special Issue [12]. The VIIRS aerosol type classification scheme was further updated using the BAOD threshold. The results indicate that the updated scheme can reliably detect changes in aerosol types under low aerosol loading conditions.

Using aerosol products, further scientific studies of atmospheric aerosol are conducted and included in this Special Issue. The seasonal characteristics and long-term variations in aerosol optical parameters in Hong Kong are analyzed using AERONET data and satellite-based observations based on the extreme-point symmetric mode decomposition (ESMD) model [13]. The interactions between aerosol loading and meteorological factors are also discussed. Another study uses Cloud–Aerosol LiDAR with Orthogonal Polarization (CALIOP) aerosol products to identify the global long-range aerosol transport pathways (the trans-Atlantic, the trans-Pacific, and the trans-Arabian Sea) [14]. Two significant paths within the range of the trans-Pacific transport pathway (aerosols from the Taklimakan Desert and aerosols from the North China Plain) are analyzed in detail. A three-stage conceptual model is further built, providing a straightforward and evident approach to exploring long-range aerosol transport pathways. To invesfigate frequently occurring severe haze pollution in northeast China, the vertical characteristics of aerosols and the causes of aerosol pollution throughout the year are analyzed using multisource data of ground-based LiDAR and Cloud–Aerosol LiDAR Pathfinder Satellite Observations (CALIPSOs) [15]. The contribution of dust, smoke, and firework aerosols are analyzed, and recommendations for pollution control policies are provided.

The effect of aerosols on atmospheric correction is also discussed. For Soil Organic Carbon (SOC) estimation, Bottom-of-Atmosphere (BOA) VNIR/SWIR reflectance retrieved from Top-Of-Atmosphere (TOA) radiance using atmospheric correction methods is needed. A thorough sensitivity study of SOC estimation in relation to aerosol optical depth and water vapor is conducted based on Earth Observing-1 Hyperion Hyperspectral data [16]. The research suggests using the FLAASH AC method to provide BOA reflectance values before SOC mapping. Another study focuses on improving the accuracy of remote sensing reflectance products in the nearshore waters of the Shandong Peninsula [17]. To achieve that goal, a monthly aerosol model based on aerosol data collected from the Mu Ping site in the coastal area of the Shandong Peninsula is developed to replace the standard model.

In summary, this Special Issue collects a series of representative studies in the research field of aerosol and atmospheric correction, mainly focusing on the improvement in aerosol identification and retrieval methods; atmospheric aerosol formation, transfer, and spatiotemporal variation; and the effect of aerosols on atmospheric correction and quantitative remote sensing. These advancements will help to continuously improve our understanding of atmospheric aerosol and the accuracy of quantitative remote sensing research. Despite the significant progress achieved, further related studies are still needed for the scientific community, policy makers, and the public to reduce evaluation uncertainty and combat the challenges faced in our society.

Author Contributions: This Editorial was prepared by S.S. and reviewed by X.G. and J.W. All authors have read and agreed to the published version of the manuscript.

Funding: This study was funded by the Future Talent Project of AIRCAS (grant number E4Z103010F), the National Natural Science Foundation of China (grant number 42005104), and China Scholarship Council (grant number CSC 202204910187).

Acknowledgments: The Guest Editors would like to thank the authors who contributed to this Special Issue and the reviewers who helped to improve the quality of this Special Issue by providing constructive feedback to the authors.

Conflicts of Interest: The authors declare no conflicts of interest.

References

1. Kaufman, Y.J.; Tanré, D.; Boucher, O. A satellite view of aerosols in the climate system. *Nature* **2002**, *419*, 215–223. [CrossRef] [PubMed]
2. Stevens, B. Rethinking the lower bound on aerosol radiative forcing. *J. Clim.* **2015**, *28*, 4794–4819. [CrossRef]
3. Wei, J.; Li, Z.; Lyapustin, A.; Wang, J.; Dubovik, O.; Schwartz, J.; Sun, L.; Li, C.; Liu, S.; Zhu, T. First close insight into global daily gapless 1 km $PM_{2.5}$ pollution, variability, and health impact. *Nat. Commun.* **2023**, *14*, 8349. [CrossRef] [PubMed]
4. Johnston, F.; Henderson, S.; Chen, Y.; Randerson, J.T.; Marlier, M.; Defries, R.; Kinney, P.; Bowman, D.M.J.S.; Brauer, M. Estimated Global Mortality Attributable to Smoke from Landscape Fires. *Environ. Health Perspect.* **2012**, *120*, 695–701. [CrossRef] [PubMed]
5. Mobley, C.D.; Werdell, J.; Franz, B.A.; Ahmad, Z.; Bailey, S.W. *Atmospheric Correction for Satellite Ocean Color Radiometry*; NASA: Washington, DC, USA, 2016.
6. Zhao, D.; Feng, L.; He, X. Global Gridded Aerosol Models Established for Atmospheric Correction over Inland and Nearshore Coastal Waters. *J. Geophys. Res. Atmos.* **2023**, *128*, e2023JD038815. [CrossRef]
7. Mannschatz, T.; Pflug, B.; Borg, E.; Feger, K.H.; Dietrich, P. Uncertainties of LAI estimation from satellite imaging due to atmospheric correction. *Remote Sens. Environ.* **2014**, *153*, 24–39. [CrossRef]
8. Li, M.; Xia, H.; Su, L.; Han, H.; Wang, X.; Yuan, J. The Detection of Desert Aerosol Incorporating Coherent Doppler Wind Lidar and Rayleigh–Mie–Raman Lidar. *Remote Sens.* **2023**, *15*, 5453. [CrossRef]
9. Fan, Y.; Sun, X.; Ti, R.; Huang, H.; Liu, X.; Yu, H. Aerosol Retrieval Study from a Particulate Observing Scanning Polarimeter Onboard Gao-Fen 5B without Prior Surface Knowledge, Based on the Optimal Estimation Method. *Remote Sens.* **2023**, *15*, 385. [CrossRef]
10. Jiang, J.; Liu, J.; Jiao, D.; Zha, Y.; Cao, S. Evaluation of MODIS DT, DB, and MAIAC Aerosol Products over Different Land Cover Types in the Yangtze River Delta of China. *Remote Sens.* **2023**, *15*, 275. [CrossRef]
11. Si, Y.; Chen, L.; Zheng, Z.; Yang, L.; Wang, F.; Xu, N.; Zhang, X. A Novel Algorithm of Haze Identification Based on FY3D/MERSI-II Remote Sensing Data. *Remote Sens.* **2023**, *15*, 438. [CrossRef]
12. Chen, Q.-X.; Huang, C.-L.; Dong, S.-K.; Lin, K.-F. Satellite-Based Background Aerosol Optical Depth Determination via Global Statistical Analysis of Multiple Lognormal Distribution. *Remote Sens.* **2024**, *16*, 1210. [CrossRef]
13. Yu, X.; Nichol, J.; Lee, K.H.; Li, J.; Wong, M.S. Analysis of Long-Term Aerosol Optical Properties Combining AERONET Sunphotometer and Satellite-Based Observations in Hong Kong. *Remote Sens.* **2022**, *14*, 5220. [CrossRef]
14. Wang, L.; Wang, W.; Lyu, B.; Zhang, J.; Han, Y.; Bai, Y.; Guo, Z. The Identification and Analysis of Long-Range Aerosol Transport Pathways with Layered Cloud-Aerosol Lidar with Orthogonal Polarization Datasets from 2006 to 2016. *Remote Sens.* **2023**, *15*, 4537. [CrossRef]
15. Duanmu, L.; Chen, W.; Guo, L.; Yuan, Y.; Yang, H.; Fu, J.; Song, G.; Xia, Z. Vertical Profiles of Aerosols Induced by Dust, Smoke, and Fireworks in the Cold Region of Northeast China. *Remote Sens.* **2024**, *16*, 1098. [CrossRef]
16. Mruthyunjaya, P.; Shetty, A.; Umesh, P.; Gomez, C. Impact of Atmospheric Correction Methods Parametrization on Soil Organic Carbon Estimation Based on Hyperion Hyperspectral Data. *Remote Sens.* **2022**, *14*, 5117. [CrossRef]
17. Shan, K.; Ma, C.; Lv, J.; Zhao, D.; Song, Q. Construction of Aerosol Model and Atmospheric Correction in the Coastal Area of Shandong Peninsula. *Remote Sens.* **2024**, *16*, 1309. [CrossRef]

remote sensing

MDPI

Technical Note

The Detection of Desert Aerosol Incorporating Coherent Doppler Wind Lidar and Rayleigh–Mie–Raman Lidar

Manyi Li [1], Haiyun Xia [1,2,3,4,*], Lian Su [1], Haobin Han [2], Xiaofei Wang [5] and Jinlong Yuan [2]

1 School of Earth and Space Science, University of Science and Technology of China, Hefei 230026, China; lmy123@mail.ustc.edu.cn (M.L.); kiwensu@mail.ustc.edu.cn (L.S.)
2 School of Atmospheric Physics, Nanjing University of Information Science and Technology, Nanjing 210044, China; hanhaobin@nuist.edu.cn (H.H.); yuanjinlong@nuist.edu.cn (J.Y.)
3 National Laboratory for Physical Sciences at the Microscale, University of Science and Technology of China, Hefei 230026, China
4 Institute of Software, Chinese Academy of Sciences, Beijing 100190, China
5 Xinjiang Uygur Autonomous Region Meteorological Service, Urumqi 830002, China; yongkangl@stu.xju.edu.cn
* Correspondence: hsia@ustc.edu.cn

Abstract: Characterization of aerosol transportation is important in order to understand regional and global climatic changes. To obtain accurate aerosol profiles and wind profiles, aerosol lidar and Doppler wind lidar are generally combined in atmospheric measurements. In this work, a method for calibration and quantitative aerosol properties using coherent Doppler wind lidar (CDWL) is adopted, and data retrieval is verified by contrasting the process with synchronous Rayleigh–Mie–Raman lidar (RMRL). The comparison was applied to field measurements in the Taklimakan desert, from 16 to 21 February 2023. Good agreements between the two lidars was found, with the determination coefficients of 0.90 and 0.89 and the root-mean-square error (RMSE) values of 0.012 and 0.013. The comparative results of continuous experiments demonstrate the ability of the CDWL to retrieve aerosol properties accurately.

Keywords: aerosol; extinction coefficient; coherent Doppler wind lidar; Raman lidar; Mie lidar

Citation: Li, M.; Xia, H.; Su, L.; Han, H.; Wang, X.; Yuan, J. The Detection of Desert Aerosol Incorporating Coherent Doppler Wind Lidar and Rayleigh–Mie–Raman Lidar. *Remote Sens.* **2023**, *15*, 5453. https://doi.org/10.3390/rs15235453

Academic Editor: Pavel Kishcha

Received: 17 October 2023
Revised: 17 November 2023
Accepted: 20 November 2023
Published: 22 November 2023

1. Introduction

The study of atmospheric aerosols, especially in the planetary boundary layer (PBL), is of great importance in understanding the vertical exchange of sensible heat (temperature), latent heat (moisture), particles, and trace gases between the surface and the lower troposphere [1–3], which has a strong influence on global climate and atmospheric composition. To observe vertical and regional transport [4,5] and conduct pollution tracing and forecasting [6], it is necessary to carry out accurate detection of meteorological parameters and aerosol properties with high temporal and spatial resolutions.

In the past several decades, lidar has been proven to be a powerful and potential tool in remote sensing of the atmosphere. Atmospheric lidars are widely used in ground-based, ball-borne, airborne, and satellite-borne detection devices [7,8]. Generally, atmospheric lidars can be classified into two categories by different atmospheric backscatter recordings: heterodyne coherent detection and direct detection. Direct detection has been employed in the majority of lidar systems, and numerous lidar applications have been applied in the detection of aerosol and clouds [9–11], the boundary layer [3,12,13], temperature [14,15], trace gas [16,17], and wind [18]. Heterodyne coherent detection detects the beat signal between the backscatter and a local oscillator laser to retrieve the Doppler shift due to moving particles. Coherent detection lidar is widely used in the detection of wind profiles, boundary layers [19,20], clear air turbulence and wind shear [21], aircraft wake vortices [22], and precipitation [23,24].

With the Doppler wind lidar technique, one of the most important indicators in the observation of atmospheric vertical exchange processes, that is, the accurate measurements of the vertical wind component, can be obtained. Using the combination of Doppler wind lidar and other types of aerosol lidars, accurate aerosol backscatter signals and wind profiles were obtained simultaneously [13]. There is no doubt that the combination of several lidars would make it more accurate, but it also increases the cost and volume, making it less conducive to airborne and spaceborne detection. Recently, many studies have been devoted to calibrating the backscatter of coherent Doppler wind lidar (CDWL) to obtain the profiles of aerosols and wind with only one lidar. To accurately retrieve the extinction coefficient of aerosol using CDWL, the effect of heterodyne efficiency needs to be considered [25,26], which represents the coupling of the mode field of the local oscillator and backscatter light. To quantify the aerosol transport and change in aerosol properties from the Sahara desert to the Caribbean, backscatter and extinction coefficient profiles were retrieved from airborne CDWL using function fitting of heterodyne efficiency, measured by changing the altitude of the aircraft [27]. Normalized backscatter power data points from horizontal detection were used to fit heterodyne efficiency, and the results of aerosol optical depth were calibrated with other devices, including a sun photometer [28], ceilometers [29], and atmospheric visibility [30]. Moreover, theoretical focus function and calibration based on liquid clouds were also adopted in the retrieval of aerosol [31–33].

Comparison and calibration are usually performed on the aerosol optical depth (AOD), due to the lack of high accuracy and resolution of the reference devices. In this work, an experimental comparison between the aerosol extinction profiles using CDWL and Rayleigh–Mie–Raman lidar (RMRL) was achieved. By applying the heterodyne efficiency from horizontal measurement, the backscatter profiles of CDWL were calibrated and used to retrieve the aerosol extinction coefficient. To compare the results of the two lidars, a data assimilation and comparison method is proposed, and the data from the Raman lidar are involved to improve the accuracy of the results. A consistency analysis of the results was conducted, and good agreement with the determination coefficients of 0.90 and 0.89 was found.

The paper is organized as follows. The involved lidar systems are described in Section 2, and Section 3 introduces the retrieval and comparison methods. To focus on the detection of aerosol optical properties, the retrieval of wind speed will not be discussed in this work. Section 4 introduces the comparison results during a 5-day vertical desert aerosol observation. Finally, Section 5 summarizes the conclusion and the outlook of future studies.

2. Instrument

2.1. Coherent Doppler Wind Lidar

In this work, an all-fiber CDWL was deployed to provide the atmospheric wind profiles and backscatter measurements. The lidar system emits a laser at a wavelength of 1548 nm, with a pulse energy of 110 μJ, a repetition frequency of 10 kHz, and a pulse full width at half maximum of 200 ns. A 100 mm diameter telescope was used as a coaxial transmitter and receiver, and then, the backscatter was coupled with the local oscillator light and finally detected by the balanced detector (BD), with a noise bandwidth of 200 MHz.

Aiming at studying the transport and sedimentation of Taklimakan desert dust, the CDWL was employed in Minfeng, Xinjiang province, China (82.69°E, 37.06°N). In response to the instability caused by the large diurnal temperature variation, the lidar system was designed with an all-fiber structure and temperature control system. The CDWL was operated in a velocity azimuth display (VAD) scanning mode during the experiment, and the elevation angle was set to 70°. The scanning range of the azimuth angle is from 0° to 360°, where 0° corresponds to the north and 90° corresponds to the east. The step of the azimuth angle is 12°, and the period of one scan is about 1 min. The radial range resolution was set to 30 m/60 m/150 m in the range of 0–3 km/3–6 km/6–15 km. In our previous work, the performance of wind measurements was validated; the standard deviations of wind speed and direction were 0.84 m/s and 9.2°, respectively [23]. The key system specifications are listed in Table 1.

Table 1. Key parameters of CDWL and RMRL systems.

	Parameter	CDWL	RMRL
Laser			
	Wavelength	1548 nm	355/532/1064 nm
	Frequency offset	80 MHz	/
	Pulse energy	110 μJ	250/350/350 mJ
	Repetition rate	10 kHz	20 Hz
	Pulse width	200 ns	8 ns
Telescope			
	Diameter	100 mm	450 mm
	Elevation angle	70°	90°
Data acquisition	Detection type	BD	PMT
	Noise bandwidth	200 MHz	0.3 nm
	Sampling rate	250 MHz	20 MHz
	Temporal resolution	1 s	60 s
	Spatial resolution	30, 60, 150 m	30 m

2.2. Rayleigh–Mie–Raman Lidar

To examine the performance of the CDWL in aerosol measurements, an RMRL was running near the CDWL simultaneously. The lidar uses an Nd: YAG power laser, which generates 20 pulses per second with a pulse energy of 250/350/350 mJ at the wavelength of 355/532/1064 nm. The backscatter signals are collected by a 450 mm diameter Cassegrain telescope and then separated into several channels using dichroic beam splitters, including the elastic channels (355, 532, and 1064 nm) and the nitrogen Raman (386 nm) channels. During the daytime, the elastic channels detect backscattered light with interferometric filters of 0.3 nm bandwidth to reduce the solar light, while the inelastic channels do not work due to the weak Raman scattering and strong background noise. The backscatter signals are detected by the photomultiplier tubes (PMT), and recorded using multichannel scaler (MCS) boards with a resolution of 30 m and 60 s.

3. Methodology

The following section introduces the analysis steps applied to the signal measured by the CDWL and RMRL, including the pre-processing of data, the retrieval algorithm of optical properties, and the data assimilation method for the different kinds of lidar.

3.1. Retrieval Algorithm of the CDWL

In this subsection, we prefer to discuss the retrieval of aerosol optical properties rather than wind profiles, for wind measurement, as discussed in many works [21,34,35], is not this paper's focus. In the CDWL measurements, the photocurrent in response to the beat signal of the atmospheric backscatter and the local oscillator can be expressed as [35]

$$i_d = RP_{Lo} + RP_s(r) + 2R[\eta_h(r)P_{Lo}P_s(r)]^{0.5}\cos(\omega_c t + \theta) + i_n, \tag{1}$$

where R is the response of the detector ($R = \eta_q q/hv$, η_q is the quantum efficiency, q is the elementary charge, h is the Planck constant, and v is the photon frequency), r is the range, and P_{Lo} and $P_s(r)$ are the power of the signal from the local oscillator and the atmospheric backscatter, respectively. $\eta_h(r)$ is the heterodyne efficiency, which represents the fraction of the total signal power matched with the local oscillator field. ω_c and θ are the frequency and phase of the IF (intermediate frequency) signal. The first two terms of Equation (1) are filtered through AC (alternating current) coupling, and the third and fourth terms are signal current i_s and noise current i_n, respectively. The wide band CNR (carrier-to-noise ratio) is defined as the ratio of signal power and noise power and can be expressed as

$$CNR(r) = \frac{\langle i_s^2 \rangle}{\langle i_n^2 \rangle} = \frac{\eta_q \eta_h(r) P_s(r)}{h\upsilon B}, \quad (2)$$

where B is the receiver noise equivalent bandwidth. Combined with the lidar equation [36], the lidar equation of coherent lidar can be expressed as:

$$\frac{CNR(r)r^2}{\eta_h(r)} = \frac{\eta_q C E_0}{h\upsilon B}\beta(r)\exp\left[-\int_0^r 2\alpha(r)\mathrm{d}z\right], \quad (3)$$

where C represents the lidar constant, E_0 is the pulse energy of the outgoing light, and $\beta(r)$ and $\alpha(r)$ are the backscatter and extinction coefficient, respectively. For a monostatic pulsed coherent lidar, due to the heterodyne efficiency caused by the telescope focus function, a distance-dependent modulation is applied to the backscatter signal, resulting in the distortion of the near-field signal. This may have little impact on wind measurements, but it will cause false values in the attenuated backscatter coefficient profiles. Using the estimation or calibration of the heterodyne efficiency, the calibrated backscatter signal can be obtained, and the backscatter and extinction coefficient of atmospheric aerosol can be further retrieved.

Similar to the processing method of overlap factor correction [37], the modulation of the heterodyne efficiency over the backscatter signal can be estimated from the horizontal measurements. To calibrate the heterodyne efficiency in the near field, an advanced horizontal measurement is performed, and the average result of the CNR is applied and fitted, as shown in the subfigure of Figure 1a. Due to the different range resolutions, the density of numerical points varies with the range. Using the ratio of the range-correct CNR and the linear fit result, the heterodyne efficiency can be obtained and used to calibrate the CNR profiles in the experiment. Figure 1a depicts four heterodyne efficiency profiles at different times, which seem to be stable within the range of 3 km due to the all-fiber structure and the stable temperature of the system.

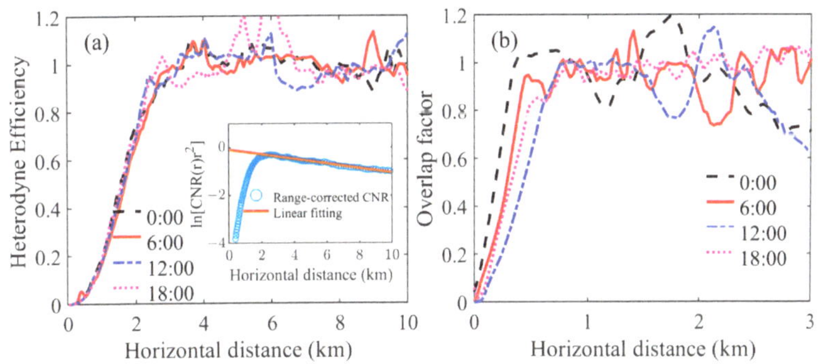

Figure 1. (**a**) The heterodyne efficiency from four horizontal experiments, which was obtained from the ratio of the range−correct CNR (blue circles) and the linear fit result (orange line), as shown in the subfigure. (**b**) The overlap factor of the 355 nm channel from three horizontal experiments, measured at 0:00, 6:00, 12:00, and 18:00 local time on 14 February 2023.

In contrast, limited by the non-coaxial telescope and the space optical structure, the RMRL system is more affected by temperature changes, which are huge in the desert. Figure 1b depicts four overlap factor profiles of the 355 nm channel, measured at 0:00, 6:00, 12:00, and 18:00 local time on 14 February 2023. The temperature variation affects the optical coupling of the space optical system significantly, resulting in an unstable overlap factor for the RMRL. The signal in the near field, mainly below 0.5 km is distorted, so only the results above 0.5 km are involved in the comparison.

As mentioned earlier, the CDWL is operated in a VAD scanning mode with an elevation angle of 70°. The CNR profiles have to be converted vertically to match the backscatter profiles of the RMRL, so an approximation is taken so that the average CNR of a circle scan is approximate to the vertical CNR profile at the projection height. Furthermore, the vertical CNR profile is interpolated to a height resolution of 30 m to match the profile of the RMRL.

3.2. Retrieval of the RMRL

In this subsection, two methods are used to retrieve the optical properties of aerosols. Generally, by assuming a constant lidar ratio between $\alpha(r)$ and $\beta(r)$, the inversion algorithm proposed by Klett and Fernald (KF method) can be used to retrieve $\alpha(r)$ and $\beta(r)$ [38,39]. This method can be used to invert the extinction and backscatter with only one elastic backscatter lidar, with the assumption of the lidar ratio and boundary value. Different assumptions of boundary value may lead to various results [40,41]. In general, the boundary value is chosen at a reference altitude where the atmosphere is relatively pure and the signal-to-noise ratio (SNR) is high enough. Sometimes, it may be difficult to find the reference height, and soft targets such as clouds are used to estimate the boundary value [42].

In addition, the Raman backscatter signals can be used alone to retrieve aerosol extinction profiles [43]. The independent aerosol extinction coefficient can be expressed as: [44,45]

$$\alpha_{aer}(\lambda_L, r) = \frac{\frac{\mathrm{d}}{\mathrm{d}r}\left[\ln \frac{n_x(r)}{r^2 P(\lambda_x, \lambda_L, r)}\right] - \alpha_{\mathrm{mol}}(\lambda_L, r) - \alpha_{\mathrm{mol}}(\lambda_x, r)}{1 + \left(\frac{\lambda_L}{\lambda_x}\right)^k}, \tag{4}$$

where the subscripts *aer* and *mol* stand for aerosol and atmospheric molecules, respectively. $n_x(r)$ is the number density of the Raman molecule; λ_X and λ_L are the wavelength of the Raman-shifted laser and the output laser, respectively. $P(\lambda_X, \lambda_L, r)$ is the backscatter power of the Raman-shifted signal, and k is the Angstrom exponent. With the combination of Mie scattering data, aerosol backscatter coefficient profiles can be obtained by:

$$\beta_{\mathrm{aer}}(\lambda_L, r) = -\beta_{\mathrm{mol}}(\lambda_L, r) + \beta_{\mathrm{mol}}(\lambda_L, r_0) \cdot \frac{P(\lambda_L, r) \cdot P(\lambda_x, \lambda_L, r_0) \cdot n_x(r)}{P(\lambda_L, r_0) \cdot P(\lambda_x, \lambda_L, r) \cdot n_x(r_0)} \cdot \frac{\exp[-\int_{r_0}^{r} \alpha_{\mathrm{aer}}(\lambda_X, r) + \alpha_{\mathrm{mol}}(\lambda_X, r)\mathrm{d}z]}{\exp[-\int_{r_0}^{r} \alpha_{\mathrm{aer}}(\lambda_L, r) + \alpha_{\mathrm{mol}}(\lambda_L, r)\mathrm{d}z]}, \tag{5}$$

where r_0 is the reference height and $P(\lambda_L, r)$ is the backscatter power of the Mie lidar signal. β_{mol} can be calculated by the atmospheric temperature and pressure from the radiosonde measurement or the atmosphere model. The two approaches present different advantages and disadvantages; for example, the KF method relies on the accurate assumption of the lidar ratio and boundary value, and the Raman method relies on the accurate assumption of the Angstrom exponent. Figure 2 depicts some retrieval results of the experiments, and the two algorithms show good consistency under correct assumptions.

To compare the data retrieval of the RMRL and CDWL, the backscatter and extinction coefficients of aerosol at the wavelengths of 355, 532, and 1064 nm had to be converted to the wavelength of 1548 nm. However, the extinction coefficient retrieved with the Raman method only uses the backscatter signal at the Raman wavelength of 386 nm, which may bring errors to the data conversion process. In this work, the two methods were combined. First, the aerosol extinction coefficient was obtained at the reference height with the Raman method using the data from 386 nm and then applied to the KF method as the boundary value. Next, with the KF method, the extinction coefficient at the wavelengths of 355, 532, and 1064 nm could be obtained using the data from the different elastic lidars. The Angstrom exponent varies with the wavelength, and an empirical relationship between aerosol extinction and wavelength can be expressed with a second-order polynomial [46,47]:

$$\ln \alpha_{aer}(\lambda) = a_0 + a_1 \ln \lambda + a_2 (\ln \lambda)^2, \tag{6}$$

where the coefficient a_2 accounts for a "curvature" often observed in Sun photometry measurements. For a special case of $a_2 = 0$, Equation (6) equals the Angstrom exponent law and $a_1 = -k$. With the profiles of the aerosol extinction coefficient from the RMRL, the profiles of the aerosol extinction and the backscatter coefficient at the wavelength of 1548 nm can be obtained, as shown in Figure 3.

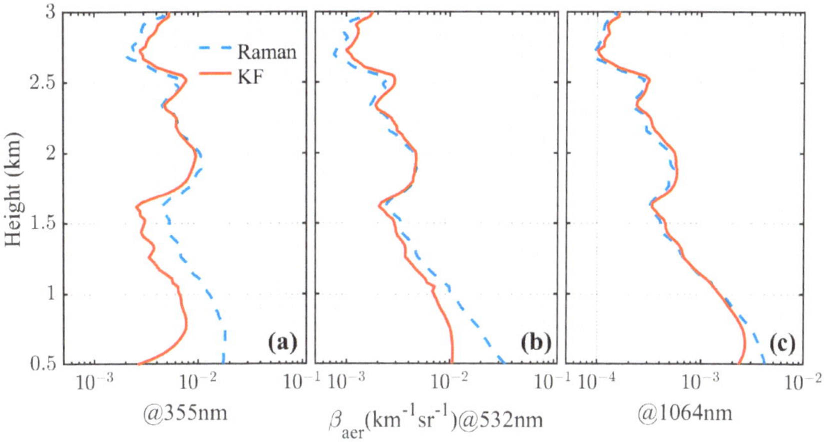

Figure 2. Comparison of retrieval using the Raman method (blue dotted line) and the KF method (red line). Aerosol backscatter coefficient β_{aer} from the data of (**a**) 355 nm, (**b**) 532 nm, and (**c**) 1064 nm lidar observed at 1:30 local time on 20 February 2023. The lidar ratios assumed for KF inversion are 30, 32, and 45. The Angstrom exponent applied in Equation (4) is set as 1.5.

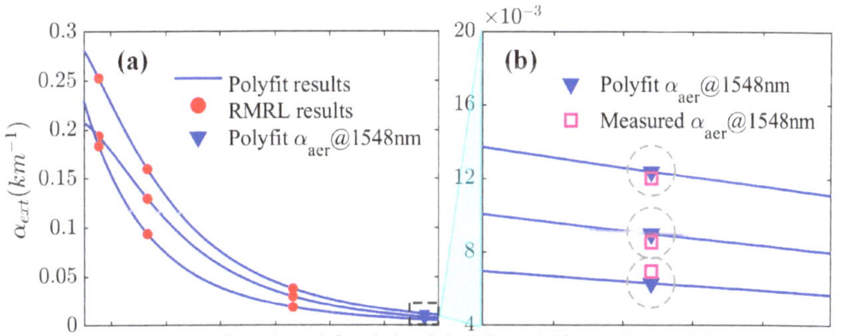

Figure 3. (**a**) Three groups of aerosol extinction coefficients at a height of 2 km retrieved from the RMRL (red dots); second-order poly fit curves using the extinction coefficients and the wavelength (blue line); aerosol extinction coefficients at the wavelength of 1548 nm derived from the poly fit curves (blue triangle); (**b**) a magnification of part of (**a**) from 1.5 to 1.6 μm; aerosol extinction coefficients retrieved from the CDWL (pink square), and the blue lines and blue triangles are the same as in (**a**), derived from the RMRL. The gray circles indicate the same group of data.

In this way, the data sets of the two lidars were assimilated to the same wavelength and their inversion results could be compared. The data processing procedure is displayed in Figure 4. The number density of Nitrogen molecules was calculated by the profile of temperature and pressure from the Atmosphere model, and the lidar ratio was set as 30/32/45/50 Sr at the wavelength of 355/532/1064/1548 nm.

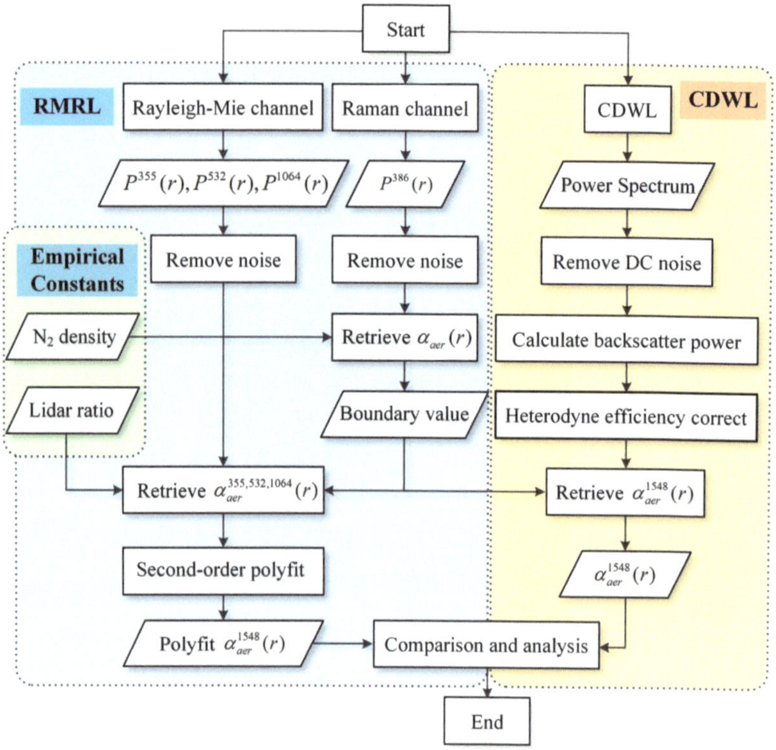

Figure 4. Overview of the data processing procedure. CDWL, coherent Doppler wind lidar; DC, direct current.

4. Experiments and Results

Field joint experiments with the CDWL and RMRL were conducted from 16 to 21 February 2023. The Raman lidar only works from 22:00 to 6:00 local time, so only the data during the nighttime were compared. Figure 5 depicts the backscatter profiles averaged over ten minutes altitude during the nighttime and daytime. Both the CDWL and RMRL demonstrate a vertical detection capability of more than 3 km during the nighttime, and it can be seen that the aerosols are mainly distributed at altitudes below 3 km. The backscatter signal at 386 nm decays smoothly because the Raman backscatter coefficient is proportional to the density of nitrogen, which is independent of the aerosol profiles. During the daytime, the Raman lidar is turned off, and the effective detection ranges of the other lidars are also reduced.

With the method in Figure 4, the extinction coefficients at various wavelengths were retrieved with the resolution of 30 m/1 min, as shown in Figure 6, which also can be used to retrieve the color ratio. Only the results from 22:00 to 6:00 local time are plotted. Figure 6d, e shows the extinction coefficients at 1548 nm using the data from the RMRL and CDWL, respectively, and the comparison results show good consistency.

It can be noticed that longitudinal stripes appear in Figure 6, which may mean the internal gravity waves strongly affect the backscatter, particularly at 0:00 on 20 February. To compare the retrieval of extinction coefficients with the two lidars in detail, the results within a height range at the same time are displayed, as shown in Figure 7. It needs to be mentioned that the results vary greatly in the near field (mainly in altitudes ranging below 0.5 km). The signal of the RMRL is uncalibrated in the range of 0–0.5 km due to the lack of a stable overlap factor, while the CDWL is calibrated with a relatively stable heterodyne efficiency. Inconsistent results appear at 6:00 local time on 18 February, the output power of the laser in the RMRL decreases, and the effective detection distance is reduced. Consequently, the measurement of the RMRL shows poor ability in aerosol retrieval, causing a large difference between the CDWL and RMRL.

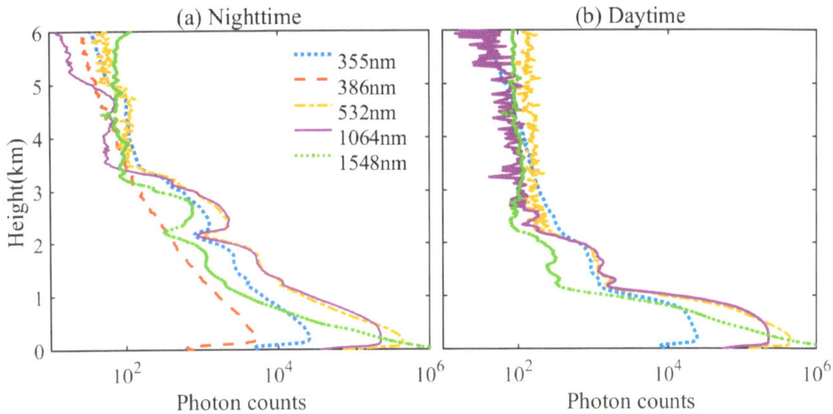

Figure 5. Ten minutes averaged altitude profiles of backscattered signals measured at (**a**) 0:00 and (**b**) 12:00 local time on 20 February. The Raman lidar does not work during the daytime. The legend indicates the wavelengths of the backscatter signals in nanometers.

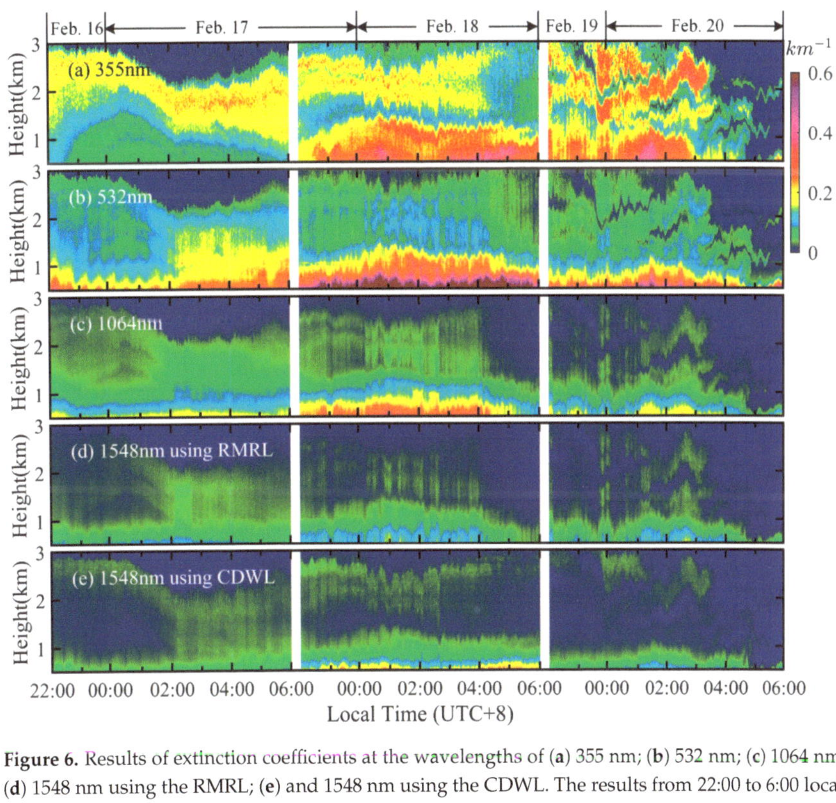

Figure 6. Results of extinction coefficients at the wavelengths of (**a**) 355 nm; (**b**) 532 nm; (**c**) 1064 nm; (**d**) 1548 nm using the RMRL; (**e**) and 1548 nm using the CDWL. The results from 22:00 to 6:00 local time on the 16, 17, 18, 19, and 20 February are displayed when the Raman lidar works routinely. The time resolution and height resolution are 1 min and 30 m, respectively.

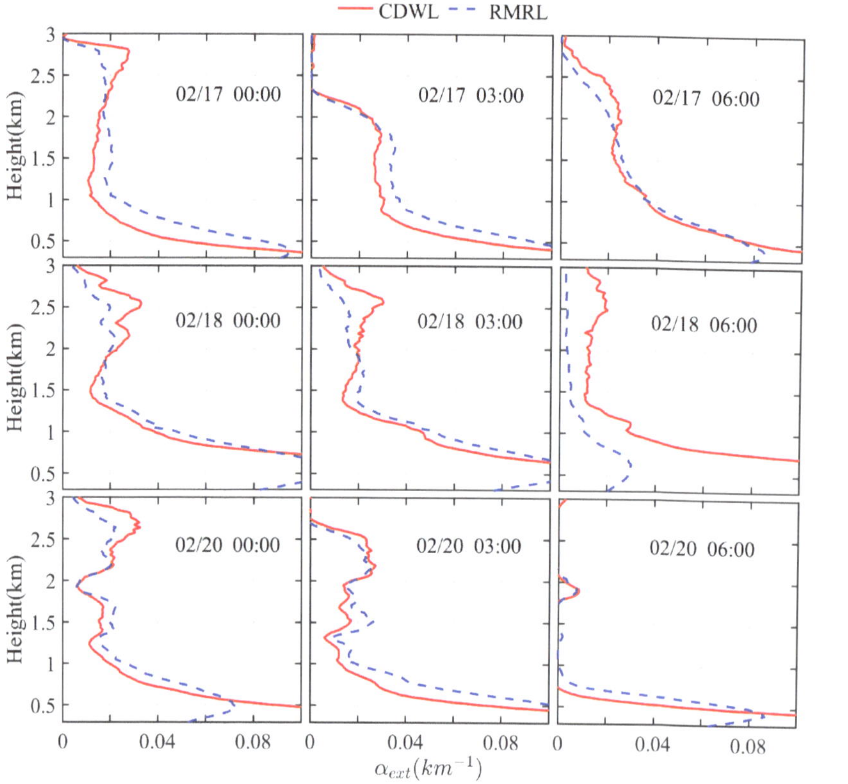

Figure 7. Comparison of the aerosol extinction coefficients with the two lidars, measured at 0:00, 3:00, and 6:00 local time on the 17, 18, and 20 February, respectively. The red lines show the results of the CDWL at 1548 nm, and the blue dotted lines show the fitting results of the RMRL at 1548 nm.

The scatter diagrams of the aerosol extinction coefficients retrieved from the CDWL and RMRL are shown in Figure 8. The results of the CDWL are plotted versus those of the RMRL, and the density of the data points is normalized and displayed in different colors. Moreover, the distributions of the bias are inserted in the bottom right corner of each panel, in which the negative bias means that the results of the CDWL are lower than that of the RMRL. In addition, due to the effects of different overlap factors, comparisons using data ranging from 0.5 to 3 km and from 0 to 0.5 km are divided into two groups. The first-order linear fitting functions of the results, the determination coefficients R, and the root-mean-square error (RMSE) are added to the upper left corner of each graph. Good performance with an R of 0.90 and 0.89 was obtained, proving the accuracy of aerosol retrieval by coherent detection after heterodyne efficiency correction. In contrast, results ranging from 0 to 0.5 km are shown in Figure 8c,d. The uncorrected overlap factor leads to the result that the aerosol extinction of the RMRL is smaller than that of the CDWL.

The continuous observation results from 16–21 February are shown in Figure 9, including the CNR, aerosol extinction coefficients, horizontal wind speed and direction, and vertical wind speed. When the Raman lidar does not work during the daytime, the boundary value is set as the interpolation of the result during the night. To ensure the accuracy of the results, the retrieval of wind speed and direction with the CNR less than −17 dB is plotted as white. In Figure 9e, the positive value of vertical velocity is defined as downward, and the upward wind speed appearing every afternoon is consistent with the results of the extinction coefficient, which also confirms the vertical transport process of aerosols. The results prove the aerosol retrieval capability of the CDWL system.

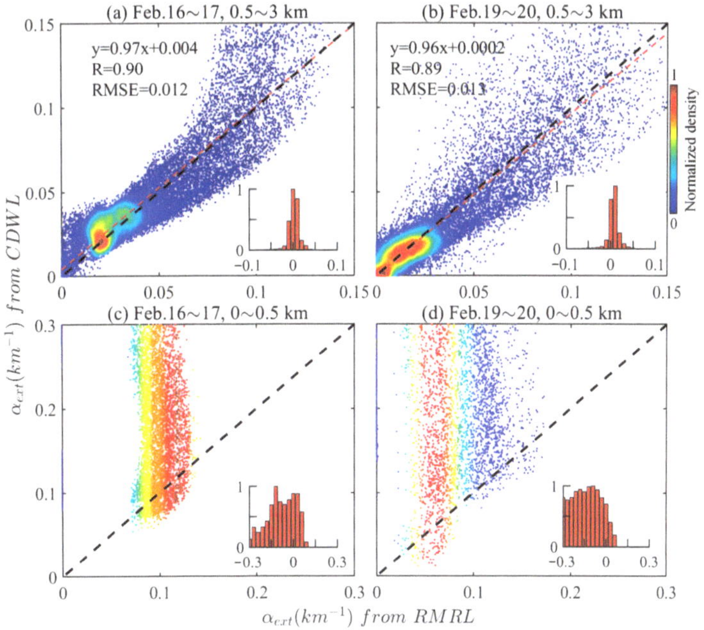

Figure 8. Scatter diagrams of the aerosol extinction coefficients derived from the RMRL and the CDWL. Results ranging from 0.5 to 3 km during the nighttime of the 17 and 20 February are compared in Figure 8 (**a**,**b**), respectively. In contrast, results ranging from 0 to 0.5 km are shown in Figure 8 (**c**,**d**). The color-shaded dots denote the normalized density. The black dashed lines denote 1:1 lines, and the red dashed lines show the linear fitting results. The distributions of y-x are inserted in the bottom right corner of each panel.

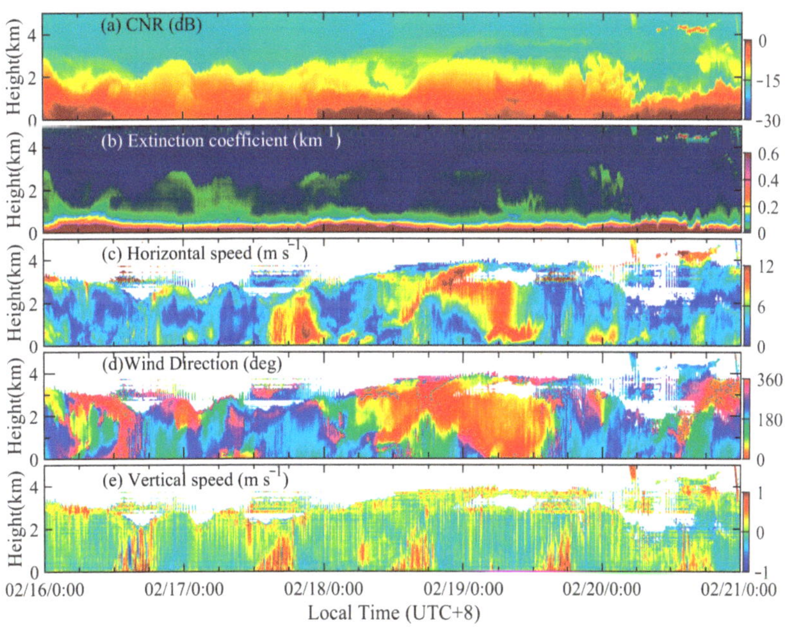

Figure 9. Continuous observation by the CDWL from the 16 to 21 February. (**a**) CNR, (**b**) aerosol extinction coefficient, (**c**) horizontal wind speed, (**d**) horizontal wind direction, and (**e**) vertical wind

speed. The CNR and extinction coefficient are averaged over a vertical scanning duration of 1 min. The positive value of vertical velocity is defined as downward.

5. Conclusions

An experimental comparison between aerosol measurements using the CDWL and the RMRL was achieved. To calibrate the aerosol retrieval, a method using boundary values from the Raman lidar was proposed, and the extinction coefficients were compared from the two lidar systems. Good agreements were achieved during the continuous field experiments, with the fitting slopes of 0.97 and 0.96, and the determination coefficients of 0.90 and 0.89. The results prove the accuracy of aerosol retrieval by coherent lidar after correction powerfully, and the aerosol detection by all-fiber wind lidar was extended, which may achieve the miniaturization and stabilization of the lidar for meteorological measurements.

Author Contributions: Conceptualization, M.L. and H.X.; methodology, M.L.; formal analysis, H.X. and M.L.; investigation, M.L.; data curation, L.S. and X.W.; writing—original draft preparation, M.L.; writing—review and editing, M.L., L.S., X.W., J.Y., H.H. and H.X.; visualization, M.L.; supervision, H.X. and H.H. All authors have read and agreed to the published version of the manuscript.

Funding: Strategic Priority Research Program of Chinese Academy of Sciences, grant number XDA22040601.

Data Availability Statement: The data underlying the results presented in this paper can be obtained from the authors upon reasonable request.

Acknowledgments: We would like to thank the Xinjiang Uygur Autonomous Region Meteorological Service for providing support with the observational data.

Conflicts of Interest: The authors declare no conflict of interest.

References

1. Kokhanovsky, A.A. *Aerosol Optics: Light Absorption and Scattering by Particles in the Atmosphere*; Springer Science & Business Media: Berlin, Germany, 2008.
2. Stocker, T. *Climate Change 2013: The Physical Science Basis: Working Group I Contribution to the Fifth Assessment Report of the Intergovernmental Panel on Climate Change*; Cambridge University Press: Cambridge, UK, 2014.
3. Caicedo, V.; Rappenglück, B.; Lefer, B.; Morris, G.; Toledo, D.; Delgado, R. Comparison of aerosol lidar retrieval methods for boundary layer height detection using ceilometer aerosol backscatter data. *Atmos. Meas. Tech.* **2017**, *10*, 1609–1622. [CrossRef]
4. Gryazin, V.; Beresnev, S. Influence of vertical wind on stratospheric aerosol transport. *Meteorol. Atmos. Phys.* **2011**, *110*, 151–162. [CrossRef]
5. Yuan, J.; Wu, Y.; Shu, Z.; Su, L.; Tang, D.; Yang, Y.; Dong, J.; Yu, S.; Zhang, Z.; Xia, H. Real-Time Synchronous 3-D Detection of Air Pollution and Wind Using a Solo Coherent Doppler Wind Lidar. *Remote Sens.* **2022**, *14*, 2809. [CrossRef]
6. Petit, J.-E.; Favez, O.; Albinet, A.; Canonaco, F. A user-friendly tool for comprehensive evaluation of the geographical origins of atmospheric pollution: Wind and trajectory analyses. *Environ. Model. Softw.* **2017**, *88*, 183–187. [CrossRef]
7. Menzies, R.T.; Tratt, D.M. Airborne CO_2 coherent lidar for measurements of atmospheric aerosol and cloud backscatter. *Appl. Opt.* **1994**, *33*, 5698–5711. [CrossRef]
8. Winker, D.M.; Vaughan, M.A.; Omar, A.; Hu, Y.; Powell, K.A.; Liu, Z.; Hunt, W.H.; Young, S.A. Overview of the CALIPSO mission and CALIOP data processing algorithms. *J. Atmos. Oceanic Technol.* **2009**, *26*, 2310–2323. [CrossRef]
9. Shimizu, A.; Nishizawa, T.; Jin, Y.; Kim, S.-W.; Wang, Z.; Batdorj, D.; Sugimoto, N. Evolution of a lidar network for tropospheric aerosol detection in East Asia. *Opt. Eng.* **2017**, *56*, 031219. [CrossRef]
10. Zhao, C.; Wang, Y.; Wang, Q.; Li, Z.; Wang, Z.; Liu, D. A new cloud and aerosol layer detection method based on micropulse lidar measurements. *J. Geophys. Res. Atmos.* **2014**, *119*, 6788–6802. [CrossRef]
11. Li, M.; Wu, Y.; Yuan, J.; Zhao, L.; Tang, D.; Dong, J.; Xia, H.; Dou, X. Stratospheric aerosol lidar with a 300 μm diameter superconducting nanowire single-photon detector at 1064 nm. *Opt. Express* **2023**, *31*, 2768–2779. [CrossRef]
12. Dang, R.; Yang, Y.; Hu, X.-M.; Wang, Z.; Zhang, S. A review of techniques for diagnosing the atmospheric boundary layer height (ABLH) using aerosol lidar data. *Remote Sens.* **2019**, *11*, 1590. [CrossRef]
13. Engelmann, R.; Wandinger, U.; Ansmann, A.; Müller, D.; Žeromskis, E.; Althausen, D.; Wehner, B. Lidar observations of the vertical aerosol flux in the planetary boundary layer. *J. Atmos. Oceanic Technol.* **2008**, *25*, 1296–1306. [CrossRef]

14. Arshinov, Y.F.; Bobrovnikov, S.M.; Zuev, V.E.; Mitev, V. Atmospheric temperature measurements using a pure rotational Raman lidar. *Appl. Opt.* **1983**, *22*, 2984–2990. [CrossRef] [PubMed]
15. Hua, D.; Uchida, M.; Kobayashi, T. Ultraviolet high-spectral-resolution Rayleigh–Mie lidar with a dual-pass Fabry–Perot etalon for measuring atmospheric temperature profiles of the troposphere. *Opt. Lett.* **2004**, *29*, 1063–1065. [CrossRef] [PubMed]
16. Yu, S.; Zhang, Z.; Xia, H.; Dou, X.; Wu, T.; Hu, Y.; Li, M.; Shangguan, M.; Wei, T.; Zhao, L. Photon-counting distributed free-space spectroscopy. *Light Sci. Appl.* **2021**, *10*, 212. [CrossRef] [PubMed]
17. Abshire, J.B.; Ramanathan, A.; Riris, H.; Mao, J.; Allan, G.R.; Hasselbrack, W.E.; Weaver, C.J.; Browell, E.V. Airborne measurements of CO_2 column concentration and range using a pulsed direct-detection IPDA lidar. *Remote Sens.* **2013**, *6*, 443–469. [CrossRef]
18. Xia, H.; Dou, X.; Sun, D.; Shu, Z.; Xue, X.; Han, Y.; Hu, D.; Han, Y.; Cheng, T. Mid-altitude wind measurements with mobile Rayleigh Doppler lidar incorporating system-level optical frequency control method. *Opt. Express* **2012**, *20*, 15286–15300. [CrossRef]
19. Tucker, S.C.; Senff, C.J.; Weickmann, A.M.; Brewer, W.A.; Banta, R.M.; Sandberg, S.P.; Law, D.C.; Hardesty, R.M. Doppler lidar estimation of mixing height using turbulence, shear, and aerosol profiles. *J. Atmos. Oceanic Technol.* **2009**, *26*, 673–688. [CrossRef]
20. Wang, L.; Qiang, W.; Xia, H.; Wei, T.; Yuan, J.; Jiang, P. Robust solution for boundary layer height detections with coherent doppler wind lidar. *Adv. Atmos. Sci.* **2021**, *38*, 1920–1928. [CrossRef]
21. Yuan, J.; Xia, H.; Wei, T.; Wang, L.; Yue, B.; Wu, Y. Identifying cloud, precipitation, windshear, and turbulence by deep analysis of the power spectrum of coherent Doppler wind lidar. *Opt. Express* **2020**, *28*, 37406–37418. [CrossRef]
22. Smalikho, I.N.; Banakh, V.; Holzäpfel, F.; Rahm, S. Method of radial velocities for the estimation of aircraft wake vortex parameters from data measured by coherent Doppler lidar. *Opt. Express* **2015**, *23*, A1194–A1207. [CrossRef]
23. Wei, T.; Xia, H.; Hu, J.; Wang, C.; Shangguan, M.; Wang, L.; Jia, M.; Dou, X. Simultaneous wind and rainfall detection by power spectrum analysis using a VAD scanning coherent Doppler lidar. *Opt. Express* **2019**, *27*, 31235–31245. [CrossRef]
24. Tridon, F.; Battaglia, A. Dual-frequency radar Doppler spectral retrieval of rain drop size distributions and entangled dynamics variables. *J. Geophys. Res. Atmos.* **2015**, *120*, 5585–5601. [CrossRef]
25. Abdelazim, S.; Santoro, D.; Arend, M.F.; Moshary, F.; Ahmed, S. Development and operational analysis of an all-fiber coherent Doppler lidar system for wind sensing and aerosol profiling. *IEEE Trans. Geosci. Remote Sens.* **2015**, *53*, 6495–6506. [CrossRef]
26. Belmonte, A. Analyzing the efficiency of a practical heterodyne lidar in the turbulent atmosphere: Telescope parameters. *Opt. Express* **2003**, *11*, 2041–2046. [CrossRef]
27. Chouza, F.; Reitebuch, O.; Groß, S.; Rahm, S.; Freudenthaler, V.; Toledano, C.; Weinzierl, B. Retrieval of aerosol backscatter and extinction from airborne coherent Doppler wind lidar measurements. *Atmos. Meas. Tech.* **2015**, *8*, 2909–2926. [CrossRef]
28. Dai, G.; Wang, X.; Sun, K.; Wu, S.; Song, X.; Li, R.; Yin, J.; Wang, X. Calibration and retrieval of aerosol optical properties measured with Coherent Doppler Lidar. *J. Atmos. Oceanic Technol.* **2021**, *38*, 1035–1045. [CrossRef]
29. Yang, S.; Preißler, J.; Wiegner, M.; von Löwis, S.; Petersen, G.N.; Parks, M.M.; Finger, D.C. Monitoring Dust Events Using Doppler Lidar and Ceilometer in Iceland. *Atmosphere* **2020**, *11*, 1294. [CrossRef]
30. Zhang, Y.; Zheng, Y.; Tan, W.; Guo, P.; Xu, Q.; Chen, S.; Lin, R.; Chen, S.; Chen, H. Two Practical Methods to Retrieve Aerosol Optical Properties from Coherent Doppler Lidar. *Remote Sens.* **2022**, *14*, 2700. [CrossRef]
31. O'Connor, E.J.; Illingworth, A.J.; Hogan, R.J. A technique for autocalibration of cloud lidar. *J. Atmos. Oceanic Technol.* **2004**, *21*, 777–786. [CrossRef]
32. Huang, T.; Yang, Y.; O'Connor, E.J.; Lolli, S.; Haywood, J.; Osborne, M.; Cheng, J.C.; Guo, J.; Yim, S.H. Influence of a weak typhoon on the vertical distribution of air pollution in Hong Kong: A perspective from a Doppler LiDAR network. *Environ. Pollut.* **2021**, *276*, 116534. [CrossRef]
33. Pentikäinen, P.; O'Connor, E.J.; Manninen, A.J.; Ortiz-Amezcua, P. Methodology for deriving the telescope focus function and its uncertainty for a heterodyne pulsed Doppler lidar. *Atmos. Meas. Tech.* **2020**, *13*, 2849–2863. [CrossRef]
34. Newsom, R.K.; Brewer, W.A.; Wilczak, J.M.; Wolfe, D.E.; Oncley, S.P.; Lundquist, J.K. Validating precision estimates in horizontal wind measurements from a Doppler lidar. *Atmos. Meas. Tech.* **2017**, *10*, 1229–1240. [CrossRef]
35. Henderson, S.W.; Gatt, P.; Rees, D.; Huffaker, R.M. Wind lidar. In *Laser Remote Sensing*; CRC Press: Boca Raton, FL, USA, 2005; pp. 487–740.
36. Fujii, T.; Fukuchi, T. *Laser Remote Sensing*; CRC Press: Boca Raton, FL, USA, 2005.
37. Berkoff, T.A.; Welton, E.J.; Campbell, J.R.; Scott, V.; Spinhirne, J.D. Investigation of overlap correction techniques for the Micro-Pulse Lidar NETwork (MPLNET). In Proceedings of the IGARSS 2003, IEEE International Geoscience and Remote Sensing Symposium, Toulouse, France, 21–25 July 2003; pp. 4395–4397.
38. Fernald, F.G. Analysis of atmospheric lidar observations: Some comments. *Appl. Opt.* **1984**, *23*, 652–653. [CrossRef]
39. Klett, J.D. Lidar inversion with variable backscatter/extinction ratios. *Appl. Opt.* **1985**, *24*, 1638–1643. [CrossRef] [PubMed]
40. Matsumoto, M.; Takeuchi, N. Effects of misestimated far-end boundary values on two common lidar inversion solutions. *Appl. Opt.* **1994**, *33*, 6451–6456. [CrossRef] [PubMed]
41. Jinhuan, Q. Sensitivity of lidar equation solution to boundary values and determination of the values. *Adv. Atmos. Sci.* **1988**, *5*, 229–241. [CrossRef]
42. Ershov, A.D.; Balin, Y.S.; Samoilova, S.V. Inversion of the lidar data in investigations of the optical characteristics of weakly turbid atmosphere. In *Ninth Joint International Symposium on Atmospheric and Ocean Optics/Atmospheric Physics. Part II: Laser Sensing and Atmospheric Physics*; SPIE: Bellingham, WA, USA, 2003; pp. 168–171. [CrossRef]

43. Gong, W.; Zhang, J.; Mao, F.; Li, J. Measurements for profiles of aerosol extinction coeffcient, backscatter coeffcient, and lidar ratio over Wuhan in China with Raman/Mie lidar. *Chin. Opt. Lett.* **2010**, *8*, 533–536. [CrossRef]
44. Melfi, S. Remote measurements of the atmosphere using Raman scattering. *Appl. Opt.* **1972**, *11*, 1605–1610. [CrossRef]
45. Whiteman, D.; Melfi, S.; Ferrare, R. Raman lidar system for the measurement of water vapor and aerosols in the Earth's atmosphere. *Appl. Opt.* **1992**, *31*, 3068–3082. [CrossRef]
46. Schuster, G.L.; Dubovik, O.; Holben, B.N. Angstrom exponent and bimodal aerosol size distributions. *J. Geophys. Res. Atmos.* **2006**, *111*, D07207. [CrossRef]
47. O'Neill, N.; Eck, T.; Holben, B.; Smirnov, A.; Dubovik, O.; Royer, A. Bimodal size distribution influences on the variation of Angstrom derivatives in spectral and optical depth space. *J. Geophys. Res. Atmos.* **2001**, *106*, 9787–9806. [CrossRef]

remote sensing

MDPI

Article

Aerosol Retrieval Study from a Particulate Observing Scanning Polarimeter Onboard Gao-Fen 5B without Prior Surface Knowledge, Based on the Optimal Estimation Method

Yizhe Fan [1,2], Xiaobing Sun [1,3,*], Rufang Ti [1], Honglian Huang [1,3], Xiao Liu [1,3] and Haixiao Yu [1,2]

1 Key Laboratory of General Optical Calibration and Characterization Technology, Hefei Institutes of Physical Science, Chinese Academy of Sciences, Hefei 230031, China
2 Science Island Branch of Graduate School, University of Science and Technology of China, Hefei 230026, China
3 Chief Studio of Agricultural Industry in Hefei, Hefei 230031, China
* Correspondence: xbsun@aiofm.ac.cn

Abstract: To meet the demand for the aerosol detection of single-angle and multi-band polarization instrument containing short-wave infrared bands, an inversion algorithm that makes full use of multi-band intensity and polarization information is proposed based on optimal estimation theory. This method uses the polarization information in the short-wave infrared band to perform surface and atmosphere decoupling without a prior information on the surface. This obtains the initial value of the aerosol, and then it uses the scalar information to obtain the final result. Moreover, the multi-band information of the instrument is used for decoupling the surface and atmospheric information, which avoids the inversion error caused by the untimely update of the surface reflectance database and the error of spatio-temporal matching. The measured data of the Particulate Observing Scanning Polarimeter (POSP) are used to test the proposed algorithm. Firstly, to verify the effectiveness of the algorithm under different surface conditions, four regions with large geographical differences (Beijing, Hefei, Baotou, and Taiwan) are selected for aerosol optical depth (AOD) inversion, and they are compared with the aerosol robotic network (AERONET) products of the nearby stations. The validation against the AERONET products produces high correlation coefficients of 0.982, 0.986, 0.718, and 0.989, respectively, which verifies the effectiveness of the algorithm in different regions. Further, we analyzed the effectiveness of the proposed algorithm under different pollution conditions. Regions with AOD >0.7 and AOD < 0.7 are screened by using the AOD products of the Moderate-Resolution Imaging Spectroradiomete (MODIS), and the AOD of the corresponding region is inverted using POSP data. It was found to be spatially consistent with the MODIS products. The correlation coefficient and root mean square error (RMSE) in the AOD high region were 0.802 and 0.217, respectively, and 0.944 and 0.022 in the AOD low region, respectively, which verified the effectiveness of the proposed algorithm under different pollution conditions.

Keywords: short-wave infrared bands; polarization; optimal estimation retrieval; aerosol optical depth; Particulate Observing Scanning Polarimeter (POSP)

Citation: Fan, Y.; Sun, X.; Ti, R.; Huang, H.; Liu, X.; Yu, H. Aerosol Retrieval Study from a Particulate Observing Scanning Polarimeter Onboard Gao-Fen 5B without Prior Surface Knowledge, Based on the Optimal Estimation Method. *Remote Sens.* **2023**, *15*, 385. https://doi.org/10.3390/rs15020385

Academic Editors: Xingfa Gu, Jing Wei and Shuaiyi Shi

Received: 5 December 2022
Revised: 3 January 2023
Accepted: 6 January 2023
Published: 8 January 2023

1. Introduction

Atmospheric aerosol, solid or liquid particles suspended in the atmosphere, including those from natural and anthropogenic sources, are an important component of the Earth's atmosphere and they play an important role in the climate system and atmospheric environment system. In the climate system, aerosol not only affects the radiative energy balance by interacting with solar and terrestrial radiation [1–3], but they also change the characteristics and lifetimes of clouds [4–8]. In terms of the environment, aerosols are considered to be the main cause of air pollution, and they have a significant negative impact on human health [9–15]. Therefore, it is of great significance to carry out the remote sensing retrieval of aerosol and particulate matter concentration.

According to different remote sensing platforms, aerosol remote sensing can be divided into ground-based, airborne, and satellite remote sensing. Satellite remote sensing has the advantages of timeliness, periodicity, and large scale, and it is an effective method for detecting the optical properties and spatial distributions of aerosols [16–20]. The inversion of aerosol properties using satellite data has been carried out for more than 40 years. According to the difference of remote sensing data and the method of surface-atmosphere decoupling, the following methods are mainly used: dark target method using the statistical relationship between the apparent reflectance of short-wave infrared channels and the surface reflectance of red and blue channels [21–23]; deep blue algorithm for bright surface area [24,25]; in the Multi-angle Imaging SpectroRadiometer (MISR) operational aerosol retrieval algorithm, a mathematical method, empirical orthogonal functions, is used to obtain the surface contribution, which gives a general solution for the surface reflectance estimation in multi-angle aerosol remote sensing [26–28]; Inversion of aerosol optical thickness using polarization information [29–34].

Although for the aerosol inversion, lookup table (LUT)-based methods are widely applied, the method turned out to be incapable of fully exploiting the information implicit in the satellite measurements [35]. So, the optimal estimation theory is introduced into atmospheric research. The optimal estimation method was applied early to ground-based observations [36]. After that Dubovik et al. developed the generalized retrieval of an aerosol and surface properties algorithm which can be applied to a multi-angle observation satellite such as Polarization and Directionality of the Earth's Reflectance (POLDER) [37]. There are some inversion framework based on optimal estimation theory that have been established [38–41]. Nevertheless, further studies based on optimal estimation theory need to be developed to apply to different satellite measurements such as single-angle multi-band polarization measurements, including short-wave infrared band.

The Particulate Observing Scanning Polarimeter (POSP) on board the Gaofen-5B satellite obtains atmospheric information in three dimensions: spectrum, observation angle and polarization through time synchronization and spatial coverage, and it obtains atmospheric parameters over the imaging area. The unique advantage of POSP is that it has nine detection channels from 380 to 2250 nm, all of which have polarization detection capabilities. Although its observation mode is single-angle observation, its advantages of multiple detection bands, wide band range and polarization detection capability will help with the inversion of aerosol optical depth (AOD).

According to the characteristics of POSP instruments, the intensity polarization joint inversion algorithm is proposed on the basis of optimal inversion theory, which makes full use of multi-band information. Firstly, the polarization reflectance in the 2250 nm band is used as the initial value of the surface polarization reflectance, and the optimal aerosol column concentration is obtained via the optimization iterative method. Then, according to the preliminary obtained AOD, the scalar signal is subjected to atmospheric correction to obtain the initial value of surface reflectance. Scalar information is used for the optimization iteration to obtain the final AOD.

This paper is divided into four parts: firstly, the characteristics of POSP data and the theory of optimization inversion framework are introduced. Secondly, the implementation details of the intensity polarization joint inversion algorithm are introduced in detail. Then, the effectiveness of the algorithm is tested in different regions and under different pollution conditions. Finally, the relevant conclusions are given.

2. Data and Optimization Estimate Framework

2.1. POSP Data Introduction

POSP is a high-precision polarization scanner developed by Anhui Institute of Optics and Fine Mechanics, Hefei Institute of Physical Sciences, Chinese Academy of Sciences, which is an onboard satellite for earth observation and can obtain multispectral and polarized radiation information on the surface and in the atmosphere. POSP adopts an on-board calibration to ensure the measurement accuracy and accuracy of the POSP operation in orbit [42]. POSP provides the first three components (I, Q, and U) of the Stokes vector corresponding to the band, and the basic parameters are shown in Table 1.

Table 1. The basic parameters of POSP.

Parameter	Value
Central wavelength/nm	380, 410, 443, 490, 670, 865, 1380, 1610, 2250
Bandwidth/nm	20, 20, 20, 20, 20, 40, 40, 60, 80
Stokes parameters	I, Q, U
Quantized digit	14 bit
Radiation calibration error	≤5%
Polarization calibration error	≤0.5%

Since the 380 nm band is greatly influenced by atmospheric molecules, the 1380 nm band is mainly used for cirrus detection. So, seven bands except for 380 nm and 1380 nm are used for AOD inversion in this paper.

2.2. Optimization Estimate Framework

The optimal inversion framework in this paper is based on the previous information analysis of aerosol [43] plus the solution of the cost function. Let x be a state vector composed of n parameters (such as aerosol volume, refractive index, particle size distribution, etc.). The state vector will be retrieved from the observation vector y containing m measured elements, such as spectral scalar and polarization reflectivity. The relationship between y and x is shown in Formula (1):

$$y = F(x) + \in \quad (1)$$

where the forward model F describes the forward model of the physical relationship between y and x. $\in$ represents the combined error of observation uncertainty and forward model uncertainty.

For satellite remote sensing, the forward model F is a nonlinear function, and the optimized solution of Equation (1) needs to go through multiple Gaussian hypothesis processes. According to the maximum likelihood method, the state vector of the t'th iteration corresponds to the solution that minimizes the quadratic cost function as follows:

$$J(x^t) = \frac{1}{2}[y - F(x^t)]^T S_\epsilon^{-1}[y - F(x^t)] + \frac{1}{2} r_a (x^t - x_a)^T S_a^{-1}(x^t - x_a) \quad (2)$$

In Formula (2), x^t represents the state vector at the t iteration, the two terms on the right side represent the constraints on the inversion results from observation and a prior estimation, respectively. The observation error covariance S_ϵ represents the uncertainty of the model and measurements. The prior error covariance S_a represents the uncertainty of the prior estimate x_a.$r_a = n/2$, where n is the number of parameters to be inverted [44].

The process of finding the minimum value of the cost function needs to go through multiple iterations, using the LM (Levenberg-Marquardt) iterative method [45]. The state vector at the (t+1)-th iteration can be expressed as:

$$x^{t+1} = x^t - \left[(r_a + \gamma)S_a^{-1} + K_t^T S_\epsilon^{-1} K_t\right]^{-1} \left\{K_t^T S_\epsilon^{-1}\left[F(x^t) - y\right] + S_a^{-1}\left[x^t - x_a\right]\right\} \quad (3)$$

In Formula (3), γ is the LM parameter, which is initially set to 10, and is adjusted according to Rodgers' update strategy after each iteration of x. K is called the weighting function matrix or Jacobian matrix, and it consists of the partial derivative of each forward model element with respect to each state vector element, i.e., $K_{i,j} = \partial F_i / \partial x_j$, $(i = 1, \ldots m; j = 1, \ldots n)$($K_{i,j}$ denotes the element in the i-th row and j-th column of matrix K, F_i is the i-th simulated measurement, and x_j is the j-th element in the state vector x).

3. Optimal Estimation Inversion Algorithm

3.1. A Priori Information on Aerosols and Surface

In the framework of optimal inversion, a priori estimation is an important factor affecting the inversion results. Reasonable priori estimation can reduce the number of inversion iterations, improve the time efficiency of the algorithm and avoid unreasonable convergence results. A priori estimation mainly comes from the a priori knowledge of state vectors and non-state vectors, such as historical meteorological data, ground measured data, etc. The influence of prior estimation on inversion is manifested in two aspects: first, the a priori estimation of the state parameter is input into the inversion program as the initial value for iteration, and the prior term that constitutes the cost function constrains the final inversion result. Secondly, the reasonable values of non-state parameters can reduce the model error and improve the accuracy of the inversion results.

3.1.1. Aerosol Model

In many studies, the bimodal lognormal distribution function is commonly used to characterize the aerosol particle size distribution, as follows:

$$n(r) = \sum_{i=1}^{2} \frac{C_i}{r\sigma_i\sqrt{2\pi}} exp\left(-\frac{(lnr - lnr_{m,i})^2}{2\sigma_i^2}\right) \tag{4}$$

In Formula (4), C denotes the volume concentration, σ is the standard deviation, and r_m is the mean radius. The bimodal log normal distribution requires six parameters to represent.

Under the assumption of a bimodal log-normal distribution, five pairs of parameters can be used to describe the aerosol model: the particle column concentration v_0^f and v_0^c, the effective radius r_{eff}^f and r_{eff}^c, the effective variance v_{eff}^f and v_{eff}^c, the real part of the refractive index m_r^f and m_r^c, and the imaginary part of the refractive index m_i^f and m_i^c. The superscripts f and c represent fine mode particles and coarse mode particles, respectively. A priori information about the aerosol models can use the results of previous studies [46], and the aerosol particle size distribution is shown in Figure 1.

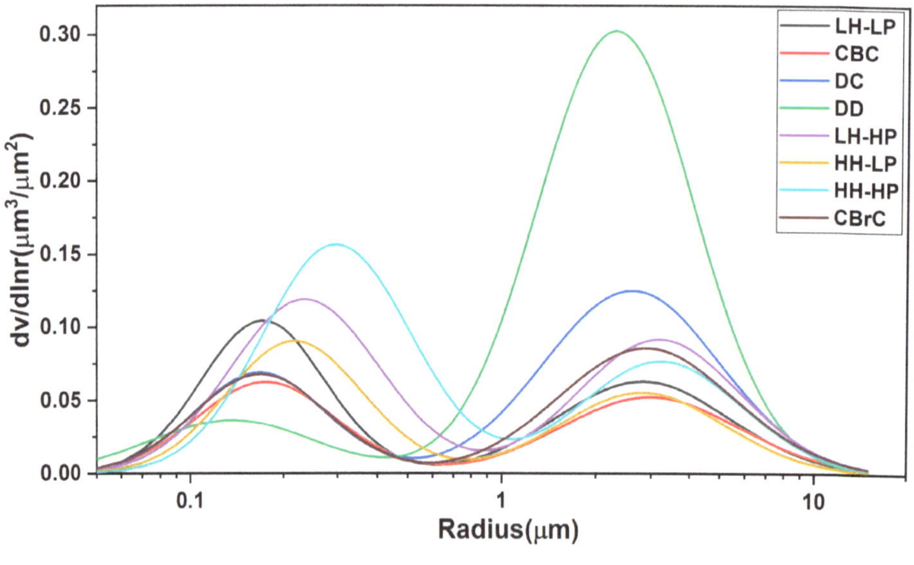

Figure 1. The size distribution of 8 aerosol models ([46]).

The legends in Figure 1 represent different types of aerosol models. LH-LP: Low Humidity-Low Pollution Aerosol, CBC: Carbonaceous-Black Carbon Aerosol, DC: Dust Urban Aerosol, DD: Sand Dust-Desert Aerosol, LH-HP: Low Humidity-High Pollution type aerosol, HH-LP: High Humidity-Low Pollution type aerosol, HH-Hp: High Humidity-High Pollution type aerosol, CBrC: Carbonaceous-Brown Carbon type aerosol.

3.1.2. Processing of A Priori Information about Surface

At present, the methods for obtaining a priori surface information mainly include the surface model method [29], the prior surface reflectivity database [24], etc. In this paper, according to the polarization reflectancenot being sensitive to band changes [47,48], the polarization reflectance of the shortwave infrared band is used as the initial value of the surface polarization reflections of other bands, and then the aerosol parameters to be retrieved are obtained according to an iterative strategy. According to the aerosol parameters obtained via polarization inversion, the atmospheric transmittance, hemispheric albedo and atmospheric path radiation are obtained using Unified Linearized Vector Radiative Transfer Model (UNL-VRTM) [47], and then the empirical value of the surface reflectance is obtained using Formula (5):

$$\rho_s = \frac{\rho - \rho_0}{(\rho - \rho_0) * S + T} \quad (5)$$

In Formula (5), ρ, ρ_0, and ρ_s are the normalized reflectances of the atmospheric top, atmospheric path radiation, and surface, respectively. S and T are the atmospheric hemisphere albedo and the total atmospheric transmittance, respectively.

3.2. Intensity Polarization Joint Optimization Inversion Algorithm

3.2.1. Satellite Observation Model

The radiation signals obtained by satellites can be characterized by the apparent reflectance. There are multiple interaction processes between the surface and the atmosphere, which are complicated by different surface reflectances and different atmospheric conditions. Therefore, the apparent reflectance model is as follows [49]:

$$R^{TOA}(\mu_s, \mu_v, \varphi) = R^{Atm}(\mu_s, \mu_v, \varphi) + \frac{T_s T_v \rho(\mu_s, \mu_v, \varphi)}{1 - s\rho(\mu_s, \mu_v, \varphi)} R^{TOA}(\mu_s, \mu_v, \varphi) = \frac{\pi I(\mu_s, \mu_v, \varphi)}{\mu_s E_0} \quad (6)$$

In Formula (6), R^{TOA} and R^{Atm} are the top-of-atmosphere (TOA) reflectance and the atmospheric path radiation, T_s and T_v are atmospheric downward and upward transmittances, respectively, ρ is the surface albedo and s is the atmospheric hemispheric albedo. μ_s, μ_v, and φ are the cosine of the solar zenith angle, the cosine of the observed zenith angle, and the relative azimuth angle. E_0 is the solar constant (the solar spectral irradiance outside the atmosphere). The surface albedo can be obtained if the apparent reflectance, atmospheric path radiation, atmospheric transmittance, and atmospheric hemispheric albedo are known.

The polarization signal obtained by the satellite is characterized by the TOA polarized reflectance. The polarization contribution caused by the multiple interactions between the surface and the atmosphere is very small. After the attenuation of the atmosphere, the contribution to the observation at the top of the atmosphere can be ignored. Therefore, the TOA polarized reflectance modeling only needs to consider the upward polarization scattering of the atmosphere and the single reflection of the surface [50], and the TOA polarized reflectance can be written as follows:

$$R_P^{TOA}(\mu_s, \mu_v, \varphi) = R_P^{Atm}(\mu_s, \mu_v, \varphi) + T_s \rho_P(\mu_s, \mu_v, \varphi) T_v R_P^{TOA}(\mu_s, \mu_v, \varphi) = \frac{\pi I_P(\mu_s, \mu_v, \varphi)}{\mu_s E_0} \quad (7)$$

In Formula (7), R_P^{TOA} is the TOA polarized reflectance, R_P^{Atm} is the polarized reflectance of the atmosphere (which can be calculated by setting the surface polarized reflectance to zero), and ρ_P is the surface polarized reflectance. $I_p = \sqrt{Q^2 + U^2}$ is the polarized radiance at the top of the atmosphere.

In summary, when using the intensity information and polarization information for the optimization solution, due to different observation models and different assumptions of the surface, the state vector x and observation vector y are shown in Table 2:

Table 2. State vector x and observation vector y settings.

Category	Polarization Setting	Intensity Setting
Observation vector	$y = \left[I_{p_{\lambda_1}}, \ldots, I_{p_{\lambda_6}}\right]^T$	$y = \left[I_{\lambda_1}, \ldots, I_{\lambda_7}\right]^T$
State vector	$x = \left[V_f, V_c, \rho_p\right]^T$	$x = \left[V_f, V_c, \rho_{\lambda_1}, \ldots, \rho_{\lambda_7}\right]^T$

In Table 2, the superscript T represents the transpose of the vector, and $\lambda_1 - \lambda_7$ are the corresponding center wavelengths of POSP: 410 nm, 443 nm, 490 nm, 670 nm, 865 nm, 1610 nm and 2250 nm, respectively. V_f and V_c are the aerosol fine-mode and coarse-mode column concentrations, respectively, and ρ represents the surface reflectance. Since it is assumed that the surface polarized reflectance is independent of the wavelength, and the TOA polarized reflectance in the 2250 nm band is used as the initial value of the surface polarized reflectance, in the polarization inversion, the state vector x has only three parameters, and the observation vector has 6 parameters. In the intensity inversion, the state vector has nine parameters and the observation vector has seven parameters.

3.2.2. Cost Function Solution Method

The optimal estimation theory provides the cost function J(x) for finding the optimal inversion result, and gives the specific form. J(x) is a nonlinear function, and Newton's iterative method is a common method for solving nonlinear functions. In this paper, the LM algorithm is used to solve the minimum value of J(x), and in the LM program module, the forward radiated transmission simulation module and the calculation module of the cost function are combined to establish the program framework for optimal iterative inversion. Under the framework of the program, the optimal estimation of the state vector is found after several iterations. The convergence condition is as follows:

$$\Delta = \frac{||x^{t+1} - x^t||_2}{||x^t||_2} \leq \varepsilon \quad (8)$$

In Formula (8), where || ||2 is the L2 norm, ε is the convergence threshold for judging the end of the iteration, which is an empirical value, and is taken as 0.1% in this paper.

In the actual operation process, when the number of iterations exceeds the set maximum number of iterations of six, or when the element value in x exceeds the boundary range, the iteration stops and an invalid value is output.

3.2.3. Algorithm Implementation

The LM program module, UNL-VRTM, and the cost function calculation module are combined to make full use of the multi-band polarization information of POSP to carry out the optimal inversion of AOD. The detailed flow chart is shown in Figure 2. After reading the POSP data, the proposed algorithm can be divided into the following two stages.

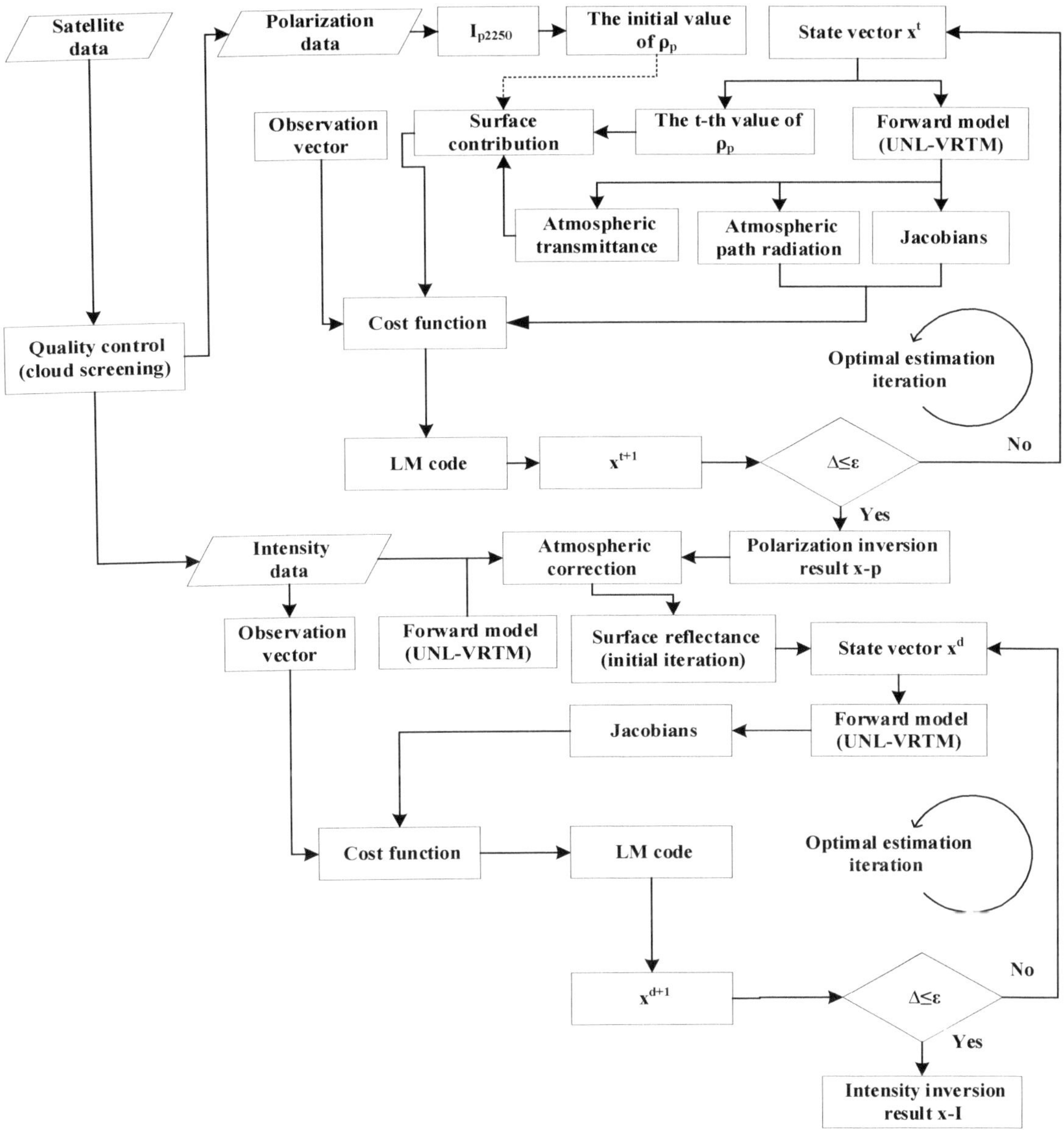

Figure 2. The flowchart of inversion algorithm.

In the first stage, the optimal inversion is performed using the polarization information to obtain the aerosol column concentration. There are three main steps at this stage: (1) Obtain the initial value of the surface polarization reflectivity of other bands through the apparent polarization reflectivity of 2250 nm; (2) Use the radiation transfer software UNL-VRTM to calculate the Jacobian matrix of the state vector; (3) Obtain a new state vector according to the update strategy of Formula (3), and then after several iterations, when the iterative convergence threshold is met, the aerosol column concentration parameter is output.

In the second stage, the final aerosol column concentration parameters are obtained by optimizing the inversion using the intensity information. First, input the aerosol column concentration parameters obtained in the previous stage into UNL-VRTM and run it, and then perform simple atmospheric correction according to Equation (5) to obtain the initial values of surface reflectance in different bands. Then, according to (2)–(3) of the first stage, the column concentration parameters of the aerosol are finally obtained.

The aerosol volume column concentration parameters are obtained via inversion according to the above process, and the spectral AOD can be further calculated by combining with the aerosol model parameters, but this calculation process does not need to be carried out separately. In the iterative process, when using the UNL-VRTM software for forward simulation, AOD can be obtained at the same time, and the calculation result can be output when the last step satisfies the iterative convergence condition.

4. Result and Discussion

4.1. Algorithm Iteration Process

According to the prior information acquisition and retrieval strategy described above, the POSP data from November 2021 to July 2022 are selected for the retrieval of AOD. An inversion case of 8 June 2022, over the Beijing site is selected to illustrate the iteration and convergence processes of the algorithm, as shown in Figure 3.

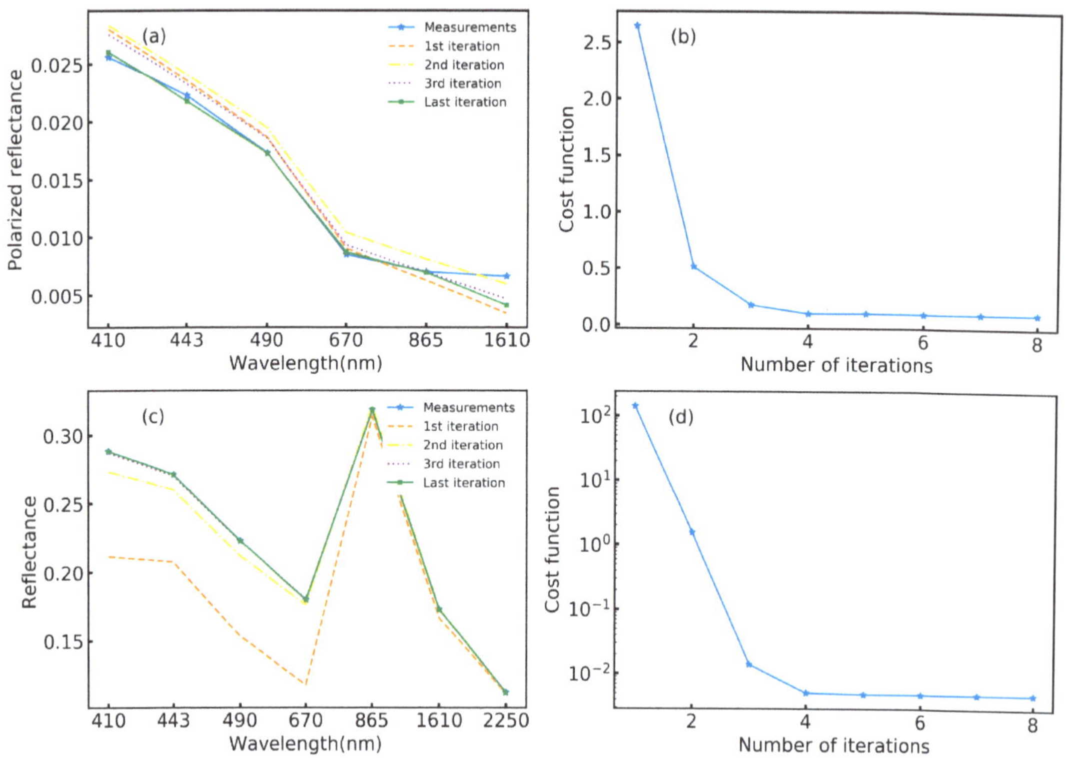

Figure 3. (**a**,**c**) Illustrations of the iterative process for TOA reflectance and TOA polarized reflectance, respectively. (**b**,**d**) The plot of the cost function as a function of iterations in polarization inversion and intensity inversion, respectively (case for 8 June 2022).

Figure 3a,c shows the fitting between the simulated values of the model and the measured values of POSP with the increase in the number of iterations in the process of inversion using polarization information and intensity information, respectively. After the last iteration, the simulated values using polarization information agree well with the observed values in the first five bands, and there is only a slight gap in the 1610 nm band. The simulated values using intensity information are in good agreement with the observed values in all bands, indicating that in single-angle observation, the intensity contains more information about the atmosphere than the polarization measurement. In Figure 3b,d of the cost function varies with the number of iterations in the process of polarization inversion and intensity inversion, respectively. It is also reflected that the intensity contains more information than polarization from (b) and (d). When convergence, the cost function value of intensity information inversion is less than that of polarization information inversion. After five iterations, the cost function almost does not change with the increase in the number of iterations, so that in order to save the time of the algorithm, the maximum number of iterations can be set to 6.

4.2. Evaluation Index of the Inversion Result

In the evaluation of the inversion results, quantitative indexes such as correlation coefficient (R), root mean square error (RMSE), expected error (EE), and the percentage of except error (Scope) are used. Taking AOD as an example, the specific definitions of each index are as follows:

$$R = \frac{\mathrm{cov}(\tau_{retrieval}, \tau_{true})}{\mathrm{std}(\tau_{retrieval}) \times \mathrm{std}(\tau_{true})} \tag{9}$$

$$RMSE = \sqrt{\frac{1}{N}\sum_{i=1}^{N}(\tau_{retrieval} - \tau_{true})^2} \tag{10}$$

$$\Delta = EE = \pm 0.15\tau_{true} \pm 0.05 \tag{11}$$

$$Scope = N_{valid}/N \times 100\% \tag{12}$$

In Formulas (9)–(12), $\tau_{retrieval}$ is the AOD result of inversion using POSP data, and τ_{true} is the data for verification. Cov() and std() represent the covariance and standard deviation calculations, respectively. Scope is the proportion of the amount of data within the range of $\tau_{true} + \Delta$ for the inversion AOD to the total data. N and N_{valid} are the amount of data and the total amount of data that fall into the error range, respectively.

4.3. Validation against Ground-Based Data

The AERONET level 2.0 category is used for ground-based validation in this paper. AERONET is a ground-based network of sun photometers that provides high-quality aerosol measurements including AOD products [36,51].

The spectral AOD are retrieved using the proposed algorithm. With regard to validation, the AOD at 670 nm is selected. Correspondingly, the AOD at 675 nm in the AERONET products is used.

In order to verify the effectiveness of the inversion algorithms in different regions, four regions (Beijing, Hefei, Baotou, and Taiwan) with considerable geographical differences were selected. Additionally these four areas have AERONET ground-based long-term observation stations, namely: Beijing area (Beijing station), Baotou area (AOE_Baotou station), Hefei area (Hefei station) and Taiwan area (Chen-Kung_Univ station), the corresponding latitude and longitude are shown in Table 3.

Table 3. AERONET data for validation.

AERONET Sites	Longitude	Latitude	Date Range
Beijing	116.3814	39.9769	2021.11–2022.7
AOE_Baotou	109.6288	40.8517	2021.11–2022.7
Hefei	117.1622	31.9047	2021.11–2022.7
Chen-Kung_Univ	120.2047	22.9934	2021.11–2022.7

The location and observation time of the AERONET sites are matched to extract the observation data corresponding to POSP from November 2021 to July 2022, as shown in Figure 4.

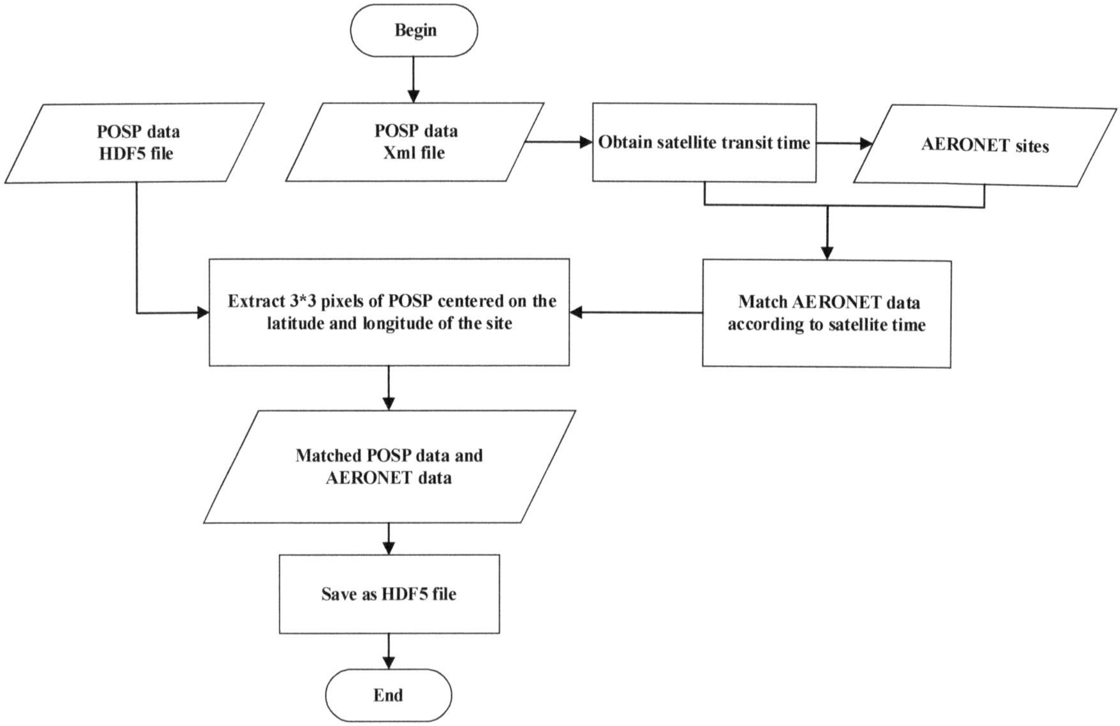

Figure 4. The flowchart of matching POSP observation data and AERONET data.

The validation against the corresponding AEROENT product is shown in Figure 5a–d are the verification results in Beijing, Baotou, Hefei and Taiwan, respectively.

It can be found from Figure 5 that the inversion results based on the algorithm proposed in this paper are basically consistent with AOD products of the AERONET sites, and the results are good. However, due to the influences of different surface types and aerosol types, the Scope is different. The order from largest to smallest is Hefei, Taiwan, Beijing, and Baotou. The reason for this may be that the surface vegetation in the Baotou area is sparse and is often considered to be a bright surface, while the Hefei site is mostly covered by evergreen vegetation, and the surface reflectance value is relatively small, so that the impact of the surface on the apparent reflectance is relatively smaller than the bright surface.

In addition, the aerosols in Baotou are mainly composed of coarse particles, so when polarization inversion is performed, the AOD obtained will have a large error, and then a large error will be introduced when the initial value of surface reflectance is provided based on the results of polarization inversion, resulting in relatively poor final inversion results.

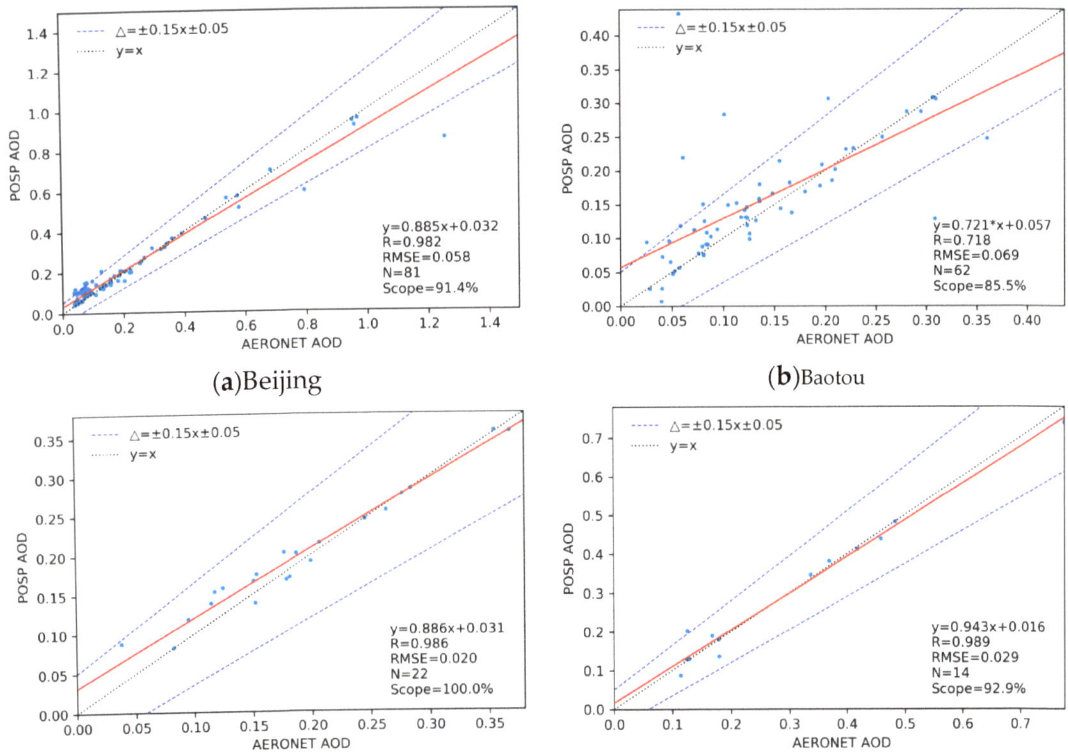

(**a**)Beijing (**b**)Baotou

(**c**)Hefei (**d**)Taiwan

Figure 5. Validation of the AOD at 670 nm with the proposed method against AOD at 675 nm in AERONET products at Beijing, Baotou, Hefei and Taiwan sites.

4.4. The Validation against MODIS Products

In order to verify the effectiveness of the intensity polarization joint inversion algorithm under different pollution conditions, MODIS's AOD products at 550 nm are first used to screen the time and longitude ranges of AOD high value (AOD > 0.7) and low value (AOD < 0.7), and then filter out the corresponding POSP observation data. The specific latitude and longitude ranges and times are shown in Table 4.

Table 4. Data information used for validation.

AOD Range	Longitude Range	Latitude Range	Date
AOD>0.7	114.6–115.9	37.5–38.5	2022.6.9
AOD<0.7	116.6–118.2	31.57–33.46	2022.5.4

The comparison between the inversion results in the high-value area and the MODIS product is shown in Figure 6. The blank area in the figure indicates no inversion results, because they are cloud pixels or no data.

Comparing the (a) and (b) figures in Figure 6, it can be found that the inversion results of POSP are consistent with the MODIS product as a whole, but that the inversion value of POSP in some areas is smaller than the AOD value in MODIS (prismatic area in the figure), and that the AOD value of POSP is greater than that of MODIS in some areas (circular area in the figure). In order to further analyze the difference between the AOD inversion results of POSP and the MODIS products, the inverted AOD and MODIS products were statistically analyzed and linear regression analysis was performed, and the results are shown in Figure 7.

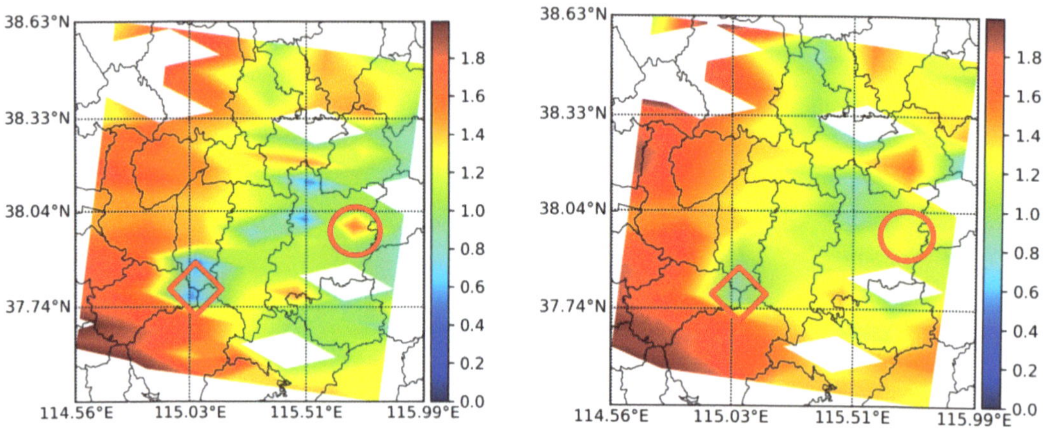

(**a**) The retrieval of AOD at 550 nm (**b**) AOD at 550 nm in MODIS products

Figure 6. POSP AOD inversion results (**a**) vs. MODIS products (**b**) (9 June 2022).

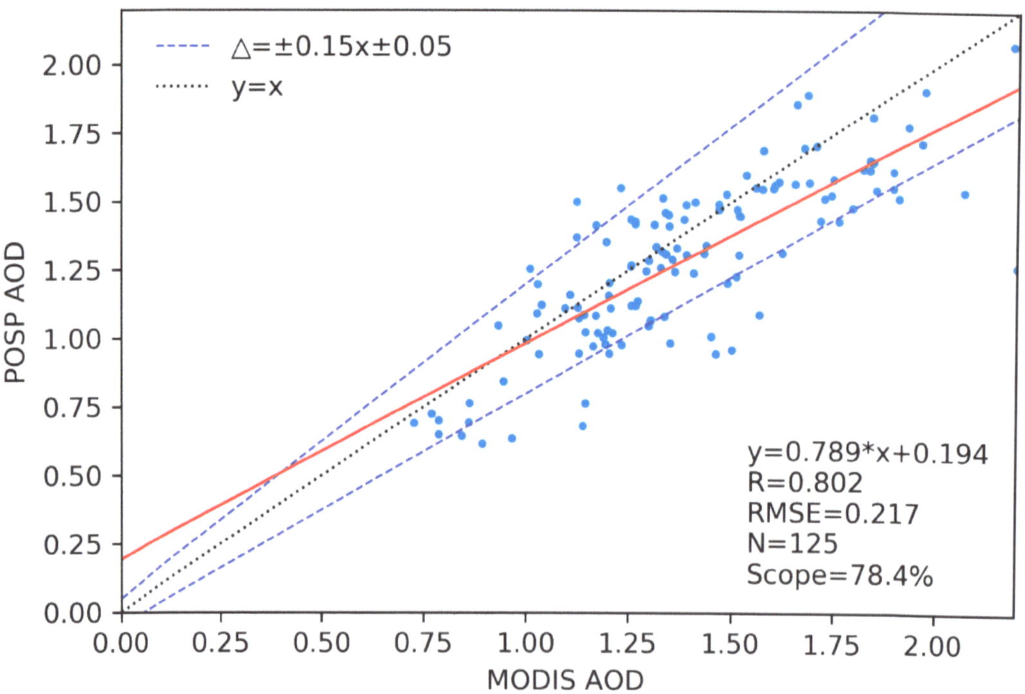

Figure 7. The scatter plot of POSP AOD inversion results against MODIS products (9 June 2022).

According to the statistical results of Figure 7, a total of 125 high-value AOD data are obtained, of which 78.4% of the data fall within the error range, and the correlation coefficient R is 0.802.

Figure 8 shows the comparison of the AOD inversion results of POSP in the low AOD value area with the MODIS product. Comparing (a) and (b), it can be found that the POSP inversion results are consistent with the MODIS product as a whole. In some areas, the inversion result of POSP is smaller than MODIS (the prismatic area in the figure). In order to further analyze the effectiveness of the inversion algorithm, the inversion results and MODIS products were statistically analyzed, and the results are shown in Figure 9.

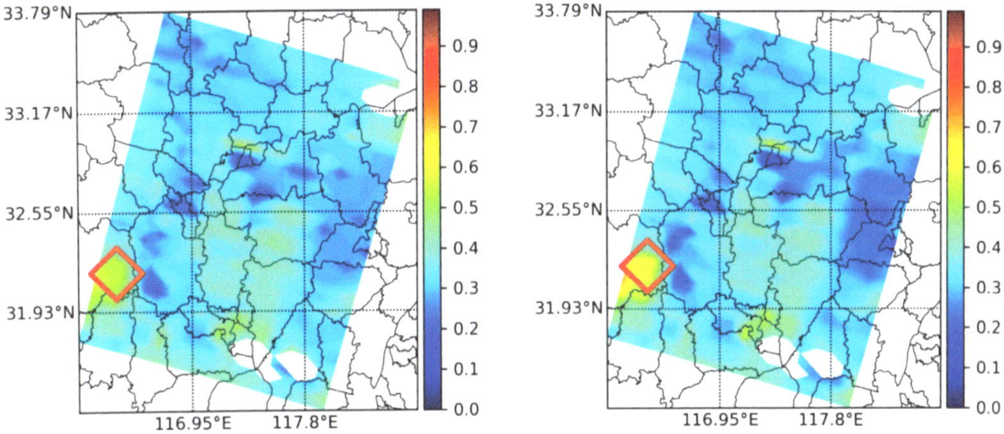

Figure 8. POSP AOD inversion results (**a**) vs. MODIS products (**b**) (4 May 2022).

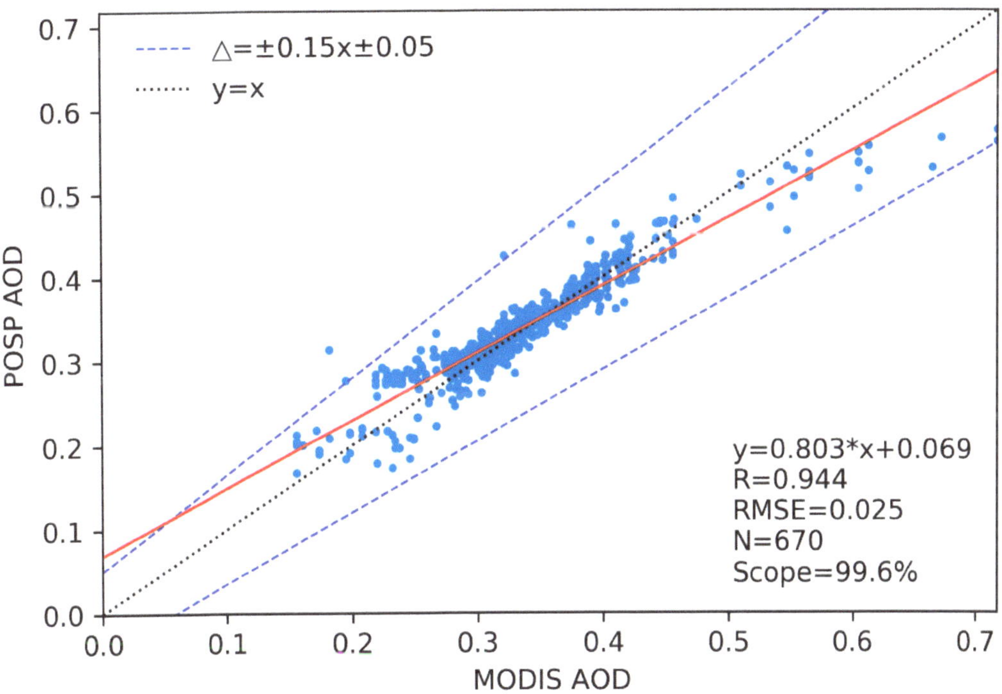

Figure 9. The scatter plot of POSP AOD inversion results against MODIS products (4 May 2022).

As can be seen from Figure 9, a total of 670 data points are obtained, 99.6% of the data are within the error range, and the correlation coefficient R is 0.944, indicating that the inversion results are in good agreement with the MODIS.

In summary, comparing the inversion results of high-value AOD and low-value AOD, it can be found that the inversion results of low-value AOD are significantly better than those in high-value AOD regions. This may be due to the assumption that the 2250 nm band is not sensitive to the atmosphere when performing polarization inversion, so that the TOA polarization reflectance of 2250 nm is used as the surface polarization reflectance of other bands. When the AOD is low, this assumption may have little effect on the inversion results, but in the high AOD value region, due to the increased probability of collision between light and particles, the 2250 nm band will carry more aerosol information, which will have a certain impact on the inversion results.

5. Conclusions

In this paper, based on the observation characteristics of POSP single-angle multi-band polarization measurements, an intensity polarization joint optimization inversion algorithm based on optimization theory is proposed, which can effectively perform ground-air decoupling without prior knowledge of the surface. Additionally, the effectiveness of the algorithm is verified in different regions and under different pollution conditions. It is found that in the vegetation coverage area (Hefei) and the urban underlying surface (Beijing and Taiwan), the correlation coefficient between the inversion AOD at 670 nm from POSP and the ground-based AOD at 675 nm from AERONET can reach more than 0.98. Good results can also be obtained on the bright surface (Baotou), the correlation coefficient can reach 0.71, and the proportion of data within the error range can reach 85.5%. In addition, the inversion results of the high-value area and the low-value area of AOD are also analyzed. It is found that the AOD at a 550 nm correlation coefficient between POSP and MODIS in the high-value area is 0.802, and in the low-value area, it is 0.994, indicating that the performance of the algorithm in the low-value area is better than that of the high-value area.

This algorithm uses the insensitivity of the short-wave infrared band to the atmosphere to obtain the initial value of the surface polarized reflectance. Therefore, in the high-value area of aerosol or under the condition coarse mode dominated aerosol, it will have a certain impact on the inversion results.

Author Contributions: Formal analysis, Y.F.; Funding acquisition, X.S.; Methodology, Y.F.; Project administration, X.S.; Supervision, X.S.; Validation, Y.F.; Writing—original draft, Y.F.; Writing—review and editing, X.S., R.T., H.H., X.L. and H.Y. All authors have read and agreed to the published version of the manuscript.

Funding: This work was supported by the Aerospace Science and Technology Innovation Application Research Project (E23Y0H555S1), the China High-resolution Earth Observation System (CHEOS) (30-Y20A010-9007-17/18), and the China Center for Resource Satellite Data and Applications Project (E13Y0J31601).

Data Availability Statement: Not applicable.

Acknowledgments: We are very grateful to Jun Wang and Xiaoguang Xu for the UNL-VRTM radiative transfer code. The data and FORTRAN codes used for the forward simulations are available on the UNL-VRTM website. We would like to thank all the relevant principal investigators and their staff for establishing and maintaining the AERONET stations used in this study. We are extremely grateful to the researchers of MODIS data products.

Conflicts of Interest: The authors declare no conflict of interest.

References

1. Boucher, O.; Anderson, T.L. General circulation model assessment of the sensitivity of direct climate forcing by anthropogenic sulfate aerosols to aerosol size and chemistry. *J. Geophys. Res. Atmos.* **1995**, *100*, 26117–26134. [CrossRef]
2. Haywood, J.; Boucher, O. Estimates of the direct and indirect radiative forcing due to tropospheric aerosols: A review. *Rev. Geophys.* **2000**, *38*, 513–543. [CrossRef]
3. Satheesh, S.; Moorthy, K.K. Radiative effects of natural aerosols: A review. *Atmospheric Environ.* **2005**, *39*, 2089–2110. [CrossRef]
4. Pincus, R.; Baker, M.B. Effect of precipitation on the albedo susceptibility of clouds in the marine boundary layer. *Nature* **1994**, *372*, 250–252. [CrossRef]
5. Tao, W.-K.; Chen, J.-P.; Li, Z.; Wang, C.; Zhang, C. Impact of aerosols on convective clouds and precipitation. *Rev. Geophys.* **2012**, *50*. [CrossRef]
6. Stevens, B.; Feingold, G. Untangling aerosol effects on clouds and precipitation in a buffered system. *Nature* **2009**, *461*, 607–613. [CrossRef] [PubMed]
7. Christensen, M.W.; Jones, W.K.; Stier, P. Aerosols enhance cloud lifetime and brightness along the stratus-to-cumulus transition. *Proc. Natl. Acad. Sci. USA* **2020**, *117*, 17591–17598. [CrossRef]
8. Fan, J.; Wang, Y.; Rosenfeld, D.; Liu, X. Review of aerosol–cloud interactions: Mechanisms, significance, and challenges. *J. Atmos Sci.* **2016**, *73*, 4221–4252. [CrossRef]
9. Seaton, A.; Godden, D.; MacNee, W.; Donaldson, K. Particulate air pollution and acute health effects. *Lancet* **1995**, *345*, 176–178. [CrossRef]
10. Kaufman, Y.J.; Tanré, D.; Boucher, O. A Satellite View of Aerosols in the Climate System. *Nature* **2002**, *419*, 215–223. [CrossRef]
11. Li, Z.; Niu, F.; Fan, J.; Liu, Y.; Rosenfeld, D.; Ding, Y. Long-term impacts of aerosols on the vertical development of clouds and precipitation. *Nat. Geosci.* **2011**, *4*, 888–894. (In English) [CrossRef]
12. Ramanathan, V.; Crutzen, P.J.; Lelieveld, J.; Mitra, A.P.; Althausen, D.; Anderson, L.M.; Andreae, M.; Cantrell, W.; Cass, G.R.; Chung, E.; et al. Indian Ocean Experiment: An integrated analysis of the climate forcing and effects of the great Indo-Asian haze. *J. Geophys. Res. Atmos.* **2001**, *106*, 28371–28398. (In English) [CrossRef]
13. Hoek, G.; Krishnan, R.M.; Beelen, R.; Peters, A.; Ostro, B.; Brunekreef, B.; Kaufman, J.D. Long-term air pollution exposure and cardio- respiratory mortality: A review. *Environ. Health* **2013**, *12*, 43. [CrossRef]
14. Zhang, Y.; Li, Z. Remote sensing of atmospheric fine particulate matter (PM2.5) mass concentration near the ground from satellite observation. *Remote Sens. Environ.* **2015**, *160*, 252–262. (In English) [CrossRef]
15. Song, C.; He, J.; Wu, L.; Jin, T.; Chen, X.; Li, R.; Ren, P.; Zhang, L.; Mao, H. Health burden attributable to ambient PM2.5 in China. *Environ. Pollut.* **2017**, *223*, 575–586. [CrossRef]
16. Hauser, A.; Oesch, D.; Foppa, N.; Wunderle, S. NOAA AVHRR derived aerosol optical depth over land. *J. Geophys. Res. Atmos.* **2005**, *110*. [CrossRef]
17. Li, C.; Lau, A.K.H.; Mao, J.; Chu, D.A. Retrieval, validation, and application of the 1-km aerosol optical depth from MODIS measurements over Hong Kong. IEEE Trans. *Geosci. Remote Sens.* **2005**, *43*, 2650–2658. [CrossRef]
18. Li, L.; Yang, J.; Wang, Y. An improved dark object method to retrieve 500m-resolution AOT (Aerosol Optical Thickness) image from MODIS data: A case study in the Pearl River Delta area, China. *ISPRS J. Photogramm. Remote Sens.* **2014**, *89*, 1–12. [CrossRef]
19. Mei, L.L.; Xue, Y.; Kokhanovsky, A.A.; von Hoyningen-Huene, W.; de Leeuw, G.; Burrows, J.P. Retrieval of aerosol optical depth over land surfaces from AVHRR data. *Atmospheric Meas. Technol.* **2014**, *7*, 2411–2420. [CrossRef]
20. Wei, J.; Huang, B.; Sun, L.; Zhang, Z.; Wang, L.; Bilal, M. A Simple and Universal Aerosol Retrieval Algorithm for Landsat Series Images over Complex Surfaces. *J. Geophys. Res. Atmos.* **2017**, *122*, 13338–13355. (In English) [CrossRef]
21. Kaufman, Y.; Wald, A.; Remer, L.; Gao, B.-C.; Li, R.-R.; Flynn, L. The MODIS 2.1-μm channel-correlation with visible reflectance for use in remote sensing of aerosol. *IEEE Trans. Geosci. Remote Sens.* **1997**, *35*, 1286–1298. [CrossRef]
22. Levy, R.C.; Remer, L.A.; Mattoo, S.; Vermote, E.F.; Kaufman, Y.J. Second-generation operational algorithm: Retrieval of aerosol properties over land from inversion of Moderate Resolution Imaging Spectroradiometer spectral reflectance. *J. Geophys. Res. Atmos.* **2007**, *112*, D13211. [CrossRef]
23. Levy, R.C.; Remer, L.A.; Kleidman, R.G.; Mattoo, S.; Ichoku, C.; Kahn, R.; Eck, T.F. Global evaluation of the Collection 5 MODIS dark-target aerosol products over land. *Atmos. Chem. Phys.* **2010**, *10*, 10399–10420. (In English) [CrossRef]
24. Hsu, N.C.; Tsay, S.-C.; King, M.D.; Herman, J.R. Aerosol Properties Over Bright-Reflecting Source Regions. *IEEE Trans. Geosci. Remote Sens.* **2004**, *42*, 557–569. (In English) [CrossRef]
25. Hsu, N.C.; Tsay, S.-C.; King, M.D.; Herman, J.R. Deep Blue Retrievals of Asian Aerosol Properties During ACE-Asia. *IEEE Trans. Geosci. Remote Sens.* **2006**, *44*, 3180–3195. (In English) [CrossRef]
26. Martonchik, J.V. Determination of aerosol optical depth and land surface directional reflectances using multiangle imagery. *J. Geophys. Res. Atmos.* **1997**, *102*, 17015–17022. [CrossRef]
27. Martonchik, J.; Diner, D.; Kahn, R.; Ackerman, T.; Verstraete, M.; Pinty, B.; Gordon, H. Techniques for the retrieval of aerosol properties over land and ocean using multiangle imaging. *IEEE Trans. Geosci. Remote Sens.* **1998**, *36*, 1212–1227. [CrossRef]
28. Diner, D.J.; Martonchik, J.V.; Kahn, R.A.; Pinty, B.; Gobron, N.; Nelson, D.L.; Holben, B.N. Using angular and spectral shape similarity constraints to improve MISR aerosol and surface retrievals over land. *Remote Sens. Environ.* **2005**, *94*, 155–171. [CrossRef]

29. Deuzé, J.L.; Bréon, F.M.; Devaux, C.; Goloub, P.; Herman, M.; Lafrance, B.; Maignan, F.; Marchand, A.; Nadal, F.; Perry, G.; et al. Remote sensing of aerosols over land surfaces from POLDER-ADEOS-1 polarized measurements. *J. Geophys. Res. Atmos.* **2001**, *106*, 4913–4926. [CrossRef]
30. Herman, M.; Deuzé, J.; Marchand, A.; Roger, B.; Lallart, P. Aerosol remote sensing from POLDER/ADEOS over the ocean: Improved retrieval using a nonspherical particle model. *J. Geophys. Res. Atmos.* **2005**, *110*. [CrossRef]
31. Tanré, D.; Bréon, F.M.; Deuzé, J.L.; Dubovik, O.; Ducos, F.; François, P.; Goloub, P.; Herman, M.; Lifermann, A.; Waquet, F. Remote sensing of aerosols by using polarized, directional and spectral measurements within the A-Train: The PARASOL mission. *Atmospheric Meas. Technol.* **2011**, *4*, 1383–1395. [CrossRef]
32. Bréon, F.; Vermeulen, A.; Descloitres, J. An evaluation of satellite aerosol products against sunphotometer measurements. *Remote Sens. Environ.* **2011**, *115*, 3102–3111. [CrossRef]
33. Wang, H.; Sun, X.; Sun, B.; Liang, T.; Li, C.; Hong, J. Retrieval of aerosol optical properties over a vegetation surface using multi-angular, multi-spectral, and polarized data. *Adv. Atmospheric Sci.* **2014**, *31*, 879–887. (In English) [CrossRef]
34. Fan, X.; Goloub, P.; Deuzé, J.-L.; Chen, H.; Zhang, W.; Tanré, D.; Li, Z. Evaluation of PARASOL aerosol retrieval over North East Asia. *Remote Sens. Environ.* **2008**, *112*, 697–707. [CrossRef]
35. Dubovik, O.; Li, Z.; Mishchenko, M.I.; Tanré, D.; Karol, Y.; Bojkov, B.; Cairns, B.; Diner, D.J.; Espinosa, W.R.; Goloub, P.; et al. Polarimetric remote sensing of atmospheric aerosols: Instruments, methodologies, results, and perspectives. *J. Quant. Spectrosc. Radiat. Transf.* **2019**, *224*, 474–511. [CrossRef]
36. Dubovik, O.; King, M.D. A flexible inversion algorithm for retrieval of aerosol optical properties from Sun and sky radiance measurements. *J. Geophys. Res. Atmos.* **2000**, *105*, 20673–20696. [CrossRef]
37. Dubovik, O.; Lapyonok, T.; Litvinov, P.; Herman, M.; Fuertes, D.; Ducos, F.; Torres, B.; Derimian, Y.; Huang, X.; Lopatin, A.; et al. GRASP: A versatile algorithm for characterizing the atmosphere. *SPIE Newsroom* **2014**, *25*, 2–1201408. [CrossRef]
38. Chen, X.; Yang, D.; Cai, Z.; Liu, Y.; Spurr, R.J.D. Aerosol Retrieval Sensitivity and Error Analysis for the Cloud and Aerosol Polarimetric Imager on Board TanSat: The Effect of Multi-Angle Measurement. *Remote Sens.* **2017**, *9*, 183. [CrossRef]
39. Hou, W.; Wang, J.; Xu, X.; Reid, J.S. An algorithm for hyperspectral remote sensing of aerosols: 2. Information content analysis for aerosol parameters and principal components of surface spectra. *J. Quant. Spectrosc. Radiat. Transf.* **2017**, *192*, 14–29. [CrossRef]
40. Hou, W.Z.; Li, Z.Q.; Zheng, F.X.; Qie, L.L. Retrieval of aerosol microphysical properties based on the optimal estimation method: Information content analysis for satellite polarimetric remote sensing measurements. In Proceedings of the ISPRS—International Archives of the Photogrammetry, Remote Sensing and Spatial Information Sciences, Beijing, China, 7–10 May 2018; pp. 533–537.
41. Zheng, F.; Li, Z.; Hou, W.; Qie, L.; Zhang, C. Aerosol retrieval study from multiangle polarimetric satellite data based on optimal estimation method. *J. Appl. Remote Sens.* **2020**, *14*, 014516. [CrossRef]
42. Yang, H.; Hong, J.; Zou, P.; Song, M.; Yang, B.; Liu, Z. Onboard Polarization Calibrators of Spaceborne Particulate Observing Scanning Polarimeter. *Acta Optics Sinica.* **2019**, *39*, 0912005. [CrossRef]
43. Fan, Y.; Sun, X.; Ti, R.; Huang, H.; Liu, X. Information analysis of aerosol and surface parameters in PSAC observation over land. *J. Infrared Millim. Waves* **2022**, *41*, 15.
44. Xu, X.; Wang, J. Retrieval of aerosol microphysical properties from AERONET photopolarimetric measurements: 1. Information content analysis. *J. Geophys. Res. Atmos.* **2015**, *120*, 7059–7078. [CrossRef]
45. Rodgers, C.D. *Inverse Methods for Atmospheric Sounding—Theory and Practice, Series on Atmospheric, Oceanic and Planetary Physics—Volume 2*; World Scientific Publishing Co.: Singapore, 2000. [CrossRef]
46. Fan, Y.; Sun, X.; Huang, H.; Ti, R.; Liu, X. The primary aerosol models and distribution characteristics over China based on the AERONET data. *J. Quant. Spectrosc. Radiat. Transf.* **2021**, *275*, 107888. [CrossRef]
47. Wang, J.; Xu, X.; Ding, S.; Zeng, J.; Spurr, R.; Liu, X.; Chance, K.; Mishchenko, M. A numerical testbed for remote sensing of aerosols, and its demonstration for evaluating retrieval synergy from a geostationary satellite constellation of GEO-CAPE and GOES-R. *J. Quant. Spectrosc. Radiat. Transf.* **2014**, *146*, 510–528. (In English) [CrossRef]
48. Waquet, F.; Léon, J.-F.; Cairns, B.; Goloub, P.; Deuzé, J.L.; Auriol, F. Analysis of the spectral and angular response of the vegetated surface polarization for the purpose of aerosol remote sensing over land. *Appl. Opt.* **2009**, *48*, 1228–1236. [CrossRef] [PubMed]
49. Vermote, E.; Tanre, D.; Deuze, J.L.; Herman, M.; Morcrette, J.J. Second Simulation of the Satellite Signal in the Solar Spectrum (6S). 6S User Guide Version 2. Appendix III: Description of the subroutines. *IEEE Trans. Geosci. Remote Sens.* **1997**, *35*, 675–686. [CrossRef]
50. Waquet, F.; Cairns, B.; Knobelspiesse, K.; Chowdhary, J.; Travis, L.D.; Schmid, B.; Mishchenko, M. Polarimetric remote sensing of aerosols over land. *J. Geophys. Res. Atmos.* **2009**, *114*, D1. [CrossRef]
51. Dubovik, O.; Smirnov, A.; Holben, B.N.; King, M.D.; Kaufman, Y.J.; Eck, T.F.; Slutsker, I. Accuracy assessments of aerosol optical properties retrieved from Aerosol Robotic Network (AERONET) Sun and sky radiance measurements. *J. Geophys. Res. Atmos.* **2000**, *105*, 9791–9806. (In English) [CrossRef]

remote sensing

Article

Evaluation of MODIS DT, DB, and MAIAC Aerosol Products over Different Land Cover Types in the Yangtze River Delta of China

Jie Jiang [1,2,*], Jiaxin Liu [1], Donglai Jiao [1,2], Yong Zha [3] and Shusheng Cao [1]

1 Department of Surveying and Geoinformatics, Nanjing University of Posts and Telecommunications, Nanjing 210023, China
2 Smart Health Big Data Analysis and Location Services Engineering Lab of Jiangsu Province, Nanjing 210023, China
3 Key Laboratory of Virtual Geographic Environment of Ministry of Education, Jiangsu Center for Collaborative Innovation in Geographical Information Resource Development and Application, College of Geographic Science, Nanjing Normal University, Nanjing 210023, China
* Correspondence: jiangj@njupt.edu.cn

Abstract: The Moderate Resolution Imaging Spectroradiometer (MODIS) aerosol optical depth (AOD) has been widely used in atmospheric environment and climate change research. Based on data of the Aerosol Robotic Network and Sun–Sky Radiometer Observation Network in the Yangtze River Delta, the retrieval accuracies of MODIS C6.1 Dark Target (DT), Deep Blue (DB), and C6.0 Multi-angle Implementation of Atmospheric Correction (MAIAC) products under different land cover types, aerosol types, and observation geometries were analyzed. About 65.64% of MAIAC AOD is within the expected error (Within EE), which is significantly higher than 41.43% for DT and 56.98% for DB. The DT product accuracy varies most obviously with the seasons, and the Within EE in winter is more than three times that in spring. The DB and MAIAC products have low accuracy in summer but high in other seasons. The accuracy of the DT product gradually decreases with the increase in urban and water land-cover proportion. After being corrected by bias and mean relative error, the DT accuracy is significantly improved, and the Within EE increases by 24.12% and 32.33%, respectively. The observation geometries and aerosol types were also examined to investigate their effects on AOD retrieval.

Keywords: Yangtze River Delta; aerosol optical depth; AERONET; SONET; MODIS; aerosol type; land cover

Citation: Jiang, J.; Liu, J.; Jiao, D.; Zha, Y.; Cao, S. Evaluation of MODIS DT, DB, and MAIAC Aerosol Products over Different Land Cover Types in the Yangtze River Delta of China. *Remote Sens.* **2023**, *15*, 275. https://doi.org/10.3390/rs15010275

Academic Editors: Michael Obland and Lunche Wang

Received: 31 October 2022
Revised: 6 December 2022
Accepted: 30 December 2022
Published: 3 January 2023

1. Introduction

Aerosols are tiny particles suspended in the atmosphere, and they are one of the main pollutants in the atmosphere [1]. This pollutant is an important factor affecting climate change, by scattering and absorbing solar radiation and changing the microphysical properties of clouds to disturb the Earth's radiation balance [2–5]. The impact of atmospheric aerosols on climate forcing is uncertain due to the relatively high spatial and temporal variability in their physical and chemical characteristics [6,7]. The observation of aerosol's optical and physical properties and the quantification of the impact of aerosol particles on the climate and environment have also attracted extensive attention. Aerosol optical depth (AOD) is the integral of the aerosol extinction coefficient in the vertical direction, and it is a key factor used to describe aerosol optical properties and determine the influence of climate effects. It is widely used in atmospheric environment monitoring, atmospheric radiation transmission, remote sensing applications, and climate change research [8–10].

Remote sensing technology is an important method for aerosol monitoring at present [11]. Satellite remote sensing has a large observation region and relatively low cost, and it can reflect the spatial and temporal variations in regional and even global atmospheric aerosol. The Moderate Resolution Imaging Spectroradiometer (MODIS) on the Terra/Aqua satellite has been widely used as a global-coverage satellite remote sensing instrument to detect aerosol optical properties [12]. However, the radiation information received by the satellite's sensor is subject to the complex influence of atmospheric scattering and surface reflection [13]. Moreover, uncertainties are introduced by surface reflectance and the aerosol model used in AOD retrieval [14,15], which results in uncertainty in the results of satellite retrieval. Therefore, the applicability of MODIS aerosol products for different regions still needs to be verified by comparison with ground observation data. East Asia is one of the regions with the largest aerosol loads in the world. Considerable research has been conducted on the applicability of MODIS AOD in East Asia, especially China, including the Yangtze River Delta (YRD), the Beijing–Tianjin–Hebei region, Pearl River Delta, and northwest China [16–18].

The land surface reflectance is high, and the aerosol types are complex compared with those in ocean aerosol retrieval, and the retrieval of aerosols over land is still an important and difficult topic [19]. The solar radiation at the top of the atmosphere (TOA) received by satellite sensors includes atmospheric scattering and surface reflection. Obtaining accurate surface reflectance is a key problem to be solved for aerosol retrieval. Land cover type also affects aerosol characteristics. The reason is that different land cover types will produce different atmospheric emissions, and this factor can change the local climate and affect the transmission of aerosols by affecting radiation transport [20]. Therefore, an increasing number of scholars focus on the influence of surface types on aerosol characteristics and AOD retrieval accuracy [21–24].

The YRD is one of the fastest-growing and most prosperous regions in China. It is considered to be a high-aerosol value area in the global context and a heavily polluted area in China, because of its dense population and industries [25]. The composition of aerosols in this region is complex, including anthropogenic aerosols such as sulfates, biomass burning aerosol, as well as natural aerosols, such as dust and sea salt transported over a long distance, and their optical properties are highly diverse [26–29]. Therefore, the applicability of the MODIS aerosol product in the YRD needs to be further analyzed to clarify the influence of different land cover types, aerosol types, and observation geometries on aerosol retrieval accuracy.

In this study, the YRD is taken as the research area, and the AOD data of the Aerosol Robotic Network (AERONET) and the Sun–Sky Radiometer Observation Network (SONET) ground-based observation sites, representing the underlying surface of forest, cropland, urban, and water, are selected as the benchmark. The aerosol retrieval accuracies of MODIS Collection 6.1 (C6.1) Dark Target (DT), Deep Blue (DB), and MODIS Collection 6 (C6) Multi-angle Implementation of Atmospheric Correction (MAIAC) AOD products under different land cover types, aerosol types, and observation geometries are studied. The accuracy of MODIS AOD is also corrected according to the land-cover proportion of urban and water areas to provide a new reference for the more accurate evaluation of the applicability of MODIS AOD products in different regions and the improvement in the retrieval method.

2. Materials and Methods

2.1. MODIS Aerosol Products

MODIS is one of the most important sensors used by the Earth Observing System (EOS) Terra/Aqua satellites for aerosol monitoring [12]. It has 36 bands covering the spectrum from 0.4 μm to 14 μm. Its spatial resolution range includes 250 m (bands 1–2), 500 m (bands 3–7), and 1 km (bands 8–36), and thus, it provides global coverage every 2 days [30]. At present, MODIS aerosol products have been developed for the DT and DB of C6.1 version [31,32]. The daily aerosol products of C6.1 Level 2 are MOD04_L2 and MYD04_L2 with a spatial resolution of 10 km, where "MOD" is used for Terra and "MYD" for Aqua. The MAIAC product has been developed for C6 [33], and its daily aerosol product is MCD19A2, with a spatial resolution of 1 km.

2.1.1. DT Products

Over vegetated and dark soiled surfaces, a proportional relationship exists between the surface reflectance in visible wavelengths (0.47 and 0.66 μm) and in the shortwave infrared wavelengths (2.12 μm), which is called the "VISvs2.12" relationship [34,35]. The DT algorithm assumes that the aerosol is transparent at 2.12 μm and that the surface reflectance equals the TOA reflectance in this band; then, the AOD is retrieved based on the selected aerosol model and look-up table (LUT). Under a dust aerosol regime, aerosol transparency is an extremely poor assumption, and the 2.12 μm channel contains information about coarse mode aerosol and the surface reflectance. Levy et al. found that a single set of "VISvs2.12" ratios is not globally applicable [13,36]. The C5 defined the "VISvs2.12" relationship as a function between the scattering angle (SA) and the Normalized Difference Vegetation Index (NDVI) based on 2.12 μm. The C6.1 product over land was corrected as follows: when the coastal proportion in the 10 km × 10 km grid was greater than 50% or the water proportion was greater than 20%, the quality assurance (QA) of AOD was dropped to 0; when the proportion of urban was more than 20%, the surface reflectance calculation scheme was modified using the MODIS land surface reflectance and land cover type products [37]. The dataset of "Optical_Depth_Land_And_Ocean" from C6.1 MOD04_L2 and MYD04_L2 was used, and it contains AOD values for the filtered, quantitatively useful retrievals over dark targets, where the QA for AOD over land is 3 (representing high-quality retrieval).

2.1.2. DB Products

Hsu et al. found that the surface reflectance of MODIS in the blue channel is still low, even in areas of high surface reflectance, such as urban and desert [38]. The DB algorithm assumes that the surface reflectance of most figures remains constant in a short time and retrieves AOD by constructing a seasonal surface reflectance database [39]. The C6 algorithm divides the land surface into three types: arid and semi-arid regions, vegetated regions, and urban/built-up and transitional regions [40]. The surface reflectance database is improved using knowledge of NDVI, SA, and season. The C6.1 product has been updated in the following aspects: radiometric calibration, heavy smoke detection, artifact correction for heterogeneous terrain, seasonal and regional aerosol model, and surface reflectance in elevated terrain [41]. The dataset of "Deep_Blue_Aerosol_Optical_Depth_550_Land_Best_Estimate" from the MODIS C6.1 product was used, where QA = 2, 3 (representing good quality retrieval). Datasets such as "Solar_Zenith", "Sensor_Zenith", "Solar_Azimuth", "Sensor_Azimuth", and "Scattering_Angle" were used to evaluate the impact of observation geometry on the accuracy of DT and DB AOD.

2.1.3. MAIAC Products

The MAIAC algorithm assumes that (1) surface reflectance slightly changes in a short period of time and that (2) AOD does not change remarkably in adjacent regions [42,43]. MAIAC retrieves AOD through time-series MODIS observations. MAIAC applies the MODIS L1B data to a fixed grid with a resolution of 1 km, analyzes the time series of observation data from the past 4 days (polar) to 16 days (equatorial) using the sliding window strategy to obtain the linear spectral regression coefficient (SRC), and calculates AOD based on SRC and surface bidirectional reflectance factors via LUT. MODIS C6 has been improved in terms of cloud mask, SRC estimation, and aerosol properties. MAIAC utilizes the minimal ratio of spectral reflectance (0.47/2.13 μm) in a 2-month period to estimate SRC at a 1 km scale, which helps to remove the occasional blockiness at the 25 km scale in the AOD and in the surface reflectance. Smoke and dust models are introduced in C6 to distinguish between fine and coarse aerosol models [33]. "Optical_Depth_055" and "AOD_QA" in the C6 product were used in this study, where "0000" for the QA of AOD was selected to represent the best quality. The datasets "cosSZA", "cosVZA", "RelAZ", and "Scattering_Angle" were selected to evaluate the impacts of observation geometry on the accuracy of MAIAC products.

2.2. Ground-Based Measurements

2.2.1. AERONET Data

AERONET is a global aerosol optical property monitoring network established by the National Aeronautics and Space Administration [44]. Through the establishment of ground-based monitoring sites in representative regions around the world, the benchmark data of global aerosol parameters have been obtained to study aerosol optical and physicochemical properties. The AERONET uses the French CIMEL CE318 Sun photometer to obtain the columnar AOD for the whole atmosphere by measuring the direct (collimated) solar radiation at different wavelengths, directions, and times. AERONET uses the spectral ed-convolution algorithm to yield fine and coarse AOD at a standard wavelength of 500 nm, from which the fine mode fraction to total AOD can be computed [45]. The CE318 Sun photometer sky radiance measurements can be inverted to produce aerosol characteristics, such as size distribution, single scattering albedo (SSA), phase functions, and the complex index of refraction. The AERONET aerosol products are classified into Level 1.0, Level 1.5, and Level 2.0 [46]. L1.0 data are unscreened and do not undergo final calibration; L1.5 data are cloud-cleared, and quality controls have been applied, but these data may not undergo final calibration; L2.0 data are automatically cloud-cleared and quality-assured, with pre-field and post-field calibration applied. The V3 Level 2.0 data (Table 1) for Taihu (TH), Qiandaohu (QDH), Hangzhou_City (HZC), Hangzhou_ZFU (HZZ), NUIST, Shouxian (SX), Hefie (HF), XuZhou-CUMT (XZ), and LA-TM in the YRD (Figure 1) from 2007 to 2018 were used in this study.

Table 1. Information of ground-based observation sites.

Site	Project	Longitude	Latitude	Data Period	Number of Matches (n)		
					DT	DB	MAIAC
TH	AEROENT	120.215	31.421	09/2005–08/2016	722	956	1512
QDH	AEROENT	119.053	29.556	08/2007–10/2008	53	53	88
HZC	AEROENT	120.157	30.290	04/2008–02/2009	69	84	103
HZZ	AEROENT	119.727	30.257	08/2007–08/2009	174	176	195
NUIST	AEROENT	118.717	32.206	09/2008–08/2010	107	137	156
SX	AEROENT	116.782	32.558	05/2008–12/2008	86	98	96
XZ	AEROENT	117.142	34.217	06/2013–05/2019	831	1135	1238
HF	AEROENT	117.162	31.905	11/2005–11/2008	86	107	118
LA-TM	AEROENT	119.440	30.324	10/2007–03/2009	93	101	112
HF	SONET	117.162	31.905	01/2013–12/2019	484	510	562
NJ	SONET	118.957	32.115	01/2013–12/2019	357	476	528
SH	SONET	121.481	31.284	01/2013–12/2019	221	370	397

2.2.2. SONET Data

SONET is a ground-based observation network organized and implemented by the Chinese Academy of Sciences in typical areas of China, including rural, urban, desert, coastal, basin, mountain, and plateau areas [47]. It obtains the physical, chemical, and optical characteristics of the total column aerosol for aerosol characteristic modeling and the authenticity inspection of satellite remote sensing products in China. SONET uses the French CIMEL's multiwavelength polarized sun–sky radiometer CE318-DP to observe the solar and sky radiation and their polarization characteristics in eight wavebands (the central wavelengths are 340, 380, 440, 500, 675, 870, 1020, and 1640 nm) and to detect the total column water vapor at 936 nm. Aerosol products are divided into three grades according to the AERONET product grades, namely, Level 1.0, Level 1.5, and Level 2.0 [48]. L1.0 data are original data; L1.5 data are cloud-cleared through the automatic cloud identification algorithm; L2.0 data are cloud-cleared, and calibration coefficient interpolation and expert identification have been applied. SONET L2.0 data (Table 1) from Hefei (HF), Nanjing (NJ), and Shanghai (SH) were adopted in this study.

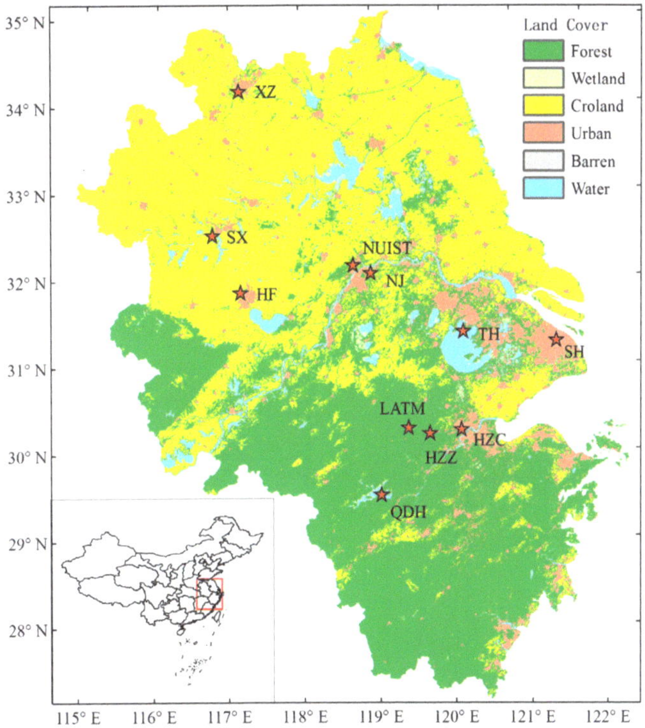

Figure 1. Locations of the AERONET and SONET sites in the Yangtze River Delta displayed on the land-cover map.

The AERONET/SONET AOD at 500 and 675 nm were interpolated into AOD at 550 nm using the Ångström exponent (AE) because MODIS AOD is at 550 nm, but AERONET/SONET AOD does not have such a wavelength.

$$\tau_{550} = \tau_{675}(550/675)^{-\alpha} \quad (1)$$

$$\alpha = -\ln(\tau_{500}/\tau_{675})/\ln(500/675) \quad (2)$$

where τ_{500} is AOD at 500 nm, τ_{675} is AOD at 675 nm, α is the AE at 500–675 nm, and τ_{550} is AOD at 550 nm obtained by interpolation.

2.3. Land Cover Data

MCD12Q1 data are the 500 m resolution land classification products formed by extracting different land cover types based on the annual MODIS Aqua and Terra data using the decision tree-supervised classification method [49]. These data can be divided into 17 land cover types. The land cover type percentages of each of the AERONET and SONET sites in the YRD corresponding to 3 × 3 pixels of MODIS L2 data within the range of 30 km × 30 km are counted in Table 2. The areas around QDH, LATM, and HZZ are dominated by forest; notably, the forest proportions of LATM and HZZ account for more than 90%. HF and SX are dominated by cropland, and the proportions account for more than 60%. The land cover types at HZC and SH are mainly urban, of which the urban area in SH accounts for more than 90%. TH is dominated by water; approximately half of it is water, and it is slightly more urban than forest. The land cover types at XZ, NJ, and NUIST are mixtures of forest, cropland, and urban. According to the land cover proportion, the ground-based observation sites are classified into five land cover types, which are forest, cropland, urban, mixed, and water.

Table 2. Percentage of land cover around AERONET/SONET sites (%).

Land Cover Type	Site	Forest	Wetland	Cropland	Urban	Barren	Water
Forest	QDH	**64.86**	9.67	0.00	1.08	0.17	24.22
	LATM	**99.92**	0	0	0.08	0	0
	HZZ	**93.47**	0.27	1.4	4.19	0	0.67
Cropland	HF	3.28	1.05	**60.33**	33.41	0.05	1.88
	SX	9.47	2.69	**63.61**	15.67	0.00	8.56
Urban	HZC	21.92	0.42	0.22	**71.22**	0.39	5.83
	SH	0.3	0.11	0	**92.93**	0.05	6.61
Mixed	XZ	12.66	0.08	43.94	42.35	0.03	0.94
	NJ	36.15	1.72	26.9	29.08	0.51	5.64
	NUIST	33.39	1.00	39.14	23.17	0.22	3.08
Water	TH	20.67	2.75	1.61	27.58	0.14	**47.25**

2.4. *Evaluation Method*

2.4.1. Spatiotemporal Matching

AERONET/SONET data represent the continuous observation of the ground sites at fixed intervals daily, whereas MODIS AOD represents the instantaneous observations with a 10 km × 10 km spatial resolution. The spatiotemporal scales of MODIS and ground-based observations differ. Therefore, the ground-based data were filtered with a temporal window of 30 min before and after satellite overpass to obtain the average of AOD. The 30 km × 30 km range of the ground site was considered a spatial window, and the average MODIS AOD under this window was calculated to form the AERONET/SONET-MODIS AOD dataset matching the temporal and spatial resolution.

2.4.2. Evaluation Method

The expected error (EE hereinafter) is usually used to evaluate the quality of AOD retrieved by satellite. A large number of experiments have been conducted by the MODIS science team; EE adopts ±(0.05 + 0.15 × AOD) for DT and DB AOD products, and ±(0.05 + 0.1 × AOD) is used for MAIAC products [33,50–52]. To uniformly compare and analyze the quality of each algorithm, the EE of ±(0.05 + 0.15 × AOD) is adopted. Above EE, Within EE and Below EE are expressed by Equations (4)–(6):

$$\mathrm{EE} = \pm(0.05 + 0.15 \times \mathrm{AOD_G}) \tag{3}$$

$$\mathrm{AOD_M} > \mathrm{AOD_G} + |\mathrm{EE}| \tag{4}$$

$$\mathrm{AOD_G} - |\mathrm{EE}| \le \mathrm{AOD_M} \le \mathrm{AOD_G} + |\mathrm{EE}| \tag{5}$$

$$\mathrm{AOD_M} < \mathrm{AOD_G} + |\mathrm{EE}| \tag{6}$$

where $\mathrm{AOD_G}$ is the observed AEROENT/SONET AOD, and $\mathrm{AOD_M}$ is the AOD retrieved from MODIS. In addition, error analysis is performed by root mean squared errors (RMSE), bias, the square of the correlation coefficient (R^2), and the slope (a) of regression analysis.

$$\mathrm{RMSE} = \sqrt{\frac{1}{n}\sum(\mathrm{AOD_M} - \mathrm{AOD_G})^2} \tag{7}$$

$$\mathrm{Bias} = \frac{1}{n}\sum(\mathrm{AOD_M} - \mathrm{AOD_G}) \tag{8}$$

$$\mathrm{R}^2 = \left(\frac{\Sigma(\mathrm{AOD_M} - \overline{\mathrm{AOD_M}})(\mathrm{AOD_G} - \overline{\mathrm{AOD_G}})}{\sqrt{\Sigma(\mathrm{AOD_M} - \overline{\mathrm{AOD_M}})^2}\sqrt{\Sigma(\mathrm{AOD_G} - \overline{\mathrm{AOD_G}})^2}}\right)^2 \tag{9}$$

3. Results

3.1. Overall Accuracy of DT, DB, and MAIAC

The accuracies of DT, DB, and MAIAC products in the YRD differ greatly (Figure 2), and the matched number and accuracy of MAIAC are significantly better than those of DT and DB. The matched number of DT is the lowest, with a value of 3239, which indicates that it is more difficult to satisfy the DT algorithm than MAIAC or DB. The DT product also has the lowest accuracy, with a Within EE of only 41.43% (Figure 2a), which is more than 24% lower than that of MAIAC, and the RMSE of DT is the highest at 0.243. In the matched data of DT, about 54.58% is Above EE, only 3.98% is Below EE, and the bias is 0.151, which implies that DT AOD in the YRD is mostly overestimated. The bias box at different AODs is shown in Figure 2d. When the AOD is less than 0.8, the mean bias exceeds the EE, and the mean bias decreases gradually with the increase in AOD. The mean bias is −0.006 when the AOD is 1.78, and this suggests that the DT AOD in the YRD is overestimated mainly in the middle and low value ranges.

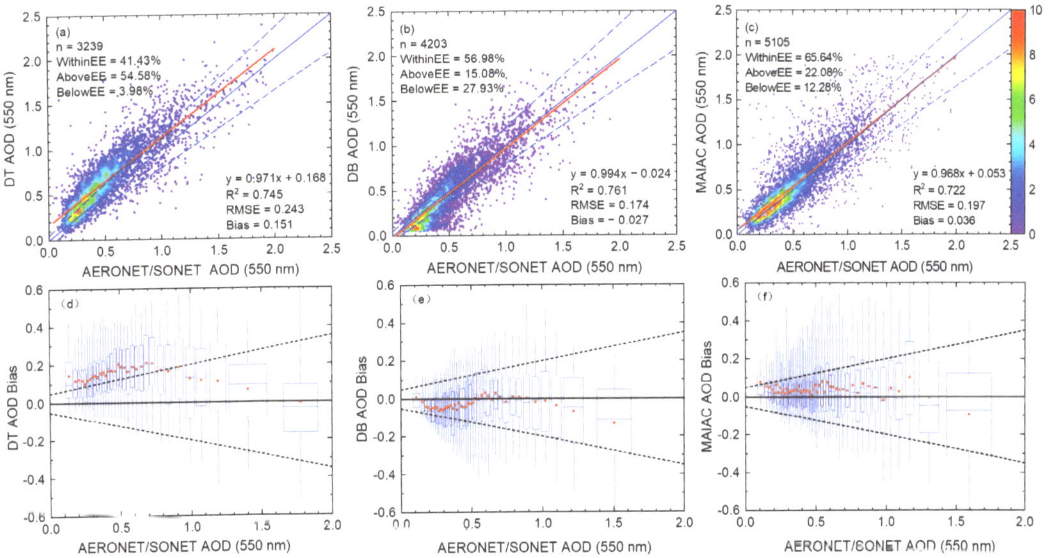

Figure 2. Overall accuracy of (**a**) DT, (**b**) DB, (**c**) MAIAC; the solid red line represents the regression line; the solid blue line represents the 1:1 reference line, and the dashed blue line represents the expected error line. The AOD bias box plot of (**d**) DT, (**e**) DB, (**f**) MAIAC products; each box is based on 100 matched data; the dashed line represents the expected error line, and the red dot represents mean bias.

A total of 4203 DB matched data have a Within EE of 56.98%, an RMSE of 0.174, and a bias of −0.027. The number and accuracy of DB matched data are better than those of DT but lower than those of MAIAC. However, 27.93% of DB AOD is Below EE, which is the highest among the three methods. The bias statistics of different AODs are shown in Figure 2e. The mean bias is less than 0 when the AOD is between 0.15 and 0.5. The mean bias is relatively small and fluctuates around 0 when the AOD is between 0.5 and 1.0. The deviation is larger and becomes less than 0 with the increase in AOD when the AOD is greater than 1.0. Therefore, DB AOD is mostly underestimated when AOD is greater than 1.0 and less than 0.5.

The MAIAC product is the best of the three algorithms. It has a matched number of 5105, a Within EE of 65.64%, an RMSE of 0.196, and a bias of 0.036, which values are obviously better than those of DT and DB. The bias statistics of different AODs are shown in Figure 2f. The mean bias of MAIAC is relatively large when the AOD is low. In particular, the mean bias is 0.079 when the AOD is 0.105, and it is the only mean bias value that exceeds the EE in the MAIAC product. Therefore, the MAIAC AOD is obviously overestimated at a low AOD. When the AOD is between 0.2 and 0.5, the mean bias is relatively low, most of which values are lower than 0.04. The mean bias slightly increases with the rise in AOD.

The retrieval accuracy values for DT, DB, and MAIAC products show discrepancy in different seasons (Table 3). The matched number is the lowest in winter, and the Within EE is the lowest in spring for the DT product. The matched number and Within EE of DB and MAIAC AOD products are the lowest in summer. The matched numbers for the three algorithms in summer are very low; among them, the matched number of DT in summer is 575, which is smaller than the 1224 in spring and the 996 in autumn, and this value for DB and MAIAC is lower in summer than in other seasons, only accounting for 12.0% and 13.6% of all matched numbers, which may be due to the rainy weather in the YRD in summer [25].

Table 3. Accuracy validation of DT, DB, and MAIAC AOD in the Yangtze River Delta in each season.

Season	Data	*n*	Above EE (%)	Below EE (%)	Within EE (%)	RMSE	Bias	R^2	a
Spring	DT	1224	75.65	1.06	**23.29**	0.279	0.228	0.789	1.018
	DB	1296	9.88	31.17	58.95	0.169	−0.052	0.772	0.974
	MAIAC	1497	21.24	17.04	61.72	0.217	0.033	0.667	1.013
Summer	DT	575	49.22	4.52	46.26	0.248	0.134	0.767	0.917
	DB	503	7.16	52.48	**40.36**	0.228	−0.116	0.806	1.011
	MAIAC	696	28.74	14.8	**56.46**	0.265	0.071	0.691	1.058
Autumn	DT	996	42.97	5.92	51.11	0.218	0.104	0.726	0.913
	DB	1073	12.86	26	**61.14**	0.149	−0.032	0.786	0.912
	MAIAC	1388	28.17	6.34	65.49	0.192	0.071	0.772	1.000
Winter	DT	444	29.51	6.98	**63.51**	0.175	0.068	0.736	0.982
	DB	1331	24.94	17.06	58.00	0.175	0.035	0.781	1.156
	MAIAC	1524	14.3	11.88	**73.82**	0.132	−0.009	0.821	0.830

The matched number of DT AOD in winter is the lowest, with a value of 444, which is mainly due to the reduction in green vegetation in winter. The Within EE of DT shows the most significant seasonal variation, with a peak of 63.51% in winter, and only 23.29% in spring, which is one-third of that in winter.

The Within EE of DB does not change considerably in spring, autumn, and winter, and the values are all close to 60%, while the lowest value is 40.36% in summer. The main reason is that the Deep Blue surface database method is mainly adopted for the surface reflection in the YRD, and the lush vegetation in summer leads to low accuracy for AOD retrieval.

The Within EE of MAIAC AOD also shows significant seasonal variation, with 56.46% in summer, which is the lowest value, 61.72% in spring, 65.49% in autumn, and 73.82% in winter, which is the highest value. The matched number is also the lowest in summer, while little change is observed in other seasons. Therefore, the quantity and accuracy of MAIAC AOD are good, except in summer.

3.2. *Influence of Land Cover Types on AOD Retrieval*

The three algorithms perform significantly differently for different land cover types (Table 4). The highest accuracy of the DT product regards the data for forest, where it obtains a Within EE of 68.75%, an RMSE of 0.163, and a bias of −0.059. The Within EE of DT for cropland with a higher vegetation proportion is 53.59%, which is obviously higher than that for mixed, urban, and water areas. The Within EE values of DT for urban and water are 18.96% and 26.04%, respectively, which are relatively low. The analysis of the relationship between the land cover proportion of urban and water (LCP_{UW}) and the Within EE at eight sites with more matched data in the YRD (Figure 3a) shows that Within

EE is significantly negatively related to LCP_{UW}, that the R^2 is 0.892, and that the accuracy of DT AOD decreases with the increase in LCP_{UW}.

Table 4. Accuracy of DT, DB, and MAIAC AOD in different land cover types in the Yangtze River Delta.

Land Cover	Data	*n*	Above EE (%)	Below EE (%)	Within EE (%)	RMSE	Bias	R^2	a
Forest	DT	320	5.94	25.31	68.75	0.163	−0.059	0.794	0.874
	DB	330	0.91	62.12	36.97	0.201	−0.151	0.826	0.910
	MAIAC	395	8.35	12.41	79.24	0.130	−0.022	0.843	0.909
Urban	DT	290	79.66	1.38	18.96	0.304	0.254	0.715	0.887
	DB	454	23.13	28.19	48.68	0.196	−0.013	0.610	0889
	MAIAC	500	18.00	19.00	63.00	0.175	−0.022	0.650	0.720
Cropland	DT	612	45.1	1.31	53.59	0.188	0.123	0.829	1.08
	DB	715	9.23	31.89	58.88	0.154	−0.042	0.808	1.059
	MAIAC	776	9.15	12.11	78.74	0.126	−0.008	0.849	0.960
Mixed1	DT	464	59.27	1.51	39.22	0.219	0.156	0.840	1.034
	DB	613	8.48	38.01	53.51	0.183	−0.078	0.794	0.998
	MAIAC	684	8.19	16.37	75.44	0.145	−0.029	0.831	0.907
Mixed2	DT	831	53.67	1.93	44.4	0.200	0.133	0.834	0.974
	DB	1135	22.73	16.3	60.97	0.172	0.025	0.791	1.035
	MAIAC	1238	17.12	9.53	73.35	0.141	0.021	0.841	0.998
Water	DT	722	72.16	1.8	26.04	0.332	0.245	0.671	0.878
	DB	956	15.69	20.4	63.91	0.164	−0.01	0.765	0.953
	MAIAC	1512	43.98	10.52	45.50	0.286	0.135	0.634	1.029

Mixed1 is NJ and NJUPT, and its forest proportion is significantly higher than that of XZ. Mixed2 is XZ, and its urban proportion is higher than those of NJ and NJUPT.

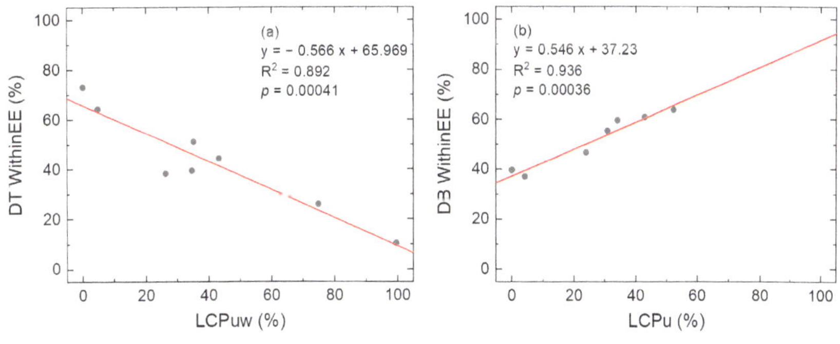

Figure 3. (**a**) Correlations between land-—cover proportions of urban and water and DT Within EE. (Three sites with the least matched data are removed, which are QDH, HZC, and SX). (**b**) Correlations between land-cover proportion of urban without water and DB Within EE (QDH, HZC, and SX with fewer matching data and SH with a high urban proportion close to 100% are removed).

The lowest accuracy of the DB product regards the data for forest, where it obtains a Within EE of only 36.97%. Different from other regions [21,53], the Within EE of DB for urban in the YRD is only 48.68%, which is lower than that for cropland, mixed, and water land cover. Among the two sites with urban cover, the Within EE at HZC is 63.10%, which is significantly higher than that at SH (45.40%). Therefore, other factors besides land cover type have a great impact on the DB product in SH. The Within EE in TH mainly dominated by water is 63.91%, which is the highest among all land cover types. The reason is that DB filters water land cover according to the MCD12C1 dataset and retrieves aerosols of other land cover types [40]. After the water land cover around TH is removed, about 52.29% of the underlying surface is urban, which value is second only to SH and HZC. After the water land cover is removed in each site, a significant linear relationship can be observed between the land cover proportion of urban without water (LCP_U) and the Within EE of DB (Figure 3b), and the Within EE of DB increases with the rise in the LCP_U.

The accuracy of MAIAC in forest, cropland, and mixed land cover types with a high vegetation proportion is relatively high. The Within EE values of Mixed1 and Mixed2 are over 70%, while the Within EE values of cropland and forest are close to 80%, which are 78.74% and 79.24%, respectively. The RMSE and bias are also relatively low, and the lowest RMSE and bias in cropland are 0.126 and −0.008, respectively. The accuracy of MAIAC in urban is low, with a Within EE of 63%. The accuracy of MAIAC in water is the lowest, with a Within EE of only 45.50%.

3.3. DT and DB AOD Correction

The bias and Mean Relative Error (MRE) (defined by Equation (10)) of DT have an obvious linear relationship with LCP_{UW} (Figure 4a,b), and the R^2 values are 0.814 and 0.932, respectively. The bias and MRE of DT are positive, except for at LATM and HZZ, with a very small LCP_{UW}. The bias and MRE of DT increase, and the accuracy decreases with the rise in LCP_{UW}. The bias and MRE of DB have an obvious linear relationship with LCP_U (Figure 4c,d). Different from those of DT, the bias and MRE of DB are all negative, except for those for XZ, with a high LCP_U. Overall, the bias and MRE of DB increase with the rise in LCP_U, and the accuracy improves.

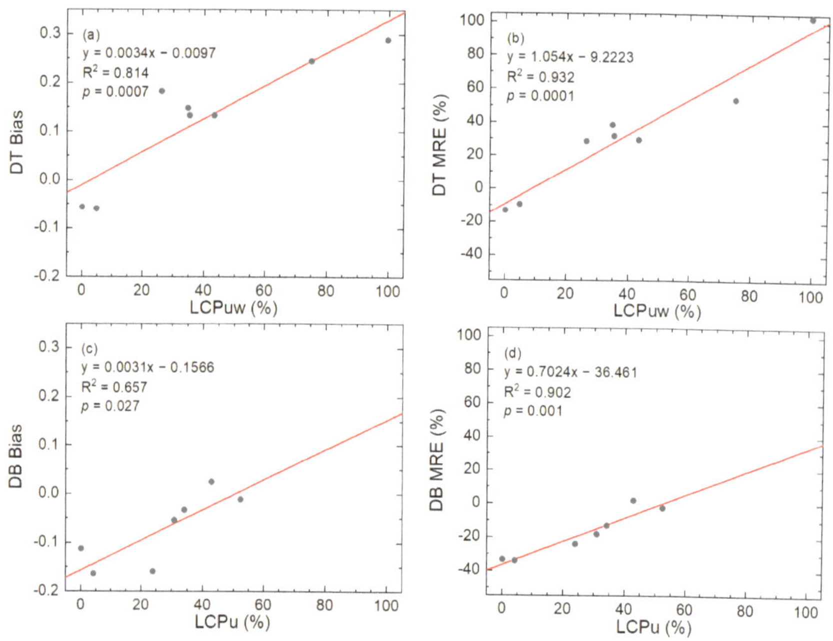

Figure 4. Correlations between land-cover proportion of urban and water and (**a**) DT bias; (**b**) DT MRE; correlations between land-cover proportion of urban without water and (**c**) DB bias; (**d**) DB MRE.

Therefore, AOD can be corrected according to LCP_{UW} and LCP_U [24]. In this study, DT and DB products are corrected according to Equations (11) and (12). If the DT product is corrected, then $Bias_{predict}$ and $MRE_{predict}$ are determined by regression equations in Figure 4a,b and LCP_{UW}. If the DB products are corrected, then $Bias_{predict}$ and $MRE_{predict}$ are determined by the regression equations in Figure 4c,d and LCP_U.

$$MRE = 100 \times \sum (AOD_M - AOD_G)/AOD_G \quad (10)$$

$$\text{AOD}_{\text{M}}^{\text{correctd}} = \text{AOD}_{\text{M}} - \text{Bias}_{\text{predict}} \tag{11}$$

$$\text{AOD}_{\text{M}}^{\text{correctd}} = \text{AOD}_{\text{M}} / \left(1 + \text{MRE}_{\text{predict}}\right) \tag{12}$$

The accuracy of DT is greatly improved when corrected by the two methods (Figure 5). In particular, the Within EE of the DT MRE-corrected product has been increased to 72.29%, but its regression coefficient is only 0.704. The RMSE and bias of the DT MRE-corrected product are slightly higher than those of the DT bias-corrected product. The change curve of the mean bias of the DT bias-corrected product (Figure 5e) is similar to that of the DT-uncorrected product, but the mean bias is significantly reduced and around zero. The mean bias of the DT MRE-corrected product (Figure 5f) gradually decreases with the increase in AOD, with a value from 0.048 to −0.327. The lower quartile limit of bias exceeds the EE when AOD is around 0.8, and the mean bias also exceeds the EE when AOD is around 1.0. Therefore, the accuracy of the DT MRE-corrected product is significantly improved in the middle and low AOD intervals. However, the error will already exceed that of DT-uncorrected products with the increase in AOD, especially when the AOD is greater than 1.0.

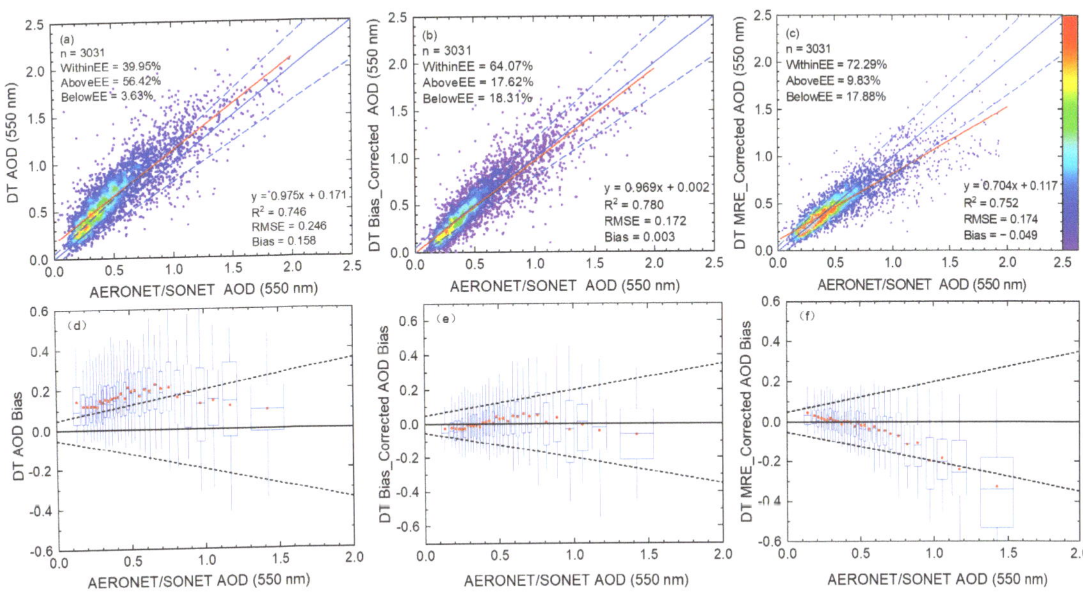

Figure 5. Accuracy of (**a**) DT, (**b**) DT bias-corrected, (**c**) DT MRE-corrected AOD; the solid red line represents the regression line; the solid blue line represents the 1:1 reference line, and the blue dashed line represents the expected error line. Bias boxplot of (**d**) DT, (**e**) DT bias-corrected, (**f**) DT MRE-corrected AOD; each box is based on 100 matching data points; the dashed line represents the expected error line, and the red dot represents the mean bias.

A comparison of the DT Within EE before and after correction at each site is shown in Figure 6a. The Within EE of LATM and HZZ with low LCP_{UW} does not change considerably after correction, but the Within EE at TH and SH with a large LCP_{UW} is significantly improved. Therefore, the DT product can be corrected effectively, especially for urban and water land cover types.

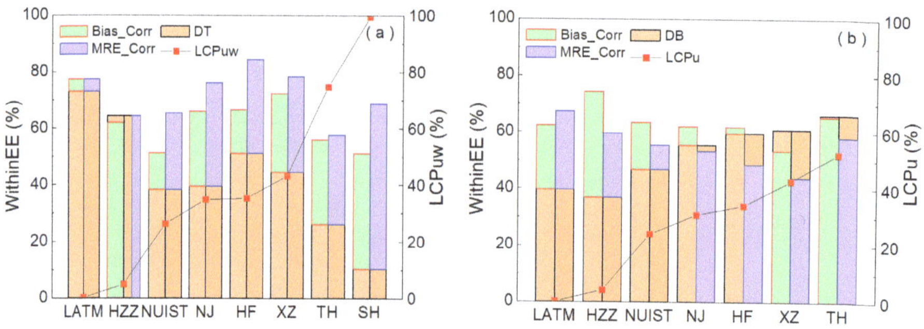

Figure 6. Accuracy comparison at each site before and after the correction of DT (**a**) and DB (**b**) products. (**a**) The orange bar represents the DT Within EE, and the green and purple bars represent the increased Within EE value following bias correction and MRE correction, respectively. The Within EE shows no change before and after MRE correction at HZZ. (**b**) The lower orange bar represents the DB Within EE, and the upper orange bar represents the decreased value after correction.

The comparison of the Within EE of DB before and after correction at each site is shown in Figure 6b. The Within EE of the DB-corrected product at LATM, HZZ, and NUIST with low LCP_U is significantly improved, but the improvement of Within EE gradually decreases with the rise in LCP_U. When the LCP_U exceeds 30%, the Within EE of the DB MRE-corrected product becomes lower than that of the uncorrected product, and the bias and MRE-corrected methods lead to a decrease in DB Within EE at HF, XZ, and TH. Therefore, only the DB product with an LCP_U less than 30% can be corrected effectively.

Table 5 shows the accuracy comparison results at LATM, HZZ, and NUIST with low LCP_U before and after the correction of DB AOD. The Within EE based on both correction methods has been significantly improved; in particular, the Within EE of the DB bias-corrected product has been increased to 67.88%, and the RMSE and bias have been significantly reduced, but the R^2 has been decreased.

Table 5. Accuracy comparison at LATM, HZZ, and NUIST before and after correction of DB AOD.

Data	*n*	Above EE (%)	Below EE (%)	Within EE (%)	RMSE	Bias	R^2	a
DB	414	2.66	56.52	40.82	0.211	−0.150	0.834	0.915
DB Bias-Corrected	414	14.49	17.63	67.88	0.155	−0.023	0.818	0.881
DB MRE-Corrected	414	21.26	18.60	60.14	0.227	0.025	0.803	1.208

3.4. Influence of Observation Geometry on AOD Retrieval

The influence of observation geometry, such as solar zenith angle (SZA), view zenith angle (VZA), scattering angle (SA), and relative azimuth angle (RAA), on the aerosol retrieval accuracy is shown in Figure 7. SZA, VZA, and SA are divided into three sections: Low (0°–30°), Moderate (30°–60°), and High (60°–90°). RAA is also divided into six sections at 30° intervals.

SZA greatly influences the accuracy of DT, DB, and MAIAC products (Figure 7a). The accuracy generally increases with the rise in SZA. The SZA values of DT, DB, and MAIAC are mainly concentrated in the Moderate and Low sections; Moderate accounts for about 60–70% of the total. The accuracy of AOD in the Low section is significantly lower than that in the Moderate section, and the Within EE in the Moderate sections of the three algorithms is about 14 percentage points higher than that in the Low section. The RMSE value of DT and DB at a Moderate SZA is about 0.05 lower than that at the Low SZA, while the difference between the RMSE values of MAIAC in the two sections is 0.101. The proportion of SZA in the High section is very low. The proportion of SZA of MAIAC in the High section, which represents the highest proportion of the High section in the three algorithms, is only 5.22%. The SZA of DT in the High section is less than 2%, but its accuracy is the highest, with a Within EE of 85.71% and an RMSE of 0.094.

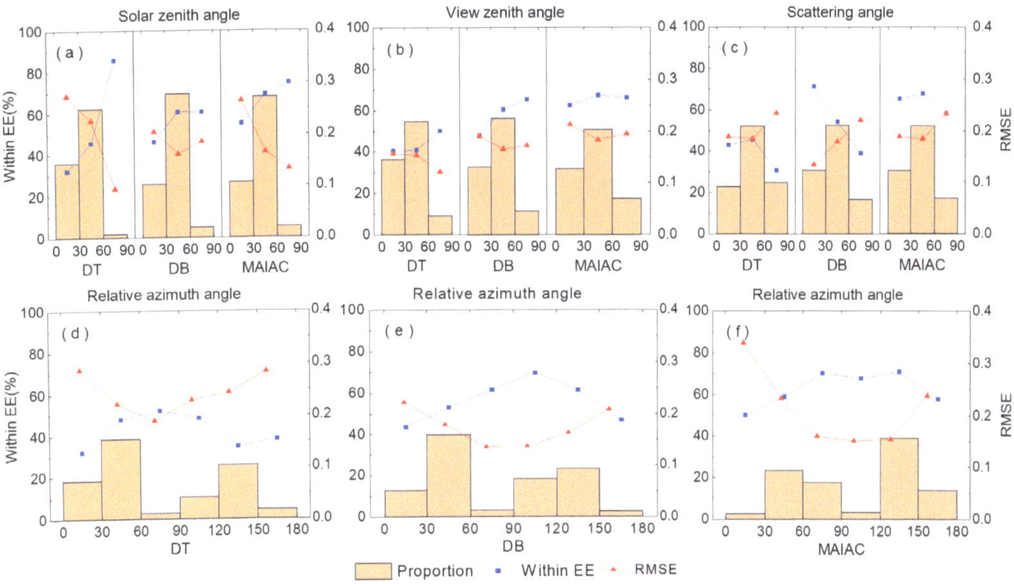

Figure 7. Influence of SZA (**a**), VZA (**b**), SA (**c**), and RAA (**d–f**) on the accuracy of DT, DB, and MAIAC products. The orange vertical bar represents the proportions of matches in each angle section.

VZA is also mainly concentrated in the Moderate and Low sections (Figure 7b). The accuracies of DT, DB, and MAIAC in the Moderate and High VZA are also higher than that in the Low VZA, but the influence of VZA is obviously smaller than that of SZA. In particular, the accuracy shows nearly no difference between the DT in the Moderate VZA and that in the Low VZA; the Within EE values are 40.74% and 40.27%, and the RMSE values of DT are 0.153 and 0.156. However, the accuracy of DT in the High section is significantly improved. The difference in the Within EE of MAIAC between Moderate and Low VZA is no more than five percentage points, and the difference in RMSE is about 0.023. The accuracy of MAIAC in High VZA is slightly reduced, and the accuracy does not change considerably with VZA. The DB product shows the largest change with VZA—the Within EE of Moderate VZA is 12.38 percentage points higher than that of Low VZA, and the RMSE is 0.026 lower.

The influence of SA is mainly reflected in that the accuracy of the three algorithms is the lowest for High SA (Figure 7c). The accuracy of the DT and MAIAC AOD varies with the SA, and the accuracy in Low SA is similar to that in Moderate SA. The difference in the Within EE between them is less than 3%, and the difference in RMSE between them is less than 0.1. Meanwhile, the accuracy of DB decreases with the rise in SA, and the accuracy difference of DB between Low and Moderate SA is large. The difference in Within EE is close to 17%, and the difference in RMSE is about 0.044.

The influence of RAA on the AOD accuracy of the three algorithms mainly manifests as high accuracy when RAA is close to vertical, but low accuracy when RAA is close to 0 or 180. The accuracy of DT in the three sections of 30°–60°, 60°–90°, and 90°–120° are better, with Within EE values of about 50%, among which the highest is 52.27% at 60°–90°, and the lowest RMSE is 0.189. The maximum Within EE of DB is 69.77% at 90°–120° RAA, and the lowest RMSE is 0.134 at 60°–90 ° RAA. MAIAC has good accuracy at 60°–90°, 90°–120°, and 120°–150° RAA, and their within EE values are close to 70%.

3.5. Influence of Aerosol Types on AOD Retrieval

Aerosol type is an important source of aerosol retrieval error. The DT algorithm establishes five aerosol models based on the global AERONET aerosol data through cluster analysis, and the appropriate aerosol model is selected according to regions and seasons. The DB algorithm uses the maximum likelihood method to calculate a mixing ratio between various dust and smoke models and then retrieves the AOD and AE. The MAIAC algorithm divides aerosols into eight types according to aerosol size distribution and the concentration ratios of coarse and fine particle columns, as well as other parameters observed by AERONET. Model 5 is mainly used for aerosol retrieval in China. The bias distributions of DT, DB, and MAIAC products under different AE and SSA are shown in Figure 8.

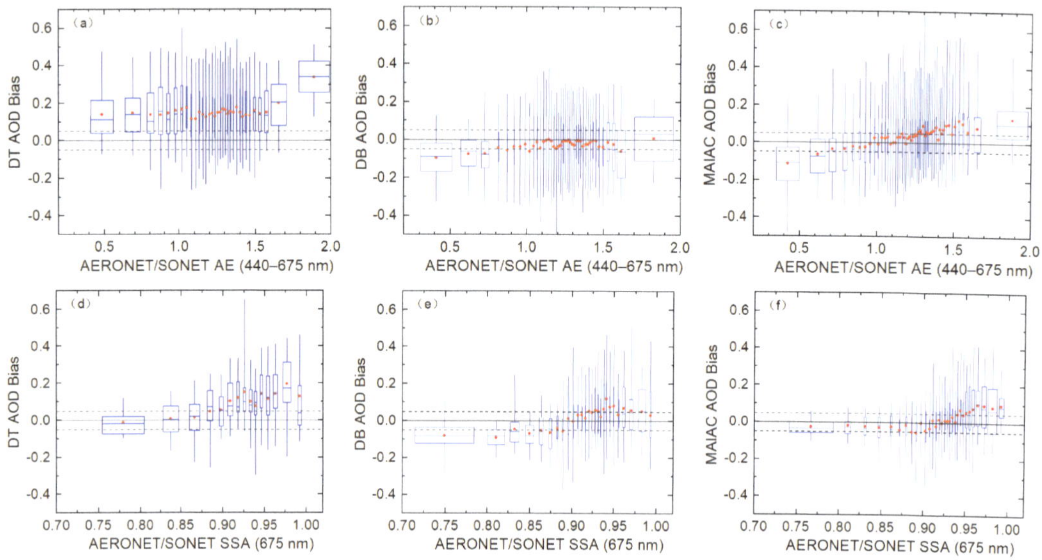

Figure 8. Accuracy of DT, DB, and MAIAC products under different AE (**a–c**) and SSA (**d–f**); each box is based on 100 matched data points, the dashed line represents the expected error line, and the red dot represents mean bias.

The bias of DT and DB does not change considerably with AE. The bias of DT is generally large and mainly positive (Figure 8a). The mean bias does not change considerably with AE, and the mean bias of DT is around 0.15 except when the AE is greater than 1.6. The bias of DB is mainly negative (Figure 8b). The mean bias of DB also does not change considerably with AE. The mean bias of DB with an AE of 0.8–1.6 is very small and does not fluctuate considerably. The bias of MAIAC increases with the rise in AE (Figure 8c). The mean bias at each interval increases linearly with AE. The mean bias is less than 0 when the AE is less than 1.0, which indicates that the AOD of MAIAC is underestimated. The AOD of MAIAC is overestimated when the AE is greater than 1.0.

The AOD bias values of three algorithms change more obviously with SSA than with AE. Overall, the mean bias increases and then flattens with the increase in SSA. The mean bias increases with SSA when it is less than 0.93. The mean bias fluctuates significantly when the SSA is greater than 0.93. The mean DB bias is negative when the SSA is lower than 0.89; otherwise, it is positive. The mean bias of DB also increases with SSA before it reaches 0.95, and then, it changes slightly. The mean bias of MAIAC increases with SSA before 0.96. The mean bias of MAIAC is negative when the SSA is lower than 0.91; otherwise, it is positive. It changes slightly when the SSA is greater than 0.96.

4. Discussion

MODIS products have been extensively evaluated and verified at both global and regional scales. Many scholars have compared the accuracy of MODIS DT, DB, and MAIAC aerosol products and found that MAIAC products have greater advantages in terms of matched number, with more than double those of the DB and DT AOD, and they show higher accuracy [17,19], which matches our findings in the YRD (Table 3). MODIS DT AOD is generally overestimated, while DB AOD is mostly underestimated [24]. MODIS products in the YRD also show a similar phenomenon. However, the Within EE of DT AOD in the YRD is only 41.43%, which is significantly lower than those in other regions [54]. There are only 320 matching data points at HZZ, LATM, and QDH for forest, accounting for 8.4% of the total matching number of DT, while SH and TH with higher urban and water proportions have 943 matching data points, which are about three times those of the forest sites. Therefore, the accuracy of DT products in the YRD is low, but the accuracy of DT products for forest in the YRD is relatively high, with a Within EE of 68.75% (Table 4). The Within EE of DT for cropland is 53.59%, which is obviously lower than that for forest, which may be due to the high urban presence in the two cropland sites, where the proportion of urban in HF is 33.41%. As can be seen from Figure 1, the southern YRD is dominated by forest, while the northern YRD is dominated by cropland. The whole YRD is covered by about 43% forest, 45% cropland, and only 10% urban and water. Therefore, the accuracy of DT in the YRD should be high. In addition, we found that the accuracy of MAIAC in the YRD in summer is significantly lower than that in other seasons, so the DT AOD in summer can be used to make up for the low accuracy of MAIAC in summer.

5. Conclusions

In this study, the overall accuracy and seasonal variations in MODIS DT, DB, and MAIAC products in the YRD were analyzed based on AERONET and SONET ground monitoring data. The influences of land cover type, observation geometry, and aerosol type on AOD retrieval accuracy were also explored. LCP_{UW} and LCP_U were used to correct DT and DB products. The conclusions are as follows.

The MAIAC product is obviously superior to the DT and DB products in matched number and accuracy. The accuracy of DT product varies most obviously with seasons, and it has the highest accuracy in winter and the lowest in spring. The accuracy of DB and MAIAC is low summer and high in other seasons.

The accuracy of MAIAC is higher in forests, cropland, and mixed land cover types with high vegetation proportions, and the accuracy is lowest in water. The accuracy of the DT product is negatively correlated with LCP_{UW} (R^2 = 0.892). The accuracy of DT AOD decreases gradually with the increase in urban and water proportion. A significant linear relationship exists between the accuracy of the DB product and the LCP_U (R^2 = 0.936), as the accuracy increases with the rise in LCP_U.

A correction method based on bias and MRE can significantly improve the accuracy of DT products, especially the accuracy at TH and SH with large LCP_{UW}. The MRE correction method works very well at low AOD, but it amplifies the error at high AOD values. The correction method also works well for DB AOD with low LCP_U, but it amplifies the error for DB AOD with high LCP_U.

The effect of SZA on accuracy is stronger than that of VZA, and the accuracy increases with the rise in SZA. The accuracy is the lowest when SA is high, but it is high when the RAA is close to vertical.

The bias of MAIAC increases with the rise in AE. The change in bias for the three algorithms with SSA is more obvious than that with AE. Notably, mean bias increases and then flattens with the rise in SSA.

Author Contributions: Conceptualization, methodology, writing, J.J.; writing, J.L. and D.J.; writing—review and editing, Y.Z. and S.C. All authors have read and agreed to the published version of the manuscript.

Funding: This research was funded by NUPTSF, grant number NY220211.

Data Availability Statement: The MODIS Collection 6.1 aerosol product (MOD04_L2 and MYD04_L2), MODIS Collection 6 aerosol product (MCD19A2) and MODIS Collection 6 land cover type product (MCD12Q1) are available at https://ladsweb.modaps.eosdis.nasa.gov/search/, accessed on 20 August 2021. The AERONET product is available at https://aeronet.gsfc.nasa.gov/, accessed on 20 August 2021.

Conflicts of Interest: The authors declare no conflict of interest.

References

1. Li, Z.Q.; Guo, J.P.; Ding, A.J.; Liao, H.; Liu, J.J.; Sun, Y.L.; Wang, T.J.; Xue, H.W.; Zhang, H.S.; Zhu, B. Aerosol and boundary-layer interactions and impact on air quality. *Natl. Sci. Rev.* **2017**, *4*, 810–833. [CrossRef]
2. Kaufman, Y.J.; Tanré, D.; Boucher, O. A satellite view of aerosols in the climate system. *Nature* **2002**, *419*, 215–223. [CrossRef]
3. Haywood, J.; Boucher, O. Estimates of the direct and indirect radiative forcing due to tropospheric aerosols: A review. *Rev. Geophys.* **2000**, *38*, 513–543. [CrossRef]
4. Rosenfeld, D.; Lohmann, U.; Raga, G.B.; O'Dowd, C.D.; Kulmala, M.; Fuzzi, S.; Reissell, A.; Andreae, M.O. Flood or drought: How do aerosols affect precipitation? *Science* **2008**, *321*, 1309–1313. [CrossRef]
5. Bellouin, N.; Quaas, J.; Gryspeerdt, E.; Kinne, S.; Stier, P.; Watson-Parris, D.; Boucher, O.; Carslaw, K.S.; Christensen, M.; Daniau, A.L.; et al. Bounding Global Aerosol Radiative Forcing of Climate Change. *Rev. Geophys.* **2020**, *58*, e2019RG000660. [CrossRef]
6. Sun, J.M.; Ariya, P.A. Atmospheric organic and bio-aerosols as cloud condensation nuclei (CCN): A review. *Atmos. Environ.* **2006**, *40*, 795–820. [CrossRef]
7. Mehta, M.; Singh, R.; Singh, A.; Singh, N.; Anshumali. Recent global aerosol optical depth variations and trends—A comparative study using MODIS and MISR level 3 datasets. *Remote Sens. Environ.* **2016**, *181*, 137–150. [CrossRef]
8. Hsu, N.C.; Herman, J.R.; Torres, O.; Holben, B.N.; Tanre, D.; Eck, T.F.; Lavenu, F. Comparisons of the TOMS aerosol index with Sun-photometer aerosol optical thickness: Results and applications. *J. Geophys. Res.* **1999**, *104*, 6269–6279. [CrossRef]
9. Butt, E.W.; Rap, A.; Schmidt, A.; Scott, C.E.; Pringle, K.J.; Reddington, C.L.; Richards, N.A.D.; Woodhouse, M.T.; Ramirez-Villegas, J.; Yang, H.; et al. The impact of residential combustion emissions on atmospheric aerosol, human health, and climate. *Atmos. Chem. Phys.* **2016**, *16*, 873–905. [CrossRef]
10. Gupta, P.; Christopher, S.A.; Wang, J.; Gehrig, R.; Lee, Y.; Kumar, N. Satellite remote sensing of particulate matter and air quality assessment over global cities. *Atmos. Environ.* **2006**, *40*, 5880–5892. [CrossRef]
11. Martin, R.V. Satellite remote sensing of surface air quality. *Atmos. Environ.* **2008**, *42*, 7823–7843. [CrossRef]
12. Remer, L.A.; Kaufman, Y.J.; Tanre, D.; Mattoo, S.; Chu, D.A.; Martins, J.V.; Li, R.R.; Ichoku, C.; Levy, R.C.; Kleidman, R.G.; et al. The MODIS aerosol algorithm, products, and validation. *J. Atmos. Sci.* **2005**, *62*, 947–973. [CrossRef]
13. Levy, R.C.; Remer, L.A.; Dubovik, O. Global aerosol optical properties and application to Moderate Resolution Imaging Spectroradiometer aerosol retrieval over land. *J. Geophys. Res. Atmos.* **2007**, *112*, 1–15. [CrossRef]
14. Li, Z.; Zhao, X.; Kahn, R.; Mishchenko, M.; Remer, L.; Lee, K.H.; Wang, M.; Laszlo, I.; Nakajima, T.; Maring, H. Uncertainties in satellite remote sensing of aerosols and impact on monitoring its long-term trend: A review and perspective. *Ann. Geophys.* **2009**, *27*, 2755–2770. [CrossRef]
15. Povey, A.C.; Grainger, R.G. Known and unknown unknowns: Uncertainty estimation in satellite remote sensing. *Atmos. Meas. Tech.* **2015**, *8*, 4699–4718. [CrossRef]
16. Xiao, Q.; Zhang, H.; Choi, M.; Li, S.; Kondragunta, S.; Kim, J.; Holben, B.; Levy, R.C.; Liu, Y. Evaluation of VIIRS, GOCI, and MODIS Collection 6 AOD retrievals against ground sunphotometer observations over East Asia. *Atmos. Chem. Phys.* **2016**, *16*, 1255–1269. [CrossRef]
17. Zhang, Z.Y.; Wu, W.L.; Fan, M.; Wei, J.; Tan, Y.H.; Wang, Q. Evaluation of MAIAC aerosol retrievals over China. *Atmos. Environ.* **2019**, *202*, 8–16. [CrossRef]
18. Li, B.; Yuan, H.; Feng, N.; Tao, S. Comparing MODIS and AERONET aerosol optical depth over China. *Int. J. Remote Sens.* **2009**, *30*, 6519–6529. [CrossRef]
19. Drury, E.; Jacob, D.J.; Wang, J.; Spurr, R.J.D.; Chance, K. Improved algorithm for MODIS satellite retrievals of aerosol optical depths over western North America. *J. Geophys. Res. Atmos.* **2008**, *113*, D16204. [CrossRef]
20. Ganzeveld, L.; Bouwman, L.; Stehfest, E.; van Vuuren, D.P.; Eickhout, B.; Lelieveld, J. Impact of future land use and land cover changes on atmospheric chemistry-climate interactions. *J. Geophys. Res. Atmos* **2010**, *115*, D23301. [CrossRef]
21. Wang, Y.; Yuan, Q.Q.; Li, T.W.; Shen, H.F.; Zheng, L.; Zhang, L.P. Evaluation and comparison of MODIS Collection 6.1 aerosol optical depth against AERONET over regions in China with multifarious underlying surfaces. *Atmos. Environ.* **2019**, *200*, 280–301. [CrossRef]
22. Liu, J.; Ding, J.; Li, L.; Li, X.; Zhang, Z.; Ran, S.; Ge, X.; Zhang, J.; Wang, J. Characteristics of aerosol optical depth over land types in central Asia. *Sci. Total Environ.* **2020**, *727*, 138676. [CrossRef] [PubMed]
23. Yang, Y.; Cermak, J.; Yang, K.; Pauli, E.; Chen, Y. Land Use and Land Cover Influence on Sentinel-2 Aerosol Optical Depth below City Scales over Beijing. *Remote Sens.* **2022**, *14*, 4677. [CrossRef]

24. Sun, K.; Gao, Y.; Qi, B.; Yu, Z.F. The Influence of Underlying Land Cover on the Accuracy of MODIS C6.1 Aerosol Products-A Case Study over the Yangtze River Delta Region of China. *Remote Sens.* **2022**, *14*, 938. [CrossRef]
25. Zhang, M.; Ma, Y.Y.; Wang, L.C.; Gong, W.; Hu, B.; Shi, Y.F. Spatial-temporal characteristics of aerosol loading over the Yangtze River Basin during 2001–2015. *Int. J. Climatol.* **2018**, *38*, 2138–2152. [CrossRef]
26. Li, L.; Chen, C.H.; Fu, J.S.; Huang, C.; Streets, D.G.; Huang, H.Y.; Zhang, G.F.; Wang, Y.; Jang, C.; Wang, H.L. Air quality and emissions in the Yangtze River Delta, China. *Atmos. Chem. Phys.* **2010**, *11*, 1621–1639. [CrossRef]
27. Ma, J.Z.; Xu, X.B.; Zhao, C.S.; Yan, P. A review of atmospheric chemistry research in China: Photochemical smog, haze pollution, and gas-aerosol interactions. *Adv. Atmos. Sci.* **2012**, *29*, 1006–1026. [CrossRef]
28. Huang, R.J.; Zhang, Y.L.; Bozzetti, C.; Ho, K.F.; Cao, J.J.; Han, Y.M.; Daellenbach, K.R.; Slowik, J.G.; Platt, S.M.; Canonaco, F.; et al. High secondary aerosol contribution to particulate pollution during haze events in China. *Nature* **2014**, *514*, 218–222. [CrossRef]
29. Li, J.; Jiang, L.; Chen, C.; Liu, D.T.; Du, S.S.; Zhang, Y.J.; Yang, Y.F.; Tang, L.L. Characteristics and Sources of Black Carbon Aerosol in a Mega-City in the Western Yangtze River Delta, China. *Atmosphere* **2020**, *11*, 315. [CrossRef]
30. Barnes, W.L.; Pagano, T.S.; Salomonson, V.V. Prelaunch characteristics of the Moderate Resolution Imaging Spectroradiometer (MODIS) on EOS-AM1. *IEEE T. Geosci. Remote* **1998**, *36*, 1088–1100. [CrossRef]
31. Levy, R.C.; Mattoo, S.; Munchak, L.A.; Remer, L.A.; Sayer, A.M.; Patadia, F.; Hsu, N.C. The Collection 6 MODIS aerosol products over land and ocean. *Atmos. Meas. Tech.* **2013**, *6*, 2989–3034. [CrossRef]
32. Tian, X.; Gao, Z. Validation and Accuracy Assessment of MODIS C6.1 Aerosol Products over the Heavy Aerosol Loading Area. *Atmosphere* **2019**, *10*, 548. [CrossRef]
33. Lyapustin, A.; Wang, Y.J.; Korkin, S.; Huang, D. MODIS Collection 6 MAIAC algorithm. *Atmos. Meas. Tech.* **2018**, *11*, 5741–5765. [CrossRef]
34. Kaufman, Y.J.; Tanre, D.; Remer, L.A.; Vermote, E.F.; Chu, A.; Holben, B.N. Operational remote sensing of tropospheric aerosol over land from EOS moderate resolution imaging spectroradiometer. *J. Geophys. Res. Atmos.* **1997**, *102*, 17051–17067. [CrossRef]
35. Remer, L.A.; Levy, R.C.; Mattoo, S.; Tanre, D.; Gupta, P.; Shi, Y.X.; Sawyer, V.; Munchak, L.A.; Zhou, Y.P.; Kim, M.; et al. The Dark Target Algorithm for Observing the Global Aerosol System: Past, Present, and Future. *Remote Sens.* **2020**, *12*, 2900. [CrossRef]
36. Levy, R.C.; Remer, L.A.; Mattoo, S.; Vermote, E.F.; Kaufman, Y.J. Second-generation operational algorithm: Retrieval of aerosol properties over land from inversion of Moderate Resolution Imaging Spectroradiometer spectral reflectance. *J. Geophys. Res.-Atmos.* **2007**, *112*, D13211. [CrossRef]
37. Gupta, P.; Levy, R.C.; Mattoo, S.; Remer, L.A.; Munchak, L.A. A surface reflectance scheme for retrieving aerosol optical depth over urban surfaces in MODIS Dark Target retrieval algorithm. *Atmos. Meas. Tech.* **2016**, *9*, 3293–3308. [CrossRef]
38. Hsu, N.C.; Tsay, S.C.; King, M.D.; Herman, J.R. Aerosol properties over bright-reflecting source regions. *IEEE T. Geosci. Remote* **2004**, *42*, 557–569. [CrossRef]
39. Hsu, N.C.; Tsay, S.C.; King, M.D.; Herman, J.R. Deep blue retrievals of Asian aerosol properties during ACE-Asia. *IEEE Trans. Geosci. Remote* **2006**, *44*, 3180–3195. [CrossRef]
40. Hsu, N.C.; Jeong, M.J.; Bettenhausen, C.; Sayer, A.M.; Hansell, R.; Seftor, C.S.; Huang, J.; Tsay, S.C. Enhanced Deep Blue aerosol retrieval algorithm: The second generation. *J. Geophys. Res. Atmos.* **2013**, *118*, 9296–9315. [CrossRef]
41. Hsu, N.C.; Lee, J.; Sayer, A.M.; Carletta, N.; Chen, S.H.; Tucker, C.J.; Holben, B.N.; Tsay, S.C. Retrieving near-global aerosol loading over land and ocean from AVHRR. *J. Geophys. Res. Atmos.* **2017**, *122*, 9968–9989. [CrossRef]
42. Lyapustin, A.; Martonchik, J.; Wang, Y.J.; Laszlo, I.; Korkin, S. Multiangle implementation of atmospheric correction (MAIAC): 1. Radiative transfer basis and look-up tables. *J. Geophys. Res. Atmos.* **2011**, *116*, D03210. [CrossRef]
43. Lyapustin, A.; Wang, Y.; Laszlo, I.; Kahn, R.; Korkin, S.; Remer, L.; Levy, R.; Reid, J.S. Multiangle implementation of atmospheric correction (MAIAC): 2. Aerosol algorithm. *J. Geophys. Res. Atmos.* **2011**, *116*, D03211. [CrossRef]
44. Holben, B.N.; Eck, T.F.; Slutsker, I.; Tanre, D.; Buis, J.P.; Setzer, A.; Vermote, E.; Reagan, J.A.; Kaufman, Y.J.; Nakajima, T.; et al. AERONET—A federated instrument network and data archive for aerosol characterization. *Remote Sens. Environ.* **1998**, *66*, 1–16. [CrossRef]
45. Dubovik, O.; King, M.D. A flexible inversion algorithm for retrieval of aerosol optical properties from Sun and sky radiance measurements. *J. Geophys. Res. Atmos.* **2000**, *105*, 20673–20696. [CrossRef]
46. Smirnov, A.; Holben, B.N.; Eck, T.F.; Dubovik, O.; Slutsker, I. Cloud-screening and quality control algorithms for the AERONET database. *Remote Sens. Environ.* **2000**, *73*, 337–349. [CrossRef]
47. Li, Z.Q. Sun-sky radiometer observation network with the extension of multi-wavelength polarization measurements. *J. Remote Sens.* **2015**, *19*, 495–519.
48. Li, Z.Q.; Xu, H.; Li, K.T.; Li, D.H.; Xie, Y.S.; Li, L.; Zhang, Y.; Gu, X.F.; Zhao, W.; Tian, Q.J.; et al. Comprehensive Study of Optical, Physical, Chemical, and Radiative Properties of Total Columnar Atmospheric Aerosols over China: An Overview of Sun–Sky Radiometer Observation Network (SONET) Measurements. *Bull. Am. Meteorol. Soc.* **2018**, *99*, 739–755. [CrossRef]
49. Sulla-Menashe, D.; Gray, J.M.; Abercrombie, S.P.; Friedl, M.A. Hierarchical mapping of annual global land cover 2001 to present: The MODIS Collection 6 Land Cover product. *Remote Sens. Environ.* **2019**, *222*, 183–194. [CrossRef]
50. Mhawish, A.; Banerjee, T.; Sorek-Hamer, M.; Lyapustin, A.; Broday, D.M.; Chatfield, R. Comparison and evaluation of MODIS Multi-angle Implementation of Atmospheric Correction (MAIAC) aerosol product over South Asia. *Remote Sens. Environ.* **2019**, *224*, 12–28. [CrossRef]

51. Sayer, A.M.; Hsu, N.C.; Bettenhausen, C.; Jeong, M.-J. Validation and uncertainty estimates for MODIS Collection 6 "Deep Blue" aerosol data. *J. Geophys. Res. Atmos.* **2013**, *118*, 7864–7872. [CrossRef]
52. Sayer, A.M.; Munchak, L.A.; Hsu, N.C.; Levy, R.C.; Bettenhausen, C.; Jeong, M.J. MODIS Collection 6 aerosol products: Comparison between Aqua's e-Deep Blue, Dark Target, and "merged" data sets, and usage recommendations. *J. Geophys. Res. Atmos.* **2014**, *119*, 13965–13989. [CrossRef]
53. Wei, J.; Li, Z.Q.; Peng, Y.R.; Sun, L. MODIS Collection 6.1 aerosol optical depth products over land and ocean: Validation and comparison. *Atmos. Environ.* **2019**, *201*, 428–440. [CrossRef]
54. Liu, N.; Zou, B.; Feng, H.H.; Wang, W.; Tang, Y.Q.; Liang, Y. Evaluation and comparison of multiangle implementation of the atmospheric correction algorithm, Dark Target, and Deep Blue aerosol products over China. *Atmos. Chem. Phys.* **2019**, *19*, 8243–8268. [CrossRef]

remote sensing

MDPI

Article

A Novel Algorithm of Haze Identification Based on FY3D/MERSI-II Remote Sensing Data

Yidan Si [1,2,3], Lin Chen [1,2,3], Zhaojun Zheng [1,2,3], Leiku Yang [4], Fu Wang [5,6], Na Xu [1,2,3] and Xingying Zhang [1,2,3,*]

1 National Satellite Meteorological Center, China Meteorological Administration, Beijing 100081, China
2 Key Laboratory of Radiometric Calibration and Validation for Environmental Satellites, National Satellite Meteorological Center (National Center for Space Weather), China Meteorological Administration, Beijing 100081, China
3 Innovation Center for FengYun Meteorological Satellite (FYSIC), Beijing 100081, China
4 School of Surveying and Land Information Engineering, Henan Polytechnic University, Jiaozuo 454003, China
5 CMA Earth System Modeling and Prediction Centre (CEMC), Beijing 100081, China
6 State Key Laboratory of Severe Weather (LaSW), Beijing 100081, China
* Correspondence: zxy@cma.gov.cn; Tel.: +86-010-68408385

Abstract: Since 2013, frequent haze pollution events in China have been attracting public attention, generating a demand to identify the haze areas using satellite observations. Many studies of haze recognition algorithms are based on observations from space-borne imagers, such as the Moderate Resolution Imaging Spectroradiometer (MODIS), the Visible Infrared Imaging Radiometer Suite (VIIRS) and the Advanced Himawari Imager (AHI). Since the haze pixels are frequently misidentified as clouds in the official cloud detection products, these algorithms mainly focus on recovering them from clouds. There are just a few studies that provide a more precise distinction between haze and clear pixels. The Medium Resolution Imaging Spectrometer II (MERSI-II), the imager aboard the FY-3D satellite, has similar bands to those of MODIS, hence, it appears to have equivalent application potential. This study proposes a novel MERSI haze mask (MHAM) algorithm to directly categorize haze pixels in addition to cloudy and clear ones. This algorithm is based on the fact that cloudy and clear pixels exhibit opposing visible channel reflectance and infrared channel brightness temperature characteristics, and clear pixels are relative brighter, and as well as this, there is a positive difference between their apparent reflectance values, at 0.865 μm and 1.64 μm, respectively, over bright surfaces. Compared with the Aqua/MODIS and MERSI-II official cloud detection products, these two datasets treat the dense aerosol loadings as certain clouds, possible clouds and possible clear pixels, and they treat distinguished light or moderate haze as possible clouds, possible clear pixels and certainly clear pixels, while the novel algorithm is capable of demonstrating the haze region's boundary in a manner that is more substantially consistent with the true color image. Using the $PM_{2.5}$ (particle matter with a diameter that is less than 2.5 μm) data monitored by the national air quality monitoring stations as the test source, the results indicated that when the ground-based $PM_{2.5} \geq 35$ μg/cm^3 is considered to be haze days, the samples with the recognition rate that is higher than 85% accounted for 72.22% of the total samples. When $PM_{2.5} \geq 50$ μg/cm^3 is considered as haze days, 83.33% of the samples had an identification rate that was higher than 85%. A cross-comparison with similar research methods showed that the method proposed in this study had better sensitivity to bright surface clear and haze areas. This study will provide a haze mask for subsequent quantitative inversion of aerosol characteristics, and it will further exert the application benefits of MERSI-II instrument aboard on FY3D satellite.

Keywords: MERSI-II; haze; heavy aerosol; cloud; bright surface; distinguish

Citation: Si, Y.; Chen, L.; Zheng, Z.; Yang, L.; Wang, F.; Xu, N.; Zhang, X. A Novel Algorithm of Haze Identification Based on FY3D/MERSI-II Remote Sensing Data. *Remote Sens.* **2023**, *15*, 438. https://doi.org/10.3390/rs15020438

Academic Editor: Carmine Serio

Received: 18 November 2022
Revised: 30 December 2022
Accepted: 30 December 2022
Published: 11 January 2023

1. Introduction

According to the explicit provisions of the World Meteorological Organization, haze is defined as a day when the average visibility is less than 10 km, the average daily relative humidity is less than 80%, and other atmospheric haze days that can lead to low visibility, such as precipitation, are excluded (WMO, 2005). Haze particles are mainly composed of extremely small particles suspended in the air, such as dust, sulfate and nitrate, which are drier than fog and wetter than dust [1]. Due to the influence of the energy consumption structure, the air pollution in China has evolved into regional and complex air pollution. In recent decades, continuous high-intensity atmospheric haze pollution has swept across central and eastern China, which has aroused the attention of domestic scholars from all walks of life [2–4].

Using the discrete ground stations that are distributed in a small area, continuous observations over a short period of time are feasible to analyze the reasons for the formation of special pollution processes and the changing trend of the pollutants. For example, the AOD increases persistently during the daytime, and the diurnal variation varies from about 15% in summer to about 45% in winter based on thirty-three months of aerosol data in Beijing [5]. In addition, a severe haze occurred on 20–27 September 2011 in the megacity of Beijing [6] and several serious haze events in Beijing between January 1 and 1 February 2013 have been observed [7]. Further, the North China Plain was analyzed by Che et al. (2014) [8] and biomass burning impacted the haze pollution in the Yangtze River Delta, China [9]. Previous studies have focused on monitoring the process of haze events and the variation characteristics of the haze formation mechanism; the distribution and variations of the haze process have seldom been analyzed over continuous spatio-temporal scales. Subsequently, there have been some studies that have discussed the spatio-temporal characteristics of haze by using the products retrieved from satellites. For example, the AOD datasets from the Collection 6 enhanced Deep Blue algorithm of MODIS aboard the Terra and Aqua satellites, the UV Aerosol Index from Ozone Monitoring Instrument (OMI) on the Aura satellite, the vertical detection of Cloud-Aerosol Lidar and Infrared Pathfinder Satellite Observations (CALIPSO) and other data were jointly used to analyze the variation characteristics and occurrence frequency of regional haze when the AOD > 1.0, which basically stabilized at 70~90 days, which was one year after 2009 [10–12]. Although the AOD data retrieved by satellites is used to analyze the characteristics of haze variation, some cloud-contaminated hazy areas have no valid AOD retrievals, which means not all of the aerosol retrieval algorithms are suitable over the heavily polluted regions [13].

Cloud detection algorithms based on satellite remote sensing data are generally designed to generate a cloud mask as an the input for the downstream cloud-relevant, environmental and continental elements retrieval [14]. Many studies revealed that the operational cloud detection products developed by MODIS, VIIRS and AHI lack the ability to classify haze aerosols, differentiate the thick aerosol layer from the cloudy pixels and they include certain uncertainties in desert and non-desert transition areas [7,15,16]. In the MOD/MYD04 aerosol retrieval algorithms, fast cloud detection has been designed to combine the standard derivation of the reflectance at 0.47 μm and 1.38 μm to distinguish the cloud edge from the reflectance identification [17]. Considering that the texture characteristic of haze is smoother than that of clouds, the variance information of a matrix composed of 3 × 3 pixels at 0.412 μm and 0.47 μm is added to the VIIRS cloud identification to avoid identifying a haze as a cloud [18].

In order to improve the discrimination of the cloud detection algorithm, many scholars have performed in-depth analyses of the characteristics of haze pixels. For example, after analyzing the spectral characteristics of cloud, fog, haze and clear pixels in visible-near infrared and infrared channels based on MODIS data, Ge et al. (2016) selected multiple sensitive channels to calculate the threshold combination of the haze distribution, and further, they designed an automatic processing algorithm for haze detection [19]. Shang et al. (2014) used MODIS secondary products for multi-day sampling to compare the difference in the cloud amount, cloud phase and cloud top pressure during clear days and heavy pollution conditions; a recognition algorithm combining the threshold values of these three parameters was established to distinguish the haze pixels from the cloud regions [20]. Based on VIIRS data, Wang et al. (2017) selected cloud, clear sky and haze areas from more than 200,000 pixels of eastern China through visual interpretation, and then, they determined the thresholds of these two variables by fitting the histograms of the standard deviation changes of the three targets at 0.412 μm and 0.47 μm, respectively, thus finally establishing a fast cloud, haze and clear sky identification algorithm for AOD inversion [15]. Ackerman (1998) believed that haze has a smaller optical thickness compared to that of clouds and is distributed on the surface. The digital elevation model (DEM) was added to retain the clear pixels above higher altitude mountains [21,22]. By analyzing the radiation distribution characteristics of 36 bands of accurate and mis-detected cloud detection points from the MODIS satellite, a cloud detection correction algorithm for mis-detected cloud products in high pollution areas was constructed, which solved the problem of cloud over-identification from the perspective of studying cloud microphysical characteristics [7]. NDVI can be used not only to identify vegetation distribution, but it can also be used to identify clouds and clear skies. Yang (2022) found that the sample with NDVI < 0.1 were treated as inland water, which is the main reason why the MODIS DT aerosol retrieval algorithm filters out haze areas [23]. Based on FY3D/MERSI-II data, Yang (2022) added a reflectance value of 2.13 μm, which is below 0.08, to this condition to recover the haze area for retrieval. This method was then adopted by the MODIS DT team to retrieve the haze area from MODIS [24].

In recent years, China's domestic satellite sensor detection technology has developed rapidly. The Fengyun series of satellites, which are resource satellites and ocean satellites, the Gaofen series of satellites, which are environmental satellites, etc., were launched successively, basically realizing the Earth observation system for multiple application fields, thus serving the business needs of various industries. As for the Fengyun satellite series, Fengyun 3 was successfully launched in May 2008 as a polar solar synchronous orbit satellite. After nearly ten years of development in FY-3A, FY-3B, FY-3C and FY-3D, the Fengyun 3 Earth Observation System has matured, and it is in the stage of stable operation [25]. In 2008, the first Moderate Resolution Spectral Imager (MERSI) on FY-3 indicated that domestic satellites could be used for the quantitative remote sensing inversion of land and ocean aerosols [26,27]. In November 2017, Fengyun 3D was once again equipped with an upgraded version of the MERSI instrument, providing a solid data source for aerosol inversion [28,29]. The MERSI load performance is comparable to that of MODIS, whose business products are relatively mature. The cloud detection products of MERSI-II are as same as those of MODIS, therefore, both lack of the ability to identify haze. For aerosol properties retrieval, the classifications of the MERSI-II operational cloud mask (MERSI-CLM) cannot meet the requirements, therefore, it is urgent to carry out research on the multi-elemental identification of cloudy, hazy and clear skies based on the MERSI load observation data at present.

It has always been a difficult problem to identify haze and clouds using satellite data because haze and clouds have similar optical and micro-physical properties, and it is difficult to distinguish them by using traditional threshold methods. Considering the effectiveness of the business products, the improved algorithm, preferably, does not rely on external data. Therefore, based on the difference of the spectral characteristics of cloud, haze and clear sky pixels, this paper proposes a parameter that is suitable for clear sky pixels on a bright surface according to the characteristics of the MERSI payload instrument. The spectral differences of many wavelengths were combined to construct a fusion identification method. Section 2 introduces the data and the algorithm principle adopted by the algorithm, and the recognition effect and accuracy evaluation of the algorithm is introduced in Section 3.

2. Materials and Methods

2.1. Study Domain

Our study area, which is bounded by latitudes of 0° and 60°N and longitudes of 60°~160°E, mostly covers China, Mongolia, Korea, Japan, Southeast Asia, India and other regions. Among them, eastern China and India are often affected by anthropogenic source emissions, resulting in the frequent occurrence of haze. In northwest China, such as in Xinjiang, Gansu and other regions, dust and bare soil cover the land, the main aerosol type is the sand dust type, the amount of anthropogenic emissions is relatively small and the atmosphere is relatively clean. The Tibetan Plateau, which is 5 km above sea level, has a typical polar plateau climate, and it is often covered by clouds. In general, the study area selected in this paper includes a haze area, a cloud area and a clear sky area, which is conducive to a more comprehensive analysis of the performance of the recognition algorithm.

2.2. Data Sources

2.2.1. FY3D/MERSI-II

The Medium Resolution Imaging Spectrometer is one of the main optical imaging payloads on the FY-3 series satellites. The second generation of MERSI (MERSI-II) has been improved in its performance compared to that of the first generation, forming a double satellite network with the FY-3C series, which greatly improves China's meteorological satellite observation capability. MERSI-II uses multiple probes (10 or 40) and provides an image of about 2900 km (cross track) × 10 km (along track) per scan. Compared with MERSI-I, MERSI-II has 19 solar reflection channels and 1 infrared emission channel, which is equipped with a total of twenty-five channels, including sixteen visible-near infrared channels, three short-wave infrared channels and six medium–long wave infrared channels. Among them, six channels have a sub-satellite point resolution of 250 m, and the remaining nineteen channels have a resolution of 1000 m. The bands used in this study are listed in Table 1. The MERSI-II L1 data used in this study can be downloaded from the website http://satellite.nsmc.org.cn/portalsite/default.aspx. (accessed on 10 April 2022).

Table 1. The wavelengths of the FY3D/MERSI-II instrument used in the study.

Band	Center Wavelength (μm)	Width (nm)	Spatial Resolution (m)
1	0.47	50	250
2	0.55	50	250
3	0.65	50	250
4	0.865	50	250
5	1.38	50	250
6	1.64	50	1000
7	2.13	50	1000
12	0.67	20	1000
15	0.865	20	1000
19	1.03	20	1000
20	3.8	180	1000
24	10.8	1000	250

The MERSI-II operational cloud detection products were also downloaded from the above website, including a granular version and daily versions. The cloud mask products using the daily scale at a resolution of 1 km × 1 km and 5 km × 5 km were released at the same time, and this study adopted the data of the finer resolution one for the cross-comparison. The parameter named CLM_DAILY_D was extracted, and the values were represented in bits and bytes for certainly the cloudy, probably cloudy, probably clear, certainly clear, coastal, sun glint and snow/ice areas. Considering that only the pixels over land were identified in this study, the pixels labeled as sun glint were treated as invalid filling values.

2.2.2. Aqua/MODIS

Two MODIS sensors were placed onboard the EOS Terra and EOS Aqua satellites. Terra/MODIS and Aqua/MODIS have been in orbit for more than 20 years, providing indispensable data sources for global and regional air quality monitoring, special weather events, urbanization development, assimilation systems and climate change, etc. [30–34]. Since FY3D is an afternoon satellite, the MYD35 cloud products with a resolution of 1 km^2 based on Aqua were selected for the evaluation, thus minimizing the inconsistency between the MERSI and MODIS mis-identification results over the time period. The meaning of the values in the dataset are as same as those of MERSI-II.

2.2.3. Auxiliary Datasets

The identification of haze pixels based on satellite remote sensing data is still in the research stage, and the corresponding official products have not released yet, therefore, there are still great difficulties in quantitative validation of haze recognition. The real-time $PM_{2.5}$ monitoring data released by the national ground air quality monitoring stations can be taken as the true ground-based observation values, which usually reflect the air quality conditions at the stations, and the larger the value is, then the more serious the pollution is. There were 1670 monitoring stations nationwide on 22 October 2020. As shown in Figure 1, most of the stations are located in eastern China, with a few in the Xinjiang and Yunnan provinces [35]. The Ministry of Environmental Protection classifies the value of $PM_{2.5}$ into six ranks, with a value below 35 $\mu g/m^3$ being excellent, a value larger than 35 $\mu g/m^3$ and smaller than 75 $\mu g/m^3$ (define as 35~75) being good, a value of 75 $\mu g/m^3$~115 $\mu g/m^3$ being classified as mild pollution, a value of 115 $\mu g/m^3$~150 $\mu g/m^3$ being classified as moderate pollution, a value of 150 $\mu g/m^3$~250 $\mu g/m^3$ being classified as severe pollution and a value of larger than 250 $\mu g/m^3$ being classified as serious pollution, which can be used as a data source to verify the identification accuracy of this study.

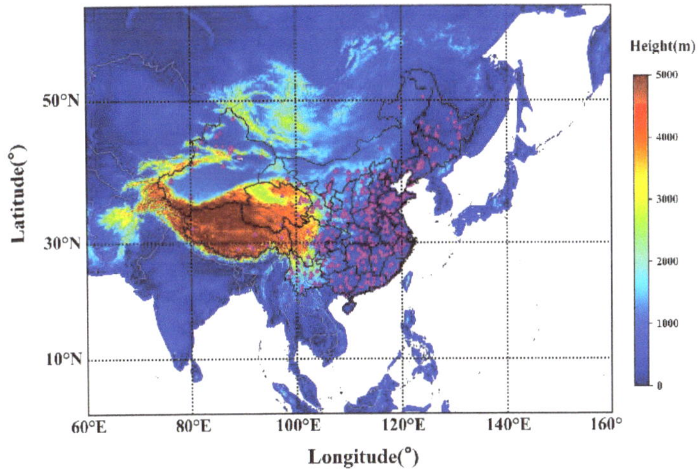

Figure 1. The study domain and overlapped digital elevation model from MERSI-II Level 1 GEO1K. The sea surface is shown in white. The national $PM_{2.5}$ monitoring stations employed by the Ministry of Environmental Protection on 22 October 2020 are depicted as the magenta-colored dots.

Additionally, Yang (2022) mentioned in his research on aerosol retrieval using MERSI remote sensing data that some haze areas could be recovered by modifying the recognition conditions of inland water bodies [23]. In this paper, the method proposed by Yang (2022) were replicated and cross-compared with the results of this study.

2.3. MHAM Algorithm

2.3.1. Algorithm Description

The design of the MHAM algorithm builds upon the spectral characteristics of different pixels, for example, the reflectance of the cloud pixels in the visible band is greater than that of the clear sky pixels, while the brightness temperature in the infrared channel is usually lower than that of the clear sky and haze pixels, and the brightness temperature difference is also higher than that of the clear sky and haze pixels. The reflectivity ratio of the near-infrared channel and the visible channel (such as the ratio of $R_{0.865}$ and $R_{0.65}$) has different values in different underlying surfaces [19,20,22]. The above characteristics were taken as the prior knowledge of this algorithm, and the adaptive improvements to the threshold selection were carried out for the MERSI instrument. Based on the selected MERSI bands, the tests were divided into two groups to distinguish the cloud, clear and haze pixels from each other. The flowchart of the algorithm is shown in Figure 2.

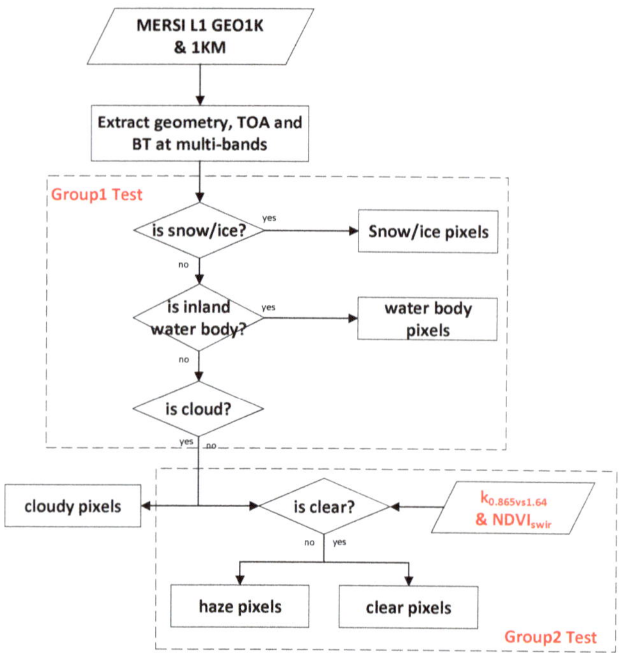

Figure 2. Flowchart of the MHAM algorithm using FY3D/MERSI-II observations.

Firstly, the DN (data number) values at different bands were extracted from the MERSI-II L1 data and converted to the apparent reflectance of the top of the atmosphere (TOA) at visible wavelengths or the brightness temperature at the infrared bands. The geolocation information and observed angles were extracted from the dataset. It must be noted that the MERSI Level 1 data at 1 km instead of those at 250 m were used in this analysis. Secondly, the ice/snow pixels were filtered out using the normalized difference snow index (NDSI), which was calculated using the apparent reflectance values at 0.55 μm ($R_{0.55}$) and 1.64 μm ($R_{1.64}$) for Landsat or $R_{0.55}$ and $R_{2.12}$ for MYD35 or $R_{0.865}$ and $R_{1.24}$ for MXD04 DT. Here, the ratio of the difference between $R_{0.55}$ and $R_{1.64}$ to the sum of $R_{0.55}$ and $R_{1.64}$ was taken as the NDSI. When both NDSI > 0.4 and $R_{0.865}$ > 0.1 were true, the pixel was treated as it was ice/snow. The inland water was identified by the combined conditions of the NDVI and $R_{2.13}$. For the remaining pixels, $R_{0.65}$, BT_{11} and the standard derivation of $R_{0.47}$ were used together for distinguishing the cloud pixels. Based on the binary tree method, the cloud and non-cloud pixels were, respectively, determined. Combined with the actual characteristics of the MERSI data, clear pixels over brighter surfaces can be capture accurately by introducing the variation between $R_{0.865}$ and $R_{1.64}$. In addition to the lower reflectance and higher brightness temperature values of the clear areas, the clear pixels and the remaining haze regions can be separated. More details about the conditions are summarized in Table 2.

Table 2. Details thresholds for MHAM algorithm using FY3D/MERSI-II observations.

Classification	Group1	Group2
ice/snow	[1] NDSI > 0.4 & $R_{0.865}$ > 0.1	/
Inland waterbody	[2] $NDVI_{nir}$ < 0.4 & $R_{2.13}$ < 0.08	/
cloud	$R_{0.65}$ > 0.45 *[defined as cloud_c1]*	/
	$R_{0.47\_std}$ > 0.0075 & $R_{0.65}$ > 0.4 *[defined as cloud_c2]*	
	BT_{11} < 250 K *[defined as cloud_c3]*	
clear	/	0 < $R_{0.65}$ < 0.2 *[defined as clear_c1]*
		$Diff_{0.865\_1.64}$ > 0 *[defined as clear_c2]*
		BT_{11} > 285 *[defined as clear_c3]*
		BT_{11}-$BT_{3.8}$ [−50, −40] *[defined as clear_c4]*
		[3] $NDVI_{swir}$ < 0.2 & 0.2 ≤ $R_{0.65}$ < 0.4 *[defined as clear_c5]*

[1] NDSI = $(R_{0.55} - R_{1.64})/(R_{0.55} + R_{1.64})$; [2] $NDVI_{nir} = (R_{0.865} - R_{0.65})/(R_{0.865} + R_{0.65})$; [3] $NDVI_{swir} = (R_{0.865} - R_{0.65})/(R_{0.865} + R_{0.65})$.

2.3.2. Selection of Spectral Characterization Based on FY3D/MERSI-II

In this analysis, the apparent reflectance at 0.65 μm ($R_{0.65}$), the brightness temperature at 11 μm (BT_{11}) and the difference between BT_{11} and $BT_{3.8}$ (BT_{11}-$BT_{3.8}$) were selected to separate the cloudy and clear skies according to different thresholds. It should be emphasized that 0.65 μm of atmospheric passage was selected in this paper, instead of using 0.47 μm of it, as was used in the MODIS aerosol algorithm. This is because the molecular scattering contribution is larger at large observation angles, and the apparent reflectance of clear sky pixels is larger than that of some haze pixels. Assuming that the solar zenith angle is 30°, the relative azimuth is 72°, the aerosol model is defined as the continental aerosol type in the mid-latitude summer and the satellite zenith angle ranges from 0 to 80°, Figure 3a,b presents the variation of molecular scattering under different aerosol loadings with the satellite zenith angles. When the observed angles are smaller than 50°, the contribution of molecular scattering changes modestly, whereas the molecular scattering increases significantly when the satellite zenith angles change from 50° to 80°. The scattering amplitude of the molecules in the red channel is less than 1/3 of that which occurs in the blue channel. The aerosol optical thickness decreases with the increase in the wavelength. As shown in Figure 3c,d, the variation trend of the aerosol scattering signal in the red band with observation angles is consistent with that of the blue band, and it is still sensitive to different aerosol contents. If 0.47 μm is used as the reflectance identification condition, it is easy to misidentify some clear sky pixels as haze ones. Therefore, it is feasible to select 0.65 μm as the identification condition in this paper.

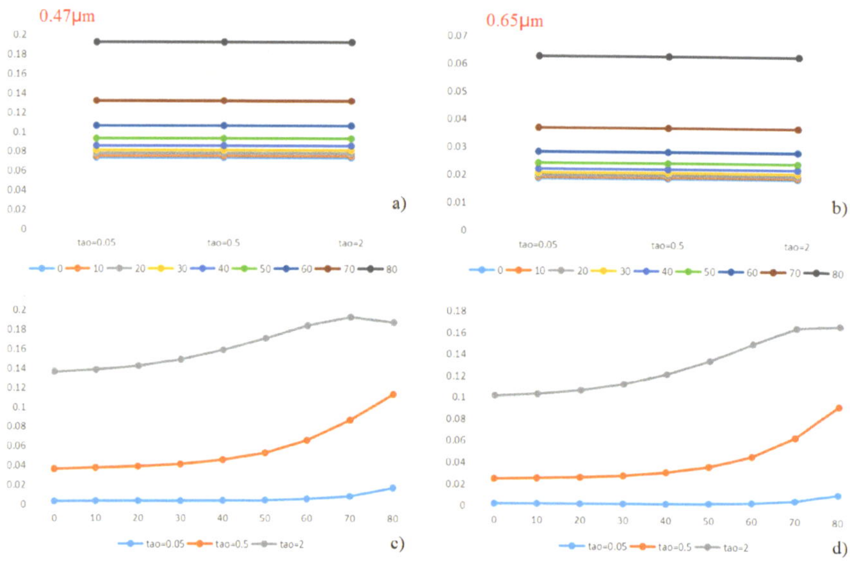

Figure 3. Variation of the apparent reflectance at two bands from molecule scattering (**a**,**b**) and aerosol scattering (**c**,**d**) as simulated by 6SV; the left panel stands for the variation in blue the channel, and the right one denotes in the red channel. Different colors in sub-panels (**a**,**b**) represent satellite zenith angles, and those in sub-panels (**c**,**d**) represent aerosol optical thickness.

2.3.3. Identification of Clear Conditions over Bright Surface

It is difficult to distinguish the clear pixels over the bright surface from the haze conditions due to them having similar spectral characteristics. The accuracy of haze detection could be heavily improved provided that the clear conditions over the bright surfaces can be accurately identified [21]. Through visual interpretation, the scenes of cloud, clear and haze regions on different underlying surfaces were selected, including spring, summer, fall and winter ones. The selected regions were kept constant to reduce the complexity of the scenes. The selected region changed only when the study region during a season was no longer pure. Details about the regions are summarized in Table 3. Inner Mongolia was selected as the clear region with a bright surface, which is defined as clear1. The area in Guangdong province was selected as clear2. The two cloud scenarios were selected to be located in Sichuan and Shaanxi, respectively. The Hebei and Henan provinces, where haze occurs frequently, were selected as the two haze scenarios. We extracted the apparent reflectance values at eight bands of each scenario, and we averaged them by region and season.

Figure 4 shows the variation trend of spectral reflectance in each scenario. A similar pattern appeared in two cloudy scenarios, but with different apparent reflectance values. The reflectance of different bands in clear2, the haze1 and haze2 showed similar variation. Surprisingly, in clear1, the apparent reflectance of the B6 and B7 channels are larger than those of the B4 and visible channels, which is contrary to that of the cloudy and haze pixels. Haze often contains absorptive components, which will attenuate the apparent reflectance after absorbing an amount of solar radiation. As we know, 1.64 μm and 2.13 μm were often used to study the characteristics of the parameters over land. For the clear pixels, 2.13 μm is only a little bit affected by the atmosphere, and the apparent reflectance is almost equal to the surface reflectance. Compared with 1.64 μm, the total apparent reflectance at 2.13 μm may be smaller than the surface reflectance with the enhancement of the atmospheric signal. As described as Section 2.3.1, visible wavelengths are susceptible to molecular scattering, therefore, the near-infrared channel (B4) and the B6 channel were selected to construct the reflectance variation for identifying the clear pixels over bright surfaces. Non-bright surface clear skies (clear2) are usually well identified when one is using a reflectance value that is below a certain threshold. If the cloud pixel coverage area can be accurately identified, clear skies and haze can be distinguished from the remaining samples.

Table 3. The MERSI-II data set used for training the identification condition with bright surfaces.

Scenarios	Selected Areas	Selected Time
clear1	42.8496°~45.3657°, 109.1090°~113.6038°	8 November 2019/4 April 2020/ 14 June 2020
	42.2122°~43.1121°, 108.6515°~109.7193°	3 December 2019
clear2	23.9992°~24.3478°, 112.7604°~113.6705°	8 November 2019/3 December 2019
	23.0716°~23.7766°, 115.0593°~115.8996°	2 December 2019
	25.3634°~25.9159°, 116.7444°~117.5885°	15 March 2020
cloud1	27.8471°~29.0374°, 103.8887°~105.2329°	8 November 2019/23 December 2019/ 3 April 2020/17 June 2020
cloud2	36.1867°~36.7887°, 108.4950°~109.3658°	8 November 2019/24 December 2019/ 1 April 2020/22 June 2020
haze1	38.0390°~39.24°, 115.2247°~116.4105°	8 November 2019/13 December 2019/ 6 April 2020/11 June 2019
haze2	33.9950°~35.4849°, 113.6284°~114.8294°	8 November 2019/13 December 2019/ 26 April 2020/18 June 2020
	34.8348°~35.8514°, 114.2112°~115.4611°	18 June 2019

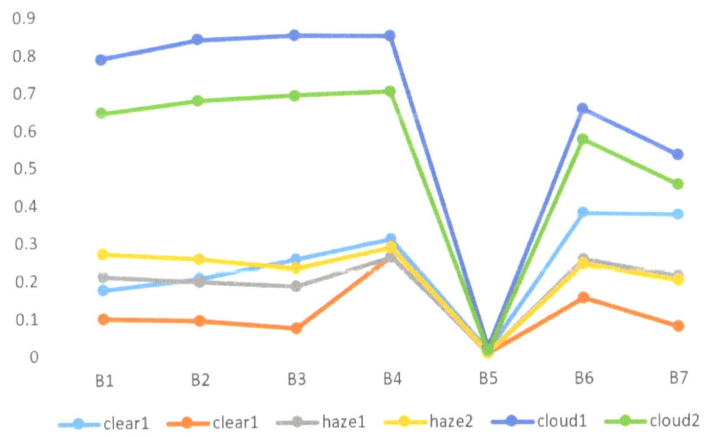

Figure 4. The spectral variation of different pixels over different underlying surfaces using MERSI-II apparent reflectance data. Different colors correspond to the scenarios in Table 2.

To confirm that whether the above parameter is reasonable, Figure 5 shows the difference of between $R_{0.865}$ and $R_{1.64}$ on 8 November 2019 and 1 December 2019, and a difference that is larger than 0 is masked in the white color. Combined with the true color maps, this parameter is sensitive to the clear sky with the bright surface. Therefore, this parameter can be added to the recognition process to further reduce the misjudgment of clear pixels over bright surfaces.

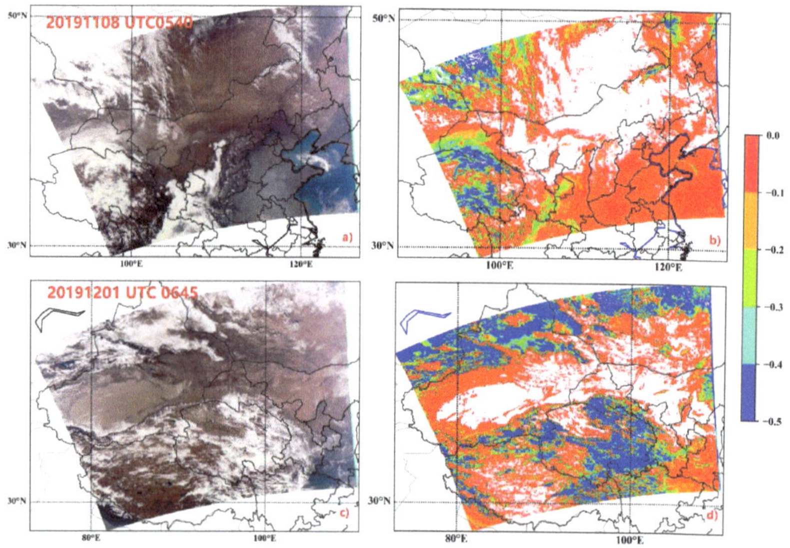

Figure 5. True color map and the distribution of the difference between $R_{0.865}$ and $R_{1.64}$. (**a**) True color map at UTC0540 on 8 November 2019; (**b**) the difference between $R_{0.865}$ and $R_{1.64}$ at UTC0540 on 8 November 2019; (**c**) True color map at UTC0645 on 1 December 2019; (**d**) the difference between $R_{0.865}$ and $R_{1.64}$ at UTC0645 on 1 December 2019.

2.3.4. Single Threshold Tests

Cloud Threshold Tests

$R_{0.65}$ is an effective indicator of clouds with large reflectance and small absorption values. BT_{11} is helpful for detecting cold clouds over inland water and certain land features. As mentioned above, the standard derivation of the reflectance at 0.47 μm usually stands for the texture patterns of the cloud edge. The cloudy regions detected using the detection conditions are shown in Figure 6. It should be addressed that Figure 6b–d intends to show that these three conditions were not independent when we were detecting cloudy pixels and the common regions distinguished by the three conditions were overwritten by the latter condition. This test of $R_{0.65}$ (defined as cloud_c1) was very robust over bright cloud layers, but it was invalid over very thin cirrus and cloud boundaries. The cloud_c2 is sensitive to the cloud edges that cloud_c1 did not detect. Part of the cloudy regions over the ocean were captured by cloud_c3, which was not detected by cloud_c1 nor cloud_c2. The complete cloudy regions exhibited in Figure 6e complement each other, showing that each criterion is necessary.

Clear Threshold Tests

As mentioned above, the red channel is less affected by molecular scattering than the blue channel ix, therefore, $R_{0.65}$ can be also used to identify the pixels over an absolutely clear sky. We assumed that a pixel is treated as clear when $R_{0.65} < 0.2$, as shown in Figure 7b. Based on the actual characteristics of the MERSI data, it is found that the difference between $R_{0.865}$ and $R_{1.64}$ can be an indicator of distinguishing the clear pixels on a bright surface. As depicted in Figure 7c, the clear sky areas in Inner Mongolia, northwest Lanzhou and part of Qinghai were accurately identified. In addition, there are few atmospheric signals in Shanxi and Hebei that had been captured by this condition. BT_{11} is also used as a clear sky identification condition. Through trial and error, when $BT_{11} > 285$ K, it is considered as a clear sky pixel, as shown in Figure 7d. It can be seen that the bright temperature difference between BT_{11} and $BT_{3.9}$, the reflectance between 0.2 and 0.4 and $NDVI_{swir} < 0.2$ are a supplement to the bright surface clear sky area. Using these three cloudy conditions and five clear sky conditions, the remaining pixels marked in white in Figure 7g were identified as haze pixels.

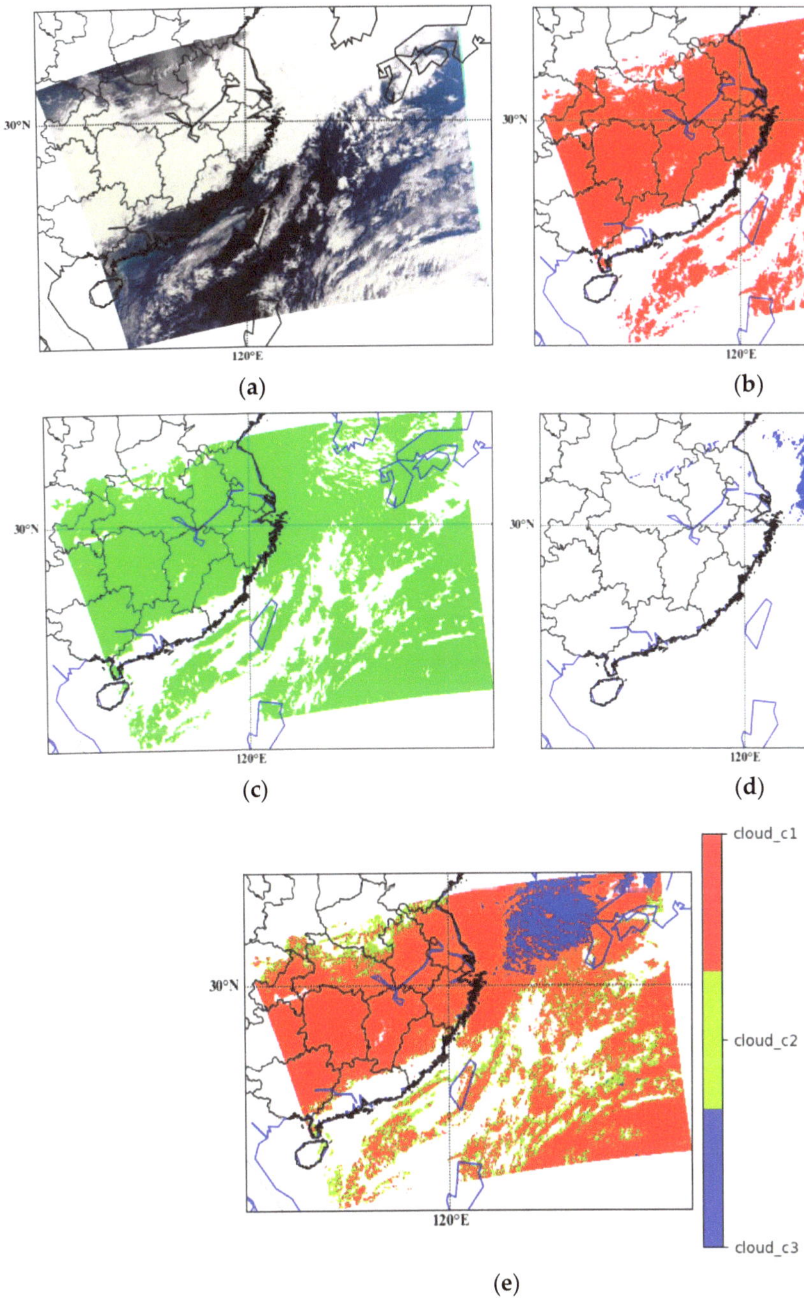

Figure 6. Test of cloud detection conditions at UTC0500 on 1 December 2019. (**a**) A true color map. (**b**) The cloudy region was detected by cloud condition 1, (**c**) the cloudy region was detected by cloud condition 2, (**d**) the cloudy region was detected by cloud condition 3 (**e**) and the final cloudy regions were detected by three conditions.

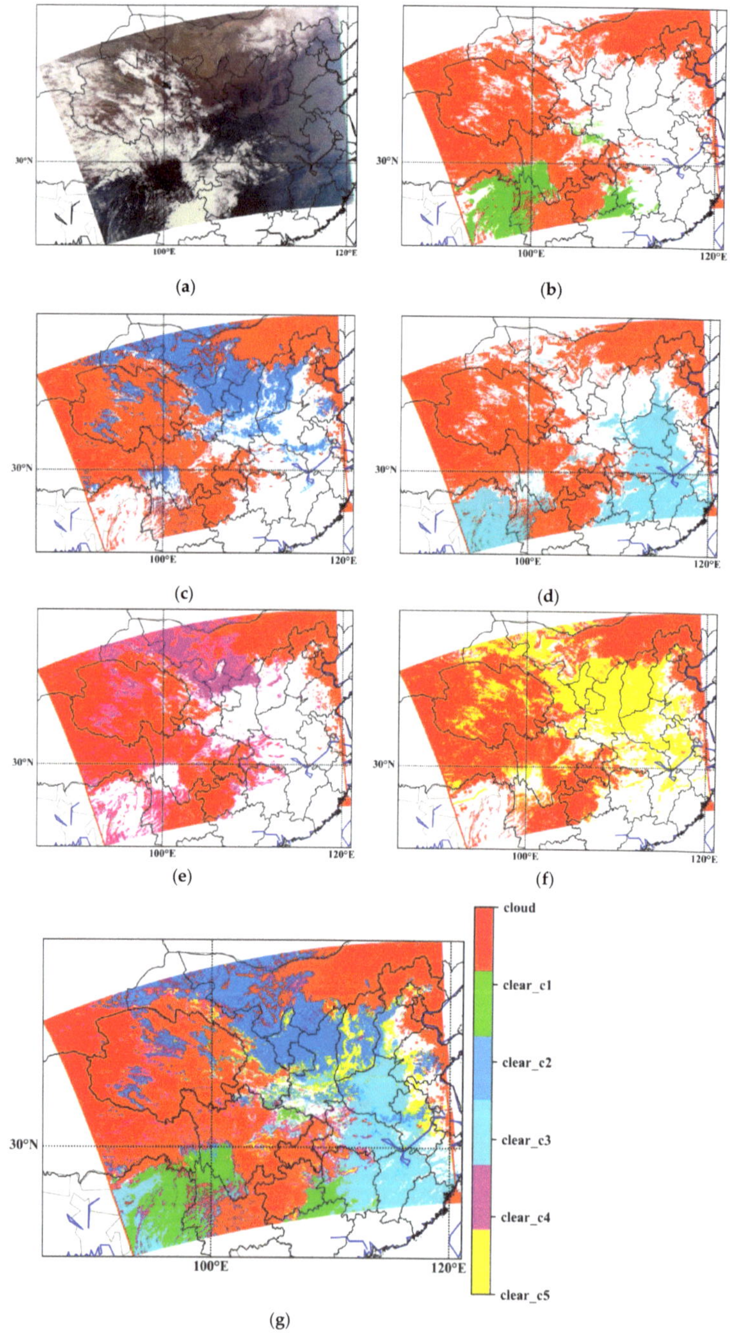

Figure 7. Test of clear detection conditions at UTC0605 on 3 December 2019. (**a**) A true color map. (**b**) The cloudy region was detected by three cloud conditions and the first clear condition, (**c**) the clear pixels were detected by clear condition 2, (**d**) the clear pixels were detected by clear condition 3, (**e**) the clear pixels were detected by clear condition 4, (**f**) the clear pixels were detected by clear condition 5 and (**g**) the final cloudy regions and clear pixels were detected by three cloudy conditions and five clear conditions.

3. Results

3.1. Distribution of Haze Identification

Two haze scenarios in winter were selected to test the performance of our proposed method, as shown in Figure 8. The first row of images is 2 December 2019, and the second row is 9 February 2020. The left panel is a true color map, and the right one is the distribution of the identified haze. On 2 December 2019, most parts of eastern China and India were shrouded by haze. Except for the clear sky area in Inner Mongolia, Xinjiang and the Qinghai-Tibet Plateau, the other parts of the whole study area were covered by cloud pixels. All of the above phenomena could be identified by the algorithm, and the boundary between the clear sky area and the cloud area could be clearly identified by the algorithm. According to the true color map, on 9 February 2020, the haze in the eastern coastal areas of China, such as Shandong and Tianjin, was relatively serious, while other areas were basically experiencing the state of light haze. In India, medium haze and light haze appeared simultaneously, and the distribution of clouds, clear skies and haze were also accurately identified. However, it is easy to miss cloud features when a thin or broken cloud covers a clear sky (e.g., South Korea and North Korea), and there is a slight error in distinguishing light haze from clear sky (e.g., Central India).

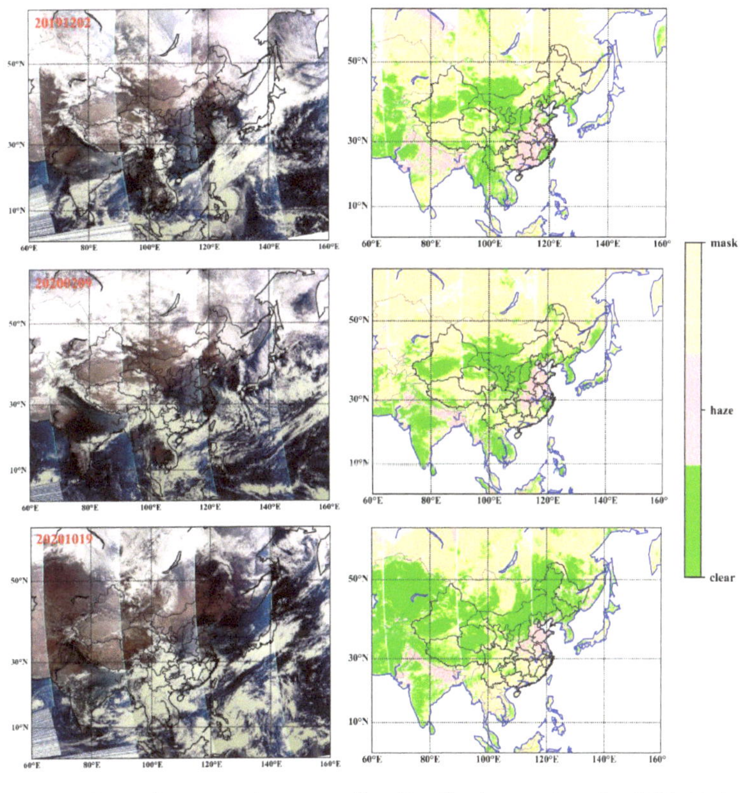

Figure 8. True color map and corresponding identification map based on MHAM algorithm, these three rows represent the distributions on 2 December 2019, 9 February 2020 and 19 October 2020, respectively.

In addition, the performance of clear pixels on a brighter surface in 19 October 2020 is further presented as the third row of images in Figure 8. Compared with the true color map, it can be clearly seen that the algorithm can accurately identify the clear sky pixels in high-altitude areas such as Inner Mongolia, northwest China, Afghanistan, Pakistan and the Qinghai-Tibet Plateau, as well as the clear sky in low-value areas of the three provinces of northeastern China. It shows that the variation of 0.865 and 1.64 μm and $NDVI_{swir}$ introduced in this algorithm are effective.

Meanwhile, the corresponding operational cloud mask with the same dates as Figure 8 derived from MYD35 and MERSI-II were also presented for a cross-comparison. As depicted in Figure 9, the cloud, probably cloud, clear and probably clear pixels are drawn in yellow, gray, green and aqua, respectively. The recognition values of the clear sky and cloud pixels are basically the same, and the difference is mainly reflected in the processing of possible cloud and possible clear sky pixels. Compared with the classifications in Figure 8, neither the MODIS nor MERSI-II operational cloud mask identified the haze regions that appeared in India and in eastern China. It can be seen that MODIS treated those as probably cloud and probably clear pixels, while the processing conditions of MERSI-II were not as strict as that of MODIS, thus most of the haze pixels were labeled as clear.

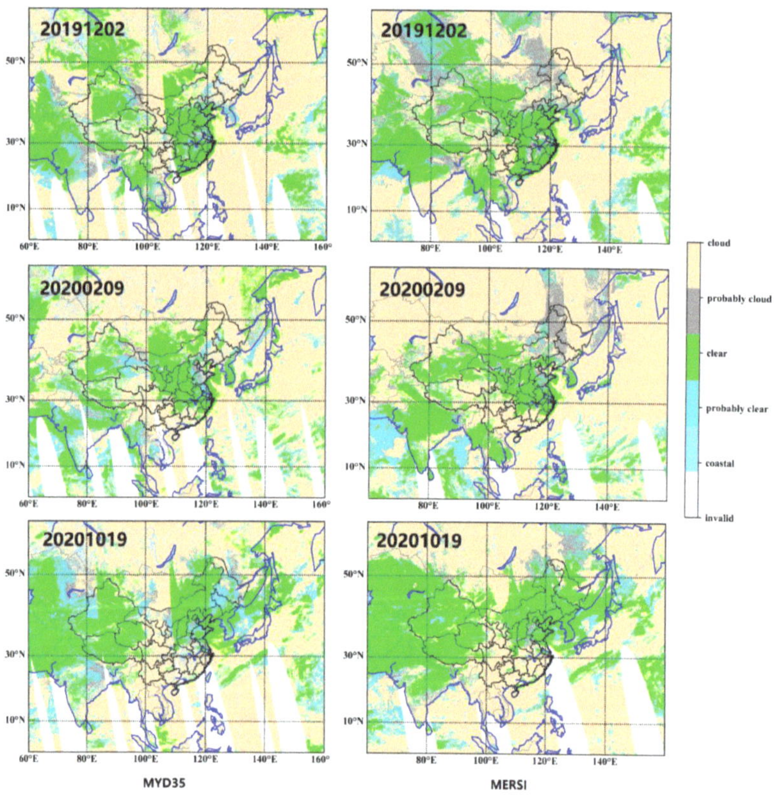

Figure 9. Aqua/MODIS cloud mask (left) and MERSI operational cloud mask (right), these three lines represent the distribution on 2 December 2019, 9 February 2020 and 19 October 2020, respectively.

3.2. Validation against $PM_{2.5}$ Measurements

As we all know, ground-based stations measure the air quality near the surface, while satellites observe the overall state in the troposphere. There are differences between these observation methods. At present, there is no unified standard, that is, when the $PM_{2.5}$ values measured by ground observation are greater than the threshold, it is consistent with the haze determined by the satellite. Therefore, this paper uses the two thresholds of 35 $\mu g/m^3$ and 50 $\mu g/m^3$ of the existing $PM_{2.5}$ levels to test the MERSI haze identification effect.

Firstly, the 18 MERSI orbits covering eastern China during the haze season were selected, which were mainly concentrated at 12–14 PM Beijing time. The national observation

$PM_{2.5}$ data during the same period were downloaded, and the samples with a $PM_{2.5}$ that was larger than 35 μg/m^3 and 50 μg/m^3 were respectively screened out for validation. Secondly, the spatial and temporal matching between the ground-based and satellite observations was performed. In the terms of the temporal scale, we chose the $PM_{2.5}$ value with the nearest hour to the MERSI orbit, for example, the transit time of MERSI is UTC0515, so we chose $PM_{2.5}$ data at 13:00 Beijing time. Spatially, taking the national station as the center, the MERSI pixels falling within a radius of 500 m were selected. If there were multiple pixels, then the nearest one was selected. Thirdly, the pixels identified by MERSI as cloud, ice/snow, inland waterbody ones were further removed from the matched samples, and only the haze and clear pixels were retained. Finally, the ratio of the number of samples identified as haze to that of the sum of the samples identified as haze and clear pixels were calculated to demonstrate the accuracy. Table 4 summarizes the test results based on the selected MERSI orbits. As it can be obviously seen in Table 4, the haze identification rate with a $PM_{2.5}$ threshold of 50 μg/m^3 is generally better than that with a $PM_{2.5}$ threshold of 35 μg/m^3. When $PM_{2.5}$ > 35 μg/m^3 was considered to be haze days, 13 samples with an identification rate that was larger than 85% accounted for 72.22% of the total samples (18), and 61.11% of the sample points (11) had an identification rate that was above 90%. When ground-based $PM_{2.5}$ > 50 μg/m^3 was considered as haze days, 83.33% and 66.67% of the samples had identification rates that were higher than 85% and 90%, respectively, indicating that the greater the ground-based $PM_{2.5}$ value is, the better the consistency is of the satellite identification of haze.

Table 4. Accuracy test of FY3D/MERSI-II haze identification effect based on ground air quality $PM_{2.5}$ monitoring data.

Date/Orbit	$PM_{2.5}$ >= 50 μg/m^3			$PM_{2.5}$ >= 35 μg/m^3		
	Haze	Clear	Hit Rate (%)	Haze	Clear	Hit Rate (%)
14 January 2020-0430	48	2	96.00	76	5	93.83
19 January 2020-0435	163	2	98.79	195	6	97.01
20 January 2020-0555	145	42	75.94	190	66	74.21
28 January 2020-0505	201	33	85.90	250	39	86.51
29 January 2020-0445	172	11	93.99	220	18	92.44
30 January 2020-0425	74	2	97.37	86	4	95.56
31 January 2020-0545	162	15	91.53	242	28	89.63
31 January 2020-0550	129	61	67.89	139	91	60.43
3 February 2020-0450	125	5	96.15	198	18	91.67
9 February 2020-0435	95	0	100	144	2	98.63
4 March 2020-0520	53	0	100	94	1	98.95
4 March 2020-0525	14	2	87.5	12	33	26.67
7 March 2020-0425	83	1	98.81	98	2	98.0
7 March 2020-0430	10	0	100	12	12	100
19 October 2020-0445	136	5	96.45	193	12	94.15
21 October 2020-0545	64	11	85.33	5	46	9.80
21 October 2020-0550	1	26	3.70	127	34	78.88
22 October 2020-0525	91	5	94.79	164	15	91.62

Taking some orbits as an example, Figure 10 shows the true color map and the MERSI identification results overlapping it with the $PM_{2.5}$ values that were observed by the national stations of UTC0505 on 20 January 2020, UTC0450 on October 19 and UTC0525 on October 22, respectively, covering the scenarios of cloudy, hazy and clear skies over different surface types. For the MERSI identification results, cloud, snow, ice and inland water were combined into masks, and they are marked with yellow. For the pixels that MERSI identifies as clouds, no matter how large the $PM_{2.5}$ observed by the state-controlled stations was, they could not be captured (these stations are not hidden here). For the pixels identified as clear skies by MERSI, most of the overlapped $PM_{2.5}$ values were smaller than 35 μg/m^3, such as the skies over Inner Mongolia and southern China on October 19. Due to its topography and large amount of anthropogenic emissions, haze events occur frequently in the North China Plain. From the true color map, it is obvious that the North China Plain was covered by haze on January 28, February 3 and October 19. This study can identify these haze areas well, and the superimposed $PM_{2.5}$ level is at least good. It should be emphasized that at UTC0525 on October 22, the haze pollution levels in the Henan, Anhui and Jiangsu Provinces were not obvious on the true color map, but these areas were identified as haze in this study. According to the observation data from the national control station, $PM_{2.5}$ was already in moderate and severe pollution. In general, the haze area and clear sky area identified by this study are consistent with the classification of the national control stations.

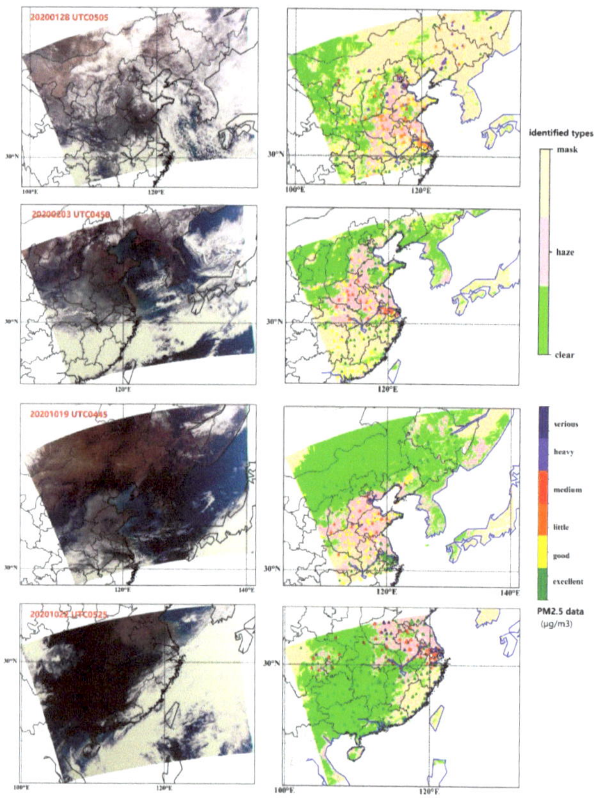

Figure 10. True color map and identified distribution overlaid with the national $PM_{2.5}$ monitoring data (unit: μg/m^3). $PM_{2.5}$ values belonging to excellent, good, little pollution, medium pollution, heavy pollution and serious pollution are marked as green, yellow, orange, red, purple and dark purple, respectively.

3.3. Cross-Comparison with Other Method

To further demonstrate the accuracy of this analysis, four MERSI orbits were selected, and the identification method proposed by Yang (2022) was implemented in this section for a cross-comparison [23]. On one hand, the pixel identification can provide the distribution of haze regions from a qualitative perspective, and on the other hand, this study's identification results can be used as a haze mask for the retrieval of aerosol quantitative data. That is to say, we should pay more attention to the pixels classified as haze and the number of the pixels that are to be retrieved. Therefore, the cloud, ice/snow and inland waterbody pixels were combined in a background mask, and the remaining pixels were used for the comparison. Figure 11 shows the true color map and the comparisons of two recognition methods on UTC0540 8 November 2019, UTC0445 29 January 2020, UTC0545 31 January 2020 and UTC0550 31 January 2020, covering clear regions over bright surfaces, haze regions over the China North Plain and clear regions in southern China, respectively. It is obviously seen that, both this study and Yang's study (2022) can identify the certainly cloudy pixels well, but there have some discrepancies between them in the identification of cloud edges, heavy haze and clear regions over bright surfaces. For example, the recognition procedure of cloud pixels in Yang's study (2022) is stricter than that of this study, as shown in 29 January 2020, when most of the areas were covered by a haze layer, and the east of Shandong Province was covered by a mixture of haze and clouds [23]. By contrast, our identification method is very sensitive to these phenomena, which is beneficial to increase the number of retrieved pixels and improve the detection ability of the MERSI instrument during haze days. At UTC0550, on 31 January 2020, the number of clear pixels in outer Mongolia detected by Yang (2022) was smaller than that of this study, which was closer to the true color map [23]. It is also found that this study is very sensitive to the identification of cloud edges/clear sky edges, such as those in the eastern part of Yunnan province shown in UTC0545 on 31 January 2020. Although it performs well in the identification of haze or clear or cloud edges, large errors will be introduced in the aerosol retrievals. This is a double-edged sword that needs to be balanced in different application scenarios.

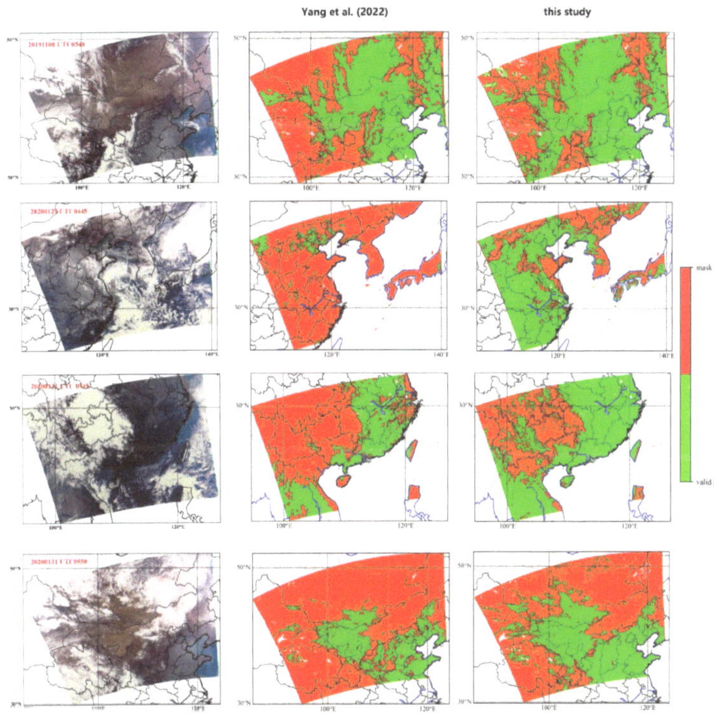

Figure 11. True color map and comparison of identified results of Yang (2022) [23] and the MHAM algorithm at UTC0540 on 8 October 2019 (the first line), UTC0445 on 29 January 2020, UTC0545 on 31 January 2020 and UTC0550 on 31 January 2020.

4. Conclusions

Cloud detection products based on FY3D/MERSI-II data have been released as operational products for many years, supplying cloud masks for the improved retrieval of atmospheric environmental products (such as aerosol and $PM_{2.5}$, etc.), and they have continental surface parameters (such as the land surface temperature). For aerosol quantitative retrieval, the cloud detection products of MERSI have limits in their haze area identification. The absolutely clear and cloudy pixels can be distinguished from each other based on the fact that they have opposite characteristics in terms of reflectance in the visible channels and brightness temperature in the infrared channels. Combined with the actual patterns of the MERSI L1 data, the positive corresponding relationship between the apparent reflectance at 0.865 μm and 1.64 μm for the clear pixels over regions with a bright surface is proposed as an identification condition. Thus, this study proposes a novel MERSI haze mask (MHAM) algorithm that divides the pixels into one of three categories: cloudy, clear or haze ones, thereby resolving the problem of distinguishing haze pixels from the clear ones over a bright surface.

In comparing the operational Aqua/MODIS and MERSI-II cloud detection products with the true color map, it reveals that these two products misclassified the heavy haze pixels as certain cloud, possible cloud and possible clear pixels, and they misclassified the pixels with light/moderate haze conditions as possible cloud, possible clear and certainly clear regions. In contrast, the novel algorithm proposed in this paper is able to capture the haze regions correctly, making it more compatible with the true color images. The real-time $PM_{2.5}$ data from the national air quality monitoring stations served as true values, and two thresholds were used to evaluate the new algorithm. The results showed that the novel algorithm has 72.22% (13 out of total 18 samples) accuracy when a $PM_{2.5}$ that was larger than 35 $\mu g/m^3$ was considered as haze conditions, while for $PM_{2.5}$ values that were larger than 50 $\mu g/m^3$, 83.33% and 66.67% of the total samples were examined with better identification rates than 85% and 90%, respectively, indicating that the greater the ground-based $PM_{2.5}$ value is, the better the consistency is in satellite haze identification. Through the cross-comparison with the method proposed by Yang (2022) [23], it was found that the novel algorithm has a better sensitivity to bright surface clear sky areas and haze areas, and there will be more valid pixels when the subsequent quantitative inversion of aerosol characteristics is performed.

Haze detection based on satellite remote sensing data is still in the research stage of recognition, and there has been no corresponding official product released so far, thus leading to great difficulties in the quantitative validation of the haze identification effect. With the definition of haze, a richer comparison between the accuracies of this identification method by using combining relative humidity and visibility data will be carried out. In addition, the haze identification method proposed in this study directly depends on the apparent reflectance and brightness temperature, that is, when the calibration coefficient is adjusted, the corresponding identification threshold also needs to be updated synchronously. We will monitor of the MERSI instrument data state and update the recognition threshold iteratively.

Author Contributions: Conceptualization, Y.S. and X.Z.; methodology, Y.S. and L.C.; software, Y.S.; validation, Y.S. and L.Y.; investigation, Y.S. and F.W.; resources, Y.S.; data curation, L.C., F.W., N.X.; writing—original draft preparation, Y.S.; writing—review and editing, L.C., Z.Z., L.Y., F.W., N.X. and X.Z.; visualization, Y.S.; supervision, L.C., N.X. and X.Z.; project administration, Y.S., F.W., L.C.; funding acquisition, Y.S. and F.W. All authors have read and agreed to the published version of the manuscript.

Funding: This work is jointly supported by the National Key Research and Development Program of China (No. 2019YFC0214601) and Beijing Municipal Natural Science Foundation (No. 8214065).

Data Availability Statement: Data sharing not applicable.

Acknowledgments: We appreciate the NASA Langley Research Center Atmospheric Sciences Data Center (https://eosweb.larc.nasa.gov/, accessed by 10 April 2022) for providing the MYD35 product.

Conflicts of Interest: The authors declare no conflict of interest.

References

1. Wu, D. A Discussion on Difference between Haze and Fog and Warning of Ash Haze Weather. *Meteorol. Mon.* **2005**, *31*, 7.
2. Gautam, R.; Hsu, N.C.; Eck, T.F.; Holben, B.N.; Janjai, S.; Jantarach, T.; Tsay, S.C.; Lau, W.K. Characterization of aerosols over the Indochina peninsula from satellite-surface observations during biomass burning pre-monsoon season. *Atmos. Environ.* **2013**, *78*, 51–59. [CrossRef]
3. Menon, S.; Hansen, J.; Nazarenko, L.; Luo, Y.F. Climate effects of black carbon aerosols in China and India. *Science* **2002**, *297*, 2250–2253. [CrossRef] [PubMed]
4. Zhang, X.Y.; Wang, F.; Wang, W.H.; Huang, F.X.; Chen, B.L.; Gao, L.; Wang, S.P.; Yan, H.H.; Ye, H.H.; Si, F.Q.; et al. The development and application of satellite remote sensing for atmospheric compositions in China. *Atmos. Res.* **2020**, *245*, 105056. [CrossRef]
5. Xia, X.A.; Chen, H.B.; Wang, P.C.; Zhang, W.X.; Goloub, P.; Chatenet, B.; Eck, T.F.; Holben, B.N. Variation of column-integrated aerosol properties in a Chinese urban region. *J. Geophys. Res.-Atmos.* **2006**, *111*. [CrossRef]
6. Liu, X.G.; Li, J.; Qu, Y.; Han, T.; Hou, L.; Gu, J.; Chen, C.; Yang, Y.; Liu, X.; Yang, T.; et al. Formation and evolution mechanism of regional haze: A case study in the megacity Beijing, China. *Atmos. Chem. Phys.* **2013**, *13*, 4501–4514. [CrossRef]
7. Zhang, X.; Wang, H.; Che, H.Z.; Tan, S.C.; Shi, G.Y.; Yao, X.P. The impact of aerosol on MODIS cloud detection and property retrieval in seriously polluted East China. *Sci. Total Environ.* **2020**, *711*, 134634. [CrossRef] [PubMed]
8. Che, H.; Xia, X.; Zhu, J.; Li, Z.; Dubovik, O.; Holben, B.; Goloub, P.; Chen, H.; Estelles, V.; Cuevas-Agullo, E.; et al. Column aerosol optical properties and aerosol radiative forcing during a serious haze-fog month over North China Plain in 2013 based on ground-based sunphotometer measurements. *Atmos. Chem. Phys.* **2014**, *14*, 2125–2138. [CrossRef]
9. Cheng, Z.; Wang, S.; Fu, X.; Watson, J.G.; Jiang, J.; Fu, Q.; Chen, C.; Xu, B.; Yu, J.; Chow, J.C.; et al. Impact of biomass burning on haze pollution in the Yangtze River delta, China: A case study in summer 2011. *Atmos. Chem. Phys.* **2014**, *14*, 4573–4585. [CrossRef]
10. Tao, M.H.; Chen, L.F.; Wang, Z.F.; Wang, J.; Tao, J.H.; Wang, X.H. Did the widespread haze pollution over China increase during the last decade? A satellite view from space. *Environ. Res. Lett.* **2016**, *11*, 054019. [CrossRef]
11. Tao, M.H.; Chen, L.F.; Xiong, X.Z.; Zhang, M.G.; Ma, P.F.; Tao, J.H.; Wang, Z.F. Formation process of the widespread extreme haze pollution over northern China in January 2013: Implications for regional air quality and climate. *Atmos. Environ.* **2014**, *98*, 417–425. [CrossRef]
12. Tao, M.H.; Li, R.; Wang, L.L.; Lan, F.; Wang, Z.F.; Tao, J.H.; Che, H.Z.; Wang, L.C.; Chen, L.F. A critical view of long-term AVHRR aerosol data record in China: Retrieval frequency and heavy pollution. *Atmos. Environ.* **2020**, *223*, 117246. [CrossRef]
13. Lim, H.; Choi, M.; Kim, J.; Kasai, Y.; Chan, P.W. AHI/Himawari-8 Yonsei Aerosol Retrieval (YAER): Algorithm, Validation and Merged Products. *Remote Sens.* **2018**, *10*, 699. [CrossRef]
14. Remer, L.A.; Mattoo, S.; Levy, R.C.; Heidinger, A.; Pierce, R.B.; Chin, M. Retrieving aerosol in a cloudy environment: Aerosol product availability as a function of spatial resolution. *Atmos. Meas. Tech.* **2012**, *5*, 1823–1840. [CrossRef]
15. Wang, Y.; Chen, L.F.; Li, S.S.; Wang, X.H.; Yu, C.; Si, Y.D.; Zhang, Z.L. Interference of Heavy Aerosol Loading on the VIIRS Aerosol Optical Depth (AOD) Retrieval Algorithm. *Remote Sens.* **2017**, *9*, 397. [CrossRef]
16. Zeng, S.; Parol, F.; Riedi, J.; Cornet, C.; Thieuleux, F. Examination of POLDER/PARASOL and MODIS/Aqua Cloud Fractions and Properties Representativeness. *J. Clim.* **2011**, *24*, 4435–4450. [CrossRef]
17. Levy, R.C.; Mattoo, S.; Munchak, L.A.; Remer, L.A.; Sayer, A.M.; Patadia, F.; Hsu, N.C. The Collection 6 MODIS aerosol products over land and ocean. *Atmos. Meas. Tech.* **2013**, *6*, 2989–3034. [CrossRef]
18. Hutchison, K.D.; Iisager, B.D.; Kopp, T.J.; Jackson, J.M. Distinguishing aerosols from clouds in global, multispectral satellite data with automated cloud classification algorithms. *J. Atmos. Ocean. Technol.* **2008**, *25*, 501–518. [CrossRef]
19. Ge, W.; Chen, L.F.; Si, Y.D.; Ge, Q.; Fan, M.; Li, S.S. Haze Spectral Analysis and Detection Algorithm Using Satellite Remote Sensing Data. *Spectrosc. Spectr. Anal.* **2016**, *36*, 3817–3824.
20. Shang, H.Z.; Chen, L.F.; Tao, J.H.; Su, L.; Jia, S.L. Synergetic Use of MODIS Cloud Parameters for Distinguishing High Aerosol Loadings from Clouds Over the North China Plain. *IEEE J. Sel. Top. Appl. Earth Obs. Remote Sens.* **2014**, *7*, 4879–4886. [CrossRef]
21. Ackerman, S.A.; Strabala, K.I.; Menzel, W.P.; Frey, R.A.; Moeller, C.C.; Gumley, L.E. Discriminating clear sky from clouds with MODIS. *J. Geophys. Res.-Atmos.* **1998**, *103*, 16. [CrossRef]
22. Shang, H.Z.; Chen, L.F.; Letu, H.S.; Zhao, M.; Li, S.S.; Bao, S.H. Development of a daytime cloud and haze detection algorithm for Himawari-8 satellite measurements over central and eastern China. *J. Geophys. Res.-Atmos.* **2017**, *122*, 3528–3543. [CrossRef]
23. Yang, L.K.; Hu, X.Q.; Wang, H.; He, X.W.; Liu, P.; Xu, N.; Yang, Z.D.; Zhang, P. Preliminary test of quantitative capability in aerosol retrieval over land from MERSI-II onboard FY3D. *Natl. Remote Sens. Bull.* **2022**, *26*, 923–940.
24. Shi, Y.X.R.; Levy, R.C.; Yang, L.K.; Remer, L.A.; Mattoo, S.; Dubovik, O. A Dark Target research aerosol algorithm for MODIS observations over eastern China: Increasing coverage while maintaining accuracy at high aerosol loading. *Atmos. Meas. Tech.* **2021**, *14*, 3449–3468. [CrossRef]

25. Yang, Z.D.; Lu, N.M.; Shi, J.M.; Zhang, P.; Dong, C.H.; Yang, J. Overview of FY-3 Payload and Ground Application System. *IEEE Trans. Geosci. Remote Sens.* **2012**, *50*, 4846–4853. [CrossRef]
26. Hu, X.Q.; Liu, J.J.; Sun, L.; Rong, Z.G.; Li, Y.; Zhang, Y.; Zheng, Z.J.; Wu, R.H.; Zhang, L.J.; Gu, X.F. Characterization of CRCS Dunhuang test site and vicarious calibration utilization for Fengyun (FY) series sensors. *Can. J. Remote Sens.* **2010**, *36*, 566–582. [CrossRef]
27. Hu, X.Q.; Sun, L.; Liu, J.J.; Ding, L.; Wang, X.H.; Li, Y.; Zhang, Y.; Xu, N.; Chen, L. Calibration for the Solar Reflective Bands of Medium Resolution Spectral Imager Onboard FY-3A. *IEEE Trans. Geosci. Remote Sens.* **2012**, *50*, 4915–4928. [CrossRef]
28. Xu, N.; Niu, X.H.; Hu, X.Q.; Wang, X.H.; Wu, R.H.; Chen, S.S.; Chen, L.; Sun, L.; Ding, L.; Yang, Z.D.; et al. Prelaunch Calibration and Radiometric Performance of the Advanced MERSI II on FengYun-3D. *IEEE Trans. Geosci. Remote Sens.* **2018**, *56*, 4866–4875. [CrossRef]
29. Chen, J.; Yao, Q.; Chen, Z.Y.; Li, M.C.; Hao, Z.Z.; Liu, C.; Zheng, W.; Xu, M.Q.; Chen, X.; Yang, J.; et al. The Fengyun-3D (FY-3D) global active fire product: Principle, methodology and validation. *Earth Syst. Sci. Data* **2022**, *14*, 3489–3508. [CrossRef]
30. Li, C.C.; Mao, J.T.; Lau, A.K.H.; Yuan, Z.B.; Wang, M.H.; Liu, X.Y. Application of MODIS satellite products to the air pollution research in Beijing. *Sci. China Ser. D Earth Sci.* **2005**, *48*, 209–219.
31. Li, S.S.; Chen, L.F.; Xiong, X.Z.; Tao, J.H.; Su, L.; Han, D.; Liu, Y. Retrieval of the Haze Optical Thickness in North China Plain Using MODIS Data. *IEEE Trans. Geosci. Remote Sens.* **2013**, *51*, 2528–2540. [CrossRef]
32. Sayer, A.M.; Hsu, N.C.; Bettenhausen, C.; Jeong, M.J. Validation and uncertainty estimates for MODIS Collection 6 "Deep Blue" aerosol data. *J. Geophys. Res.-Atmos.* **2013**, *118*, 7864–7872. [CrossRef]
33. Xiao, Q.; Zhang, H.; Choi, M.; Li, S.; Kondragunta, S.; Kim, J.; Holben, B.; Levy, R.C.; Liu, Y. Evaluation of VIIRS, GOCI, and MODIS Collection 6AOD retrievals against ground sunphotometer observations over East Asia. *Atmos. Chem. Phys.* **2016**, *16*, 1255–1269. [CrossRef]
34. You, W.; Zang, Z.L.; Pan, X.B.; Zhang, L.F.; Chen, D. Estimating PM2.5 in Xi'an, China using aerosol optical depth: A comparison between the MODIS and MISR retrieval models. *Sci. Total Environ.* **2015**, *505*, 1156–1165. [CrossRef] [PubMed]
35. Wang, Z.; Li, R.Y.; Chen, Z.Y.; Yao, Q.; Gao, B.B.; Xu, M.Q.; Yang, L.; Li, M.C.; Zhou, C.H. The estimation of hourly PM2.5 concentrations across China based on a Spatial and Temporal Weighted Continuous Deep Neural Network (STWC-DNN). *ISPRS J. Photogramm. Remote Sens.* **2022**, *190*, 38–55. [CrossRef]

remote sensing

MDPI

Article

Satellite-Based Background Aerosol Optical Depth Determination via Global Statistical Analysis of Multiple Lognormal Distribution

Qi-Xiang Chen [1,*], Chun-Lin Huang [1,2], Shi-Kui Dong [1] and Kai-Feng Lin [3]

1 School of Energy Science and Engineering, Harbin Institute of Technology, Harbin 150001, China; 20b902059@stu.hit.edu.cn (C.-L.H.); dongsk@hit.edu.cn (S.-K.D.)

2 Key Laboratory of Middle Atmosphere and Global Environment Observation, Institute of Atmospheric Physics, Chinese Academy of Sciences, Beijing 100029, China

3 School of Chemistry and Chemical Engineering, Harbin Institute of Technology, Harbin 150001, China; linkaifeng@hit.edu.cn

* Correspondence: atlas.chen@hit.edu.cn

Abstract: Determining background aerosol optical depth threshold value (BAOD) is critical to aerosol type identification and air pollution control. This study presents a statistical method to select the best BAOD threshold value using the VIIRS DB AOD products at 1 × 1 degree resolution from 2012 to 2019 as a major testbed. A series of multiple lognormal distributions with 1 to 5 peaks are firstly applied to fit the AOD histogram at each grid point, and the distribution with the highest correlation coefficient (R) gives preliminary estimations of BAOD, which is defined as either the intersection point of the first two normal distribution curves when having multiple peaks, or the midpoint between the peak AOD and the first AOD with non-zero probability when the mono peak is the best fit. Then, the lowest 1st to 100th percentile AOD distributions are compared with the preliminary BAOD distribution on a global scale. The final BAOD is obtained from the best cutoff percentile AOD distributions with the lowest bias compared with preliminary BAOD. Results show that the lowest 30th percentile AOD is the best estimation of BAOD for different AOD datasets and different seasons. Analysis of aerosol chemical information from MERRA-2 further supports this selection. Based on the BAOD, we updated the VIIRS aerosol type classification scheme, and the results show that the updated scheme is able to achieve reliable detection of aerosol type change in low aerosol loading conditions.

Keywords: aerosol type; background aerosol; threshold value; AOD; VIIRS; MERRA-2

Citation: Chen, Q.-X.; Huang, C.-L.; Dong, S.-K.; Lin, K.-F. Satellite-Based Background Aerosol Optical Depth Determination via Global Statistical Analysis of Multiple Lognormal Distribution. *Remote Sens.* **2024**, *16*, 1210. https://doi.org/10.3390/rs16071210

Academic Editors: Xingfa Gu, Jing Wei and Shuaiyi Shi

Received: 3 March 2024
Revised: 28 March 2024
Accepted: 28 March 2024
Published: 29 March 2024

1. Introduction

Atmospheric aerosols play a key role in the earth-ocean-atmosphere system [1]. As a direct effect, they absorb and scatter the incoming short-wave solar radiation and the long-wave earth radiation; as an indirect effect, they modify the cloud formation processes by acting as the cloud condensation nuclei [2]. Aerosol particles can be emitted from various sources, such as the ocean, desert, wildfires, and urban/industrial emissions. Retrieving aerosol type information helps to characterize regional aerosol optical properties, track aerosol transport paths, and provide composition information for air quality research [3].

Background aerosol optical depth (BAOD) is a parameter characterizing the light extinction ability of local aerosol particles and represents the long-term stable level of aerosol concentrations from local emissions, rather than the long-range transport pollutants (i.e., dust storm), abrupt emission events (i.e., volcanic eruptions, wildfires), or unexpected aerosol accumulations due to adverse weather condition (i.e., haze). For a single site, BAOD could simply be regarded as a certain low level of AOD, but when applied to a large regional or global scale, BAOD should vary with locations; for example, BAOD in urban areas with anthropogenic emissions is supposed to be different from (generally higher than) BAOD at remote continent or grassland. Extracting BAOD from the total aerosol loading is important for us and model simulation to separate anthropogenic effects from natural variability and evaluate climate effects of specific aerosol types, e.g., urban/industrial, biomass burning, dust, etc. [4,5]. BAOD is also a key factor for some satellite retrieval algorithms performing atmospheric corrections of surface reflectance to remove aerosol contributions from satellite signals [6,7]. Thus, determining the BAOD threshold value is fundamental for both aerosol-climate effect assessment and satellite remote sensing correction.

Typical aerosols usually have certain AOD ranges. For example, background aerosols like clean continental and maritime aerosols often have an AOD value from 0 to 0.2, while anthropogenic and biomass-burning aerosols have much higher AOD, which ranges from 0.3 to 1.0, or even over 2 in some extremely polluted situations [8]. AOD is quite useful in identifying background aerosols, which are characterized by relatively low aerosol loading emitted from local sources [9,10]. Many previous studies have determined the BAOD threshold value when identifying aerosol types (e.g., urban industry, dust) from clean or background types. For example, Zhang, Kondragunta, Laszlo, Liu, Remer, Huang, Superczynski, and Ciren [6] used the lowest 5th percentile AOD as the estimate of the background AOD based on global AERONET observations; Xia, et al. [11] set the background level AOD_{550} as less than 0.1 based on 40 ground-based observations across China; Petrenko, et al. [12] selected the peak histogram AOD as the background threshold for the month before the burning season; Chen, et al. [13] chose AOD_{440} less than 0.2 as the background threshold over an urban area in Northeast China. These BAOD threshold values are generally determined by the researcher's understanding of local emission and aerosol transport as well as their own experience and preference. There are also aerosol-type identification methods without the use of AOD [14,15], such as Angstrom Exponent and Single scattering albedo (AE-SSA) [16], fine mode fraction, and single scattering albedo (FMF-SSA) [17], and scattering Angstrom Exponent and absorbing Angstrom Exponent (SAE-AAE) [18]. However, it should be noted that the widely used aerosol properties, such as SSA and SAE from sun-photometer observations, are only available when AOD > 0.4. Therefore, these methods often miss the classification at AOD < 0.4 when ground-based sun-photometer products are used, and an AOD of 0.4 could be considered the hidden BAOD threshold value.

In addition to these ground-based studies, satellite sensors like VIIRS also provide global products with background aerosol information [19]. In the VIIRS aerosol classification algorithm, aerosols are divided into eight types, in which AOD less than 0.3 is identified as background type [3]. Different from ground-based observation, the selection of 0.3 as the BAOD threshold for VIIRS is more like a compromise to the limitation of the sensor and algorithm because the accuracy of AE is really limited when AOD is lower than 0.3. Aerosol type cannot be accurately identified, and thus, a 'background' type is used to represent these low-loading aerosol situations [20]. Zheng, Che, Xia, Wang, Yang, Chen, Wang, Zhao, Li, Zhang, Gui, Yang, Liang and Zhang [15] identified clean continent aerosols with AOD < 0.25 based on MODIS data. Chen, et al. [21] used an AOD threshold value of 0.2 to identify clean aerosols over China based on Himawari-8 observations, and the clean aerosol is identical to the background aerosol type. Whether for ground or satellite aerosol-type classification, the background AOD threshold value varies substantially from 0.1 to 0.3 and has not yet been properly defined. The classification of aerosol types, however, is rather sensitive to the selection of the BAOD threshold value [5,15,22]. Such a large variation in BAOD threshold value will cause uncertainty in the classification of aerosol types and decrease the comparability of similar studies [23–25]. Establishing a more reasonable and statistically based BAOD standard will help to unify the incoherence in aerosol-type studies, lower the level of difficulty in data comparisons, reduce the uncertainty in aerosol-type retrievals, and improve the quantification of the aerosol climate effects.

This study presents a reference map of global BAOD threshold values using a statistical method based on long-term satellite AOD observations from VIIRS, MODIS, and MISR. The statistical base of the BAOD determination involves multiple lognormal fitting processes for AOD histograms within each grid and an optimized process to search the best cutoff percentile of the lowest AOD on a global scale. With the help of the MERRA-2 dataset, we discussed the rationality of the BAOD from the chemical view, and we further discussed the implications of the BAOD threshold value for aerosol-type classification using VIIRS observations as an example.

2. Materials and Methods

2.1. Satellite AOD Data

VIIRS is a scanning radiometer onboard the Suomi National Polar-orbiting Partnership satellite launched in October 2011 [19]. It provides multi-spectral data twice per day, once during the daytime and once during the nighttime [26]. Aerosol properties are retrieved only during the daytime measurements at 1:30 P.M. local time. In this study, we choose VIIRS data as a testbed for our statistical method because of the following reasons: (1) VIIRS orbital swath is broader (by ~50%) than Moderate Resolution Imaging Spectroradiometer (MODIS), and it has no data gaps at equatorial regions and also twice or more sampling per day at the middle to high latitudes, which means VIIRS provides a broader spatiotemporal coverage of AOD measurement. (2) Since the aging of MODIS (especially Terra, launched in 1999) and Multi-angle imaging spectroradiometer (MISR, onboard) sensors, their data accuracy is getting unstable, and lots of efforts are needed to ensure the data is suitable for climate studies [3]. As a new generation sensor designed to continue the EOS-era data records, VIIRS will become the major sensor replacing MODIS soon. Thus, we choose the VIIRS DB AOD product as our major testbed. The VIIRS data used in this work is VIIRS V1 AOD from the Deep Blue Level 3 Daily aerosol data in the Level 1 Atmosphere (Archive Set 5110) collection (AERDB_D3_VIIRS_SNPP). The spatial resolution of the data is 1 × 1 degree, and the temporal coverage is from March 2012 to December 2019. Also, the latest MODIS C6.1 Level 3 Daily AOD product from Dark Target (DT), Deep Blue (DB), and their fusion (DTB) (MOD08_D3 for Terra and MYD08_D3 for Aqua) and MISR Version 23 Level 3 Global AOD product (MIL3DAEN) are also included to analyze the influence of spatial-temporal coverage of AOD on the determination of BAOD threshold value. VIIRS and MODIS data can be found at NASA's Level-1 and Atmosphere Archive and Distribution System Distributed Active Archive Center (https://ladsweb.modaps.eosdis.nasa.gov/, accessed on 25 March 2024), and MISR data can be found at the Atmospheric Science Data Center (https://eosweb.larc.nasa.gov/project/misr/misr_table, accessed on 25 March 2024). The time ranges for MODIS and MISR AOD are from January 2012 to December 2019.

2.2. AERONET Data

Aerosol Robotic Network (AERONET) is a ground-based remote sensing network that provides long-term and continuous aerosol properties around the world [27]. The AERONET collaboration provides globally distributed observations of spectral AOD, inversion products, and precipitable water. These products are often used in aerosol-related research and satellite data validations [28]. The AERONET data are publicly accessible at the AERONET Team website at NASA Goddard Space Flight Center (http://aeronet.gsfc.nasa.gov/, accessed on 25 March 2024). In this work, Level 2.0 daily AODs at 440 nm and 670 nm in the AEROSOL OPTICAL DEPTH (V3)-SOLAR collection are used to obtain AOD_{550} using an Angstrom Exponential relation. The interpolated AOD is regarded as the ground truth value to validate the VIIRS AOD.

2.3. MERRA-2 Aerosol Chemical Data

The Modern-Era Retrospective Analysis for Research and Applications, version 2 (MERRA-2), was developed by the NASA Global Modeling and Assimilation Office (GMAO) [29]. There are two primary objectives of developing MERRA-2: one is to assimilate data from NASA's Earth Observation System (EOS) and demonstrate its usefulness for climate study, and the other is to improve the representation of the atmospheric hydrological cycle in reanalysis models compared to previous ones. MERRA-2 reanalysis data has a resolution of 0.625° longitude by 0.5° latitude on a global scale, and it provides hourly column mass density of black carbon (BC), organic carbon (OC), dust, SO_4, sea salt, SO_2, and Dimethyl sulfide (DMS) [30]. MERRA-2 data are available at MDISC, managed by the NASA Goddard Earth Sciences (GES) Data and Information Services Center (DISC) (https://disc.gsfc.nasa.gov/datasets?project=MERRA-2, accessed on 25 March 2024). Since this study focuses on background AOD over land, only column mass densities of BC, OC, dust, SO_4, and SS are taken into account. The time period of MERRA-2 data used in this study is between 2018 and 2019 (two years).

2.4. Multiple Lognormal Distribution Fitting for AOD Histogram

This study determines BAOD values by utilizing multiple lognormal fitting processes for AOD histograms within each grid. Multiple models in AOD histogram can be a representation of aerosol species [31,32]. The AOD histogram for each grid is created using a log10 AOD space ranging from −2 to 1 with an interval of 0.1 (a total of 31 bins with AOD ranging from 0.01 to 10). In each grid, five lognormal distributions with 1 to 5 modes are fitted to the histogram. P(logτ) denotes the fitted AOD histogram for the five lognormal distributions. Function F is used to evaluate the fitting performance of the five lognormal distributions. The lognormal mode number with the lowest F value is considered the best estimation. Specifically, the target function is the square of the absolute bias between the AOD histogram and the fitted one.

$$P(\log \tau) = \sum_{i=1}^{m} \frac{P_i}{\sigma_i \sqrt{2\pi}} \times \exp(-\frac{\log(\tau^2/\mu_i^2)}{2\sigma_i^2}) \tag{1}$$

$$F = \sum_{j=1}^{n} \left[P_{fit}(\log \tau_j) - P_{hist}(\log \tau_j)\right]^2 \tag{2}$$

where τ represents AOD, m denotes the number of lognormal modes ranging from 1 to 5, P(log τ) is the probability density of m lognormal with modal probability density P_i, geometric mean μ_i, and standard deviation σ_i, F is the target function, j represents the j-th AOD bin, P_{fit} represents the fitted probability density function, and P_{hist} is the reference probability density function from the AOD histogram.

The least squares method is used for fitting. The main idea of the least squares method is to minimize the sum of the squares of the errors (also known as residuals) of the true and predicted values by determining the unknown parameters (usually a matrix of parameters). The loss function F can be treated as a multivariate function, and F is a partial derivative of the parameters, respectively, so that the partial derivatives are equal to 0. Thus, a system of nonlinear equations is obtained, and then the iterative method is used to solve for the unknowns (i.e., P_i, μ_i, and σ_i).

$$\frac{\partial F}{\partial P_i} = \frac{\partial F}{\partial \sigma_i} = \frac{\partial F}{\partial \mu_i} = 0 \tag{3}$$

$$\min f(\mathbf{x}) = \sum_{i=1}^{n} \left[P_{hist,i} - P_{fit,i}\right]^2 \tag{4}$$

$$F(x + dx) = F(x) + dxF'(x) + O\left(\alpha^2\right) \approx F(x) + dxF'(x) \tag{5}$$

where x is the unknowns (i.e., P_i, μ_i, and σ_i), dx is the perturbation, and F' represents the derivative.

An initial guess pool for each AOD histogram is constructed before fitting. In this pool, the highest point of the mono-mode, the turning points for the multi-modes (second, third, and later peak points), and the points where the slope changes significantly make up the initial guess of mode locations and their corresponding probability densities serve as the initial guess of mode heights. The initial standard deviations are 0.5, 0.45, 0.3, 0.35, and 0.3 for 1 to 5 modes, respectively. The boundary of each mode is set to be the same as > 0, $-2 < < 1$, and $0.05 < < 1$. If any of the three parameters in each mode exceeds the boundaries, the fit is rejected. For example, if a negative P_i value is observed in a bimodal distribution, this fit is rejected, and we turn to fit the trimodal distribution. To avoid redundant modes, we calculate the specific value of each mode on the log10 AOD space. If less than two positive values are detected, this mode is deleted. For instance, if a trimodal fit is performed and the third mode generates only two positive values in the entire log10 AOD space, with all other values being zero, we delete the mode and degrade the trimodal distribution to a bimodal distribution. Sites with fewer than 60 observations are avoided during each fitting process because they are statistically insignificant and may cause unstable or unrealistic fitting results.

The best fit goes to the lognormal distribution with the lowest F values. The preliminary BAOD is either the intersection point of the first two normal distribution curves when having multiple modes, or the midpoint between the peak AOD and the first AOD with 1/10 peak probability density values when the mono normal distribution is the best fit (Figure 1).

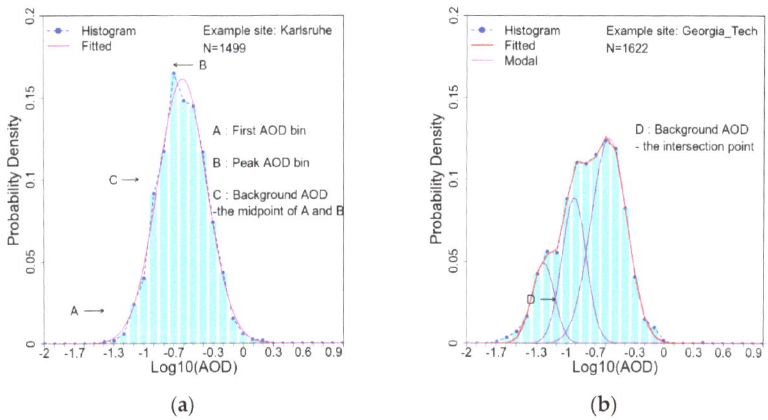

(a) (b)

Figure 1. Determination of background AOD threshold value for different AOD histogram situations. (**a**) fitted mono mode, and (**b**) fitted multiple modes. Here, two AERONET AOD histograms from Karlsruhe and Georgia Tech are used as an example to illustrate the determination of BAOD threshold value for the mono- and multi-lognormal distributions.

2.5. Determination of the Best Cutoff Percentile for BAOD

Since the preliminary BAOD is obtained from statistical methods, the results will be inevitably influenced by factors such as the difference in the observation number in each grid and the data accuracy. Thus, the preliminary BAOD may not be smooth, and some outliers may appear. To avoid the outliers, we create 100 guesses with the lowest 1st to 100th AOD as reference BAODs and select the best cutoff percentile for the final BAOD threshold value. To determine the best estimation of BAOD, we have two target functions: an absolute bias (R) between the preliminary and guess BAOD and the other is a relative bias (Q).

$$R = \sum_{lon=-180}^{180} \sum_{lat=-90}^{90} \left(\mathrm{BAOD}_{lon,lat}^{k} - \mathrm{BAOD}_{lon,lat}^{pre}\right)^2, \tag{6}$$

$$Q = \sum_{lon=-180}^{180} \sum_{lat=-90}^{90} \left(1 - \frac{\mathrm{BAOD}_{lon,lat}^{k}}{\mathrm{BAOD}_{lon,lat}^{pre}}\right)^2, \tag{7}$$

where $BAOD^{k}_{lon,lat}$ represents the AOD of the lowest kth percentile, and $BAOD^{pre}_{lon,lat}$ is the preliminary BAOD; R and Q are the k-th absolute and relative bias, respectively. The best cutoff percentile is chosen as the percentile with the lowest R and Q, and the results based on VIIRS data are presented in Section 3.3.

2.6. IDW Interpolation

To evaluate the rationality of the reference BAOD obtained from VIIRS AOD using the methods described in Sections 2.4 and 2.5, a ground-based BAOD map from global AERONET sites is obtained using an Inverse Distance Weight (IDW) interpolation [6]. The IDW interpolation is a deterministic method for multivariate interpolation with a known scattered set of points. The values at unknown points are calculated with a weighted average of the values available at the known points. A general form of finding an interpolated value y at a given point x based on samples $y_i = y(x_i)$ for i = 1,2, ... ,N using IDW is an interpolating function:

$$y(x) = \begin{cases} \frac{\sum_{i=1}^{N} w_i(x) y_i}{\sum_{i=1}^{N} w_i(x)}, if\ d(x, x_i) \neq 0 \\ y_i, if\ d(x, x_i) = 0 \end{cases}, \tag{8}$$

where $w_i = 1/d(x, x_i)^p$ is the IDW weighting function, x denotes an interpolated point, x_i is an interpolating (known) point, d is the distance from the known point x_i to the unknown point x, N is the total number of known points, and p is the power parameter [33]. Specifically, x_i denotes the BAOD values from the i-th AERONET site using the same multiple lognormal fitting processes described in Section 2.4, and x denotes the same grid point as that in the VIIRS BAOD map.

3. Results

3.1. VIIRS DB AOD Comparison with AERONET AOD

The temporal coverage of the VIIRS DB AOD measurement is shown in Figure 2a. VIIRS DB AOD product has a high temporal coverage between 10 N and 50 N and between −10 N and −50 N, while a much lower temporal coverage is observed between the two belts. Such a temporal coverage difference is probably due to the high occurrence of clouds near the equator [34]. The scattering circles in Figure 2a illustrate the 697 AERONET sites used in evaluating the VIIRS DB AOD products. The vertical color bar illustrates the magnitude of the correlation between AERONET and VIIRS DB AOD, and the horizontal color bar indicates the VIIRS DB AOD observation frequency. Since AERONET provides a point observation with a sampling frequency of 5–15 min while satellite provides a regional snapshot at a single time, a spatiotemporal matchup scheme is applied. To be specific, AERONET data is firstly averaged within a time window of ±30 min from the satellite overpass, and then a spatial average is conducted if more than one AERONET observation is available in the $1° \times 1°$ grid cell.

Figure 2b shows the overall statistics based on 189,939 common observations from paired AERONET and VIIRS observations. VIIRS DB AOD generally performs well in most AERONET sites and has an overall correlation of 0.862 with AERONET AOD, and 74.2% of VIIRS DB AODs fell in the estimated error (EE) envelopes of $\pm(0.05 + 15\%)$. The VIIRS DB data set was previously validated by Sayer et al. [35] on an instantaneous basis. Compared with the results from Sayer et al. [35], a similar RMSE (0.125) and median bias (0.005) values were obtained. Their correlation R (0.82) is slightly higher, while the Estimated Error envelope (EE) within the ratio (0.80) is a little bit lower. We also observe a regional dependency on the correlation, which is in agreement with the results from Sayer et al. [35]. The correlation coefficients tend to be less than 0.4 over low AOD areas like southwest Africa and Australia, as well as highly elevated terrain such as the Rocky Mountains and the Tibetan Plateau in western China, while they are close to 1 over other regions with high aerosol loading or low elevated terrain [3]. Figure 2d shows the VIIRS DB AOD performance over the low AOD regime. High EEs with a ratio >60% are observed for low AOD bins (0–0.3), which indicates that most of the observations fall into the EE envelope of ±(0.05 + 15%).

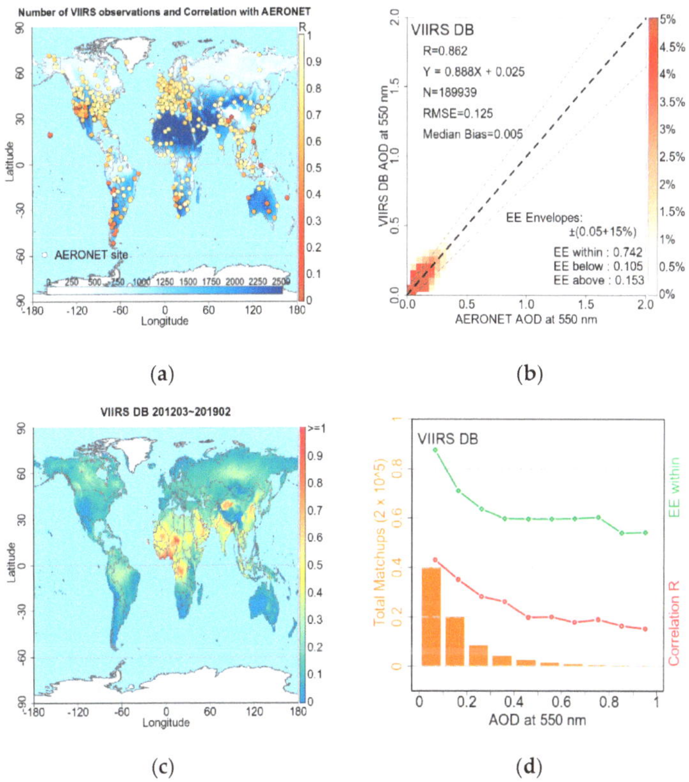

Figure 2. (**a**) Map of the available VIIRS DB AOD observations from March 2012 to December 2019 and the correlations with AERONET measurements, circles denote the involved AERONET sites, the vertical color bar denotes the magnitude of the correlation, and the horizontal color bar denotes the VIIRS DB AOD observation number during the study period; (**b**) the overall correlation between VIIRS DB and AERONET AOD. EE represents estimated error; the 1:1 line is shown in black dotted line; the gray dotted lines indicate ±(0.05 + 15%) of the AERONET AOD; and (**c**) Annual mean DB AOD map at 550 nm from VIIRS measurements during March 2012 to February 2019. (**d**) performance of VIIRS DB for different AOD bins.

A high correlation of 0.862 between VIIRS and AERONET AOD indicates that VIIRS AOD can accurately represent the daily variation of aerosol loadings globally. The low fitting intercept of 0.025 suggests that VIIRS generally overestimates AOD by 0.025, which is higher than the AERONET uncertainty of 0.01 but still accurate for satellite observations. However, it is important to note that such overestimation on VIIRS AOD would be passed to BAOD threshold values. AOD observations fall within the expected error, which means a maximum overestimation or underestimation of around 0.1 when AERONET AOD is lower than 0.3. Low R or EE ratios are observed in southwest Africa and Australia, as well as highly elevated terrain such as the Rocky Mountains and the Tibetan Plateau in western China. This can lead to an imprecise estimation of the absolute BAOD threshold value, and BAOD from satellite observations in these regions needs further examination before usage.

3.2. Global AOD Distribution

Figure 3 shows the map of VIIRS DB AOD at different lowest percentiles. The lowest 50th percentile (the median), 25th and 75th percentile (50% of the data), 16th and 83rd percentile (one-standard deviation, 68% of the data), and 3rd and 97th percentile (two-standard deviation, 95%) of AOD are selected to show the global AOD characteristics. The AOD map shown in Figure 3a with the lowest 3rd percentile could be regarded as the cleanest aerosol condition for different regions. The lowest 3rd percentile AOD in most parts of the world is lower than 0.1; however, a relatively high AOD above 0.2 is observed in the northern part of South America, North Africa, and India, indicating a high aerosol loading base. With the increase of percentile, a relatively fast increase of AOD is observed in the Middle West Africa, Middle East, India, and East Asia (Figure 3a–e), which indicates these regions are the major aerosol sources around the world. From Figure 3f, 83% of AOD in most parts of North and South America, South Africa, Europe, Australia, and North Asia are below 0.3, which means an overall clean aerosol condition over these regions. However, from the lowest 97th percentile AOD map (Figure 3g), extremely high AOD events (AOD > 1) appear in some of these clean regions. The extreme AODs found in the northern part of North America and the tropical region of South America are mostly from biomass-burning smoke [3]. Those extreme AODs observed in East Asia are partly due to the heavy anthropogenic emissions and partly due to transported dust from the desert region and occasionally biomass-burning smoke from Siberia and Southeast Asia [36]. Also, 97% of AODs are found always below 0.3 over the southwest part of North and South America, the southern part of South Africa, North Europe, and the Non-desert region of Middle Asia (e.g., Tibetan Plateau) and West Australia, indicating a year-round clean aerosol condition.

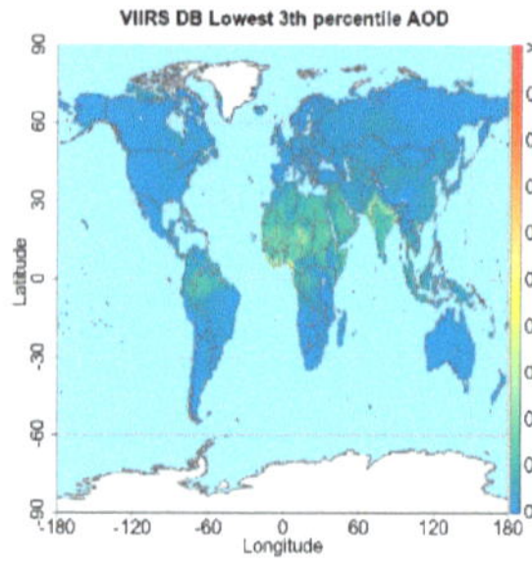

(a)

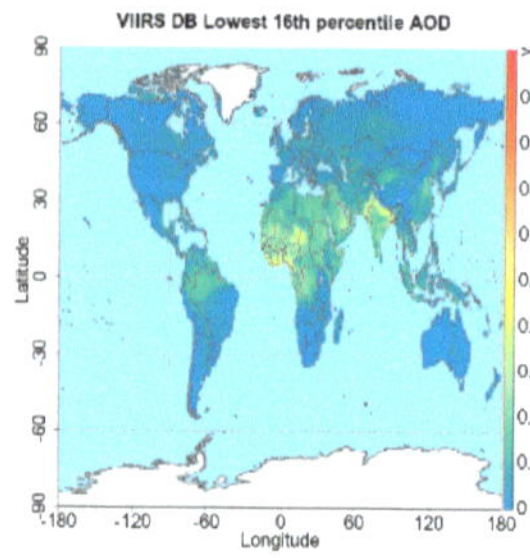

(b)

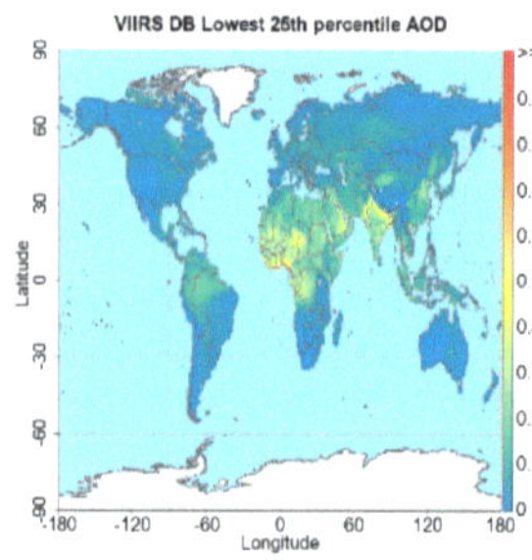

(c)

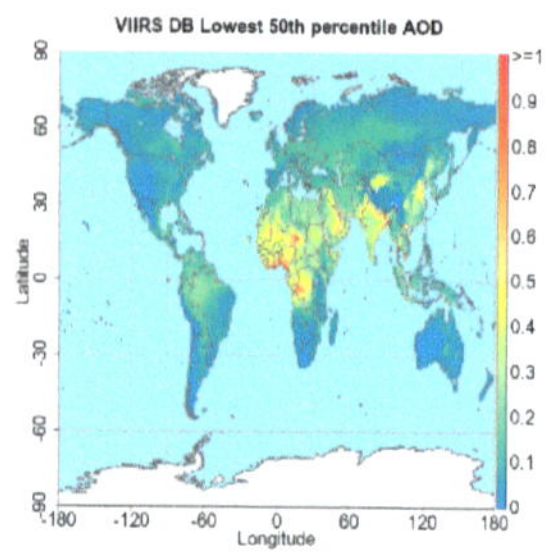

(d)

Figure 3. *Cont.*

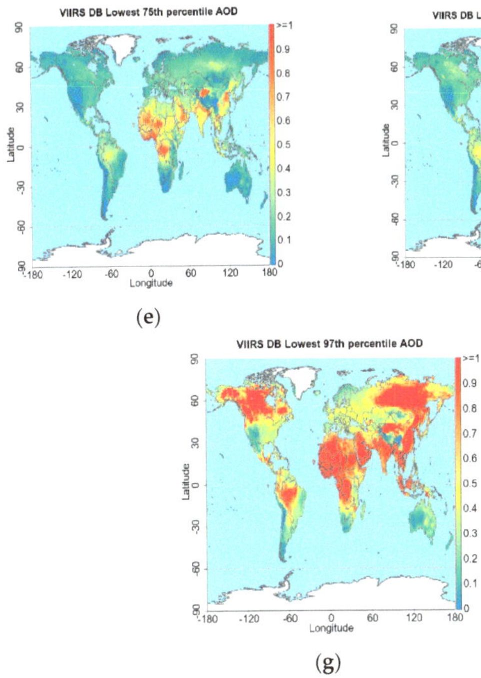

Figure 3. VIIRS DB AOD map at different lowest percentile, from (**a–g**) 3rd, 16th, 25th, 50th, 75th, 83rd, and 97th percentile.

3.3. Preliminary BAOD from Multiple Lognormal Fitting

Figure 4a shows the global multiple lognormal fitting correlation map. Most parts of the world have a very high fitting correlation of over 0.98. Although relatively low fitting correlations are observed over the high latitude northern hemisphere (50 N–80 N), the Tibetan Plateau, and Midwest Australia, it does not hinder the determination of BAOD threshold over these regions, and a generally 'smooth' BAOD map is finally obtained despite some abrupt points with abnormal BAOD values. The quadruple and triple lognormal modes are found to appear most frequently around the world and their combined fraction is around 80%. This high frequency of occurrence of multiple AOD modes indicates that aerosol types tend to vary among three or four kinds in most parts of the world [37].

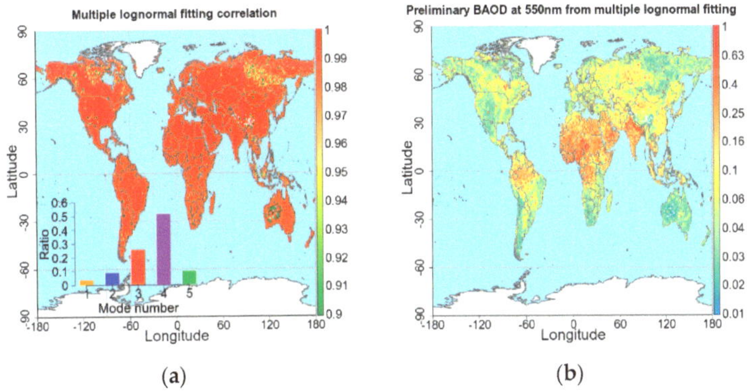

Figure 4. (**a**) Multiple lognormal fitting correlation map and fitted ratios for different lognormal modes, and (**b**) preliminary BAOD map at 550 nm.

The preliminary BAOD from multiple lognormal fitting is shown in Figure 4b. A distinct high BAOD is found over the Indo-Gangetic Plain, Arabian Peninsula, and North and Mid Africa since the annual mean AOD over these regions is generally over 0.4. Southeast Asia also has a relatively high annual mean AOD, but the preliminary BAOD there is much lower than in areas like the Indo-Gangetic Plain and the Arabian Peninsula. Such difference is due to the fact that aerosol loading over Southeast Asia varies in a wide range, and a number of AOD measurements fall in low AOD bins; thus, the corresponding AOD histogram has a potential mode in low AOD bins [38]. While in high BAOD regions like the Indo-Gangetic Plain and the northern part of South America, most AOD observations fall in relatively high AOD bins, leading to the AOD histogram moving towards the high AOD end. High BAOD values are also found in the eastern part of North America, for these areas have relatively large industrial emissions [4].

3.4. Best Cutoff Percentile of the Lowest AOD for Background Aerosol

Figure 5a shows the variations of BAOD bias (R and Q) with an increasing percentile of the lowest AOD. The lowest R and Q appear in the 9th and 30th percentile, respectively. Considering that the global mean BAOD from preliminary multiple lognormal fitting is 0.113, we select the 30th percentile of the lowest AOD as the best estimation of BAOD because its global mean (0.105) is very close to the preliminary mean BAOD of 0.113, while the 9th percentile is only about half (0.063). The final BAOD map (Figure 4b) shows similar patterns to those of the preliminary BAOD. North and Mid Africa, the Arabian Peninsula, North and Mid Africa, and the northern part of South America are still characterized by high BAOD threshold values. The eastern part of America has a slightly higher BAOD than the western part. Compared with Figure 4b, wider coverage of BAOD over 0.1 is observed over Europe and West Asia, and a distinct increase of BAOD from 0.1 to 0.4 is found in Southeast Asia. Such an increase is generally acceptable for the background aerosol loading because the annual mean AOD (especially in Southeast Asia, Figure 4b) is still much larger (nearly twice) than the lowest 30th percentile AOD.

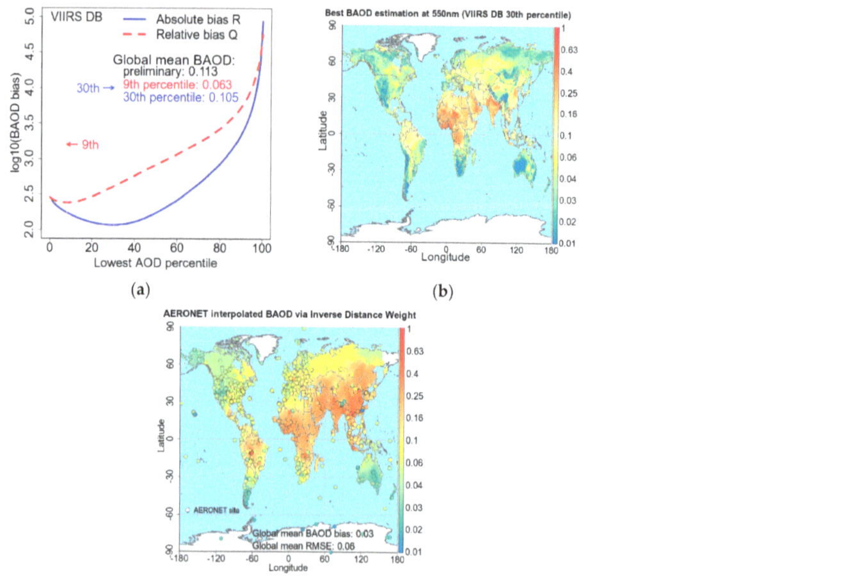

(c)

Figure 5. (**a**) Variation of the BAOD bias with lowest AOD percentiles at 550 nm from VIIRS, (**b**) global map of the 30th percentile of the lowest AOD at 550 nm as the background threshold value, (**c**) The AERONET interpolated BAOD from inverse distance weight interpolation.

To evaluate the rationality of the BAOD map in Figure 5b, the lowest 30th percentile of AERONET AOD is selected as AERONET BAOD value, and then the IDW interpolation method is applied to get a ground-based BAOD map. Here, the distance is set as 25 degrees, and the power parameter is set as 2 in the IDW interpolation. Figure 5c shows the ground-based interpolated BAOD map. We then calculate the global mean bias and RMSE between the lowest 30th percentile of VIIRS AOD and the IDW interpolated AOD from AERONET. The global mean BAOD bias and RMSE are 0.03 and 0.06, respectively. Similarly, high BAOD pools are observed over South America, North Africa, and North Asia, while low BAOD pools are over North America, Europe, and Australia. The interpolated AERONET BAOD map generally agrees well with the VIIRS BAOD map, indicating the rationality of the reference BAOD in Figure 4b. The differences between VIIRS BAOD and the BAOD obtained from AERONET mainly come from two aspects. One is that the interpolation process itself generates bias. Since the number of ground-based sites is limited, most regions are short of reliable observations. Since aerosols are of high spatiotemporal variation, the interpolated one may not be able to fully describe the variation of spatial aerosol conditions compared to satellite observations. The other is that, as a statistical method, the multiple lognormal fitting results may vary due to different inputs. Although high consistency between VIIRS and AERONET AOD is observed, a lot of observations outside the EE envelope (Figure 2b) will lead to changes in BAOD values and even outliers.

4. Discussion

4.1. Variation of BAOD to Different Satellite Datasets and Seasons

Due to the inconsistency of spatiotemporal coverage, systematic bias, and retrieval algorithm between different sensors, the best estimation of the BAOD threshold value may vary with different AOD products. To assess how the reference BAOD could vary with different satellite AOD inputs, we apply the same BAOD determination process to MISR and Terra- and Aqua-MODIS datasets. Figure 6a–g shows the overall fitting results from Terra-MODIS, Aqua-MODIS, and MISR AOD inputs. Similar to VIIRS, the best estimation of BAOD threshold value for MISR, Terra-MODIS, and Aqua-MODIS is selected between the 28th and 33rd percentile of their lowest AOD. Since the first derivatives of the absolute bias R are all around 0 in the vicinity of their best-fitting percentile, the difference is negligible between the 30th and 28th/33rd percentile of the lowest AOD being the best BAOD. Thus, the reference BAOD obtained from the current study varies little with different satellite AOD products, and the use of the 30th percentile of the lowest AOD as background AOD threshold value should be reasonable for various satellite sensors.

Aerosol loading and type usually vary with seasons, and such a seasonal variation may also have an impact on the determination of the best BAOD threshold value. Following the same data processing, the monthly mean BAOD threshold value from different sensors is shown in Table 1. Due to the least data limitation ($N > 60$), the seasonal BAOD threshold value from MISR is not available here. In general, the monthly average best percentile is selected between 26.6 and 30.7. The best percentile shows an insignificant seasonal variation, which tends to be low in June and July while high in January and December. The largest difference to the reference 30th percentile is observed from the Terra-MODIS DB product, where the 22nd percentile in June is found to be the best estimation. Figure 6h shows the corresponding bias curve, and the use of the 30th percentile is acceptable because the difference in log bias between the 30th and 22nd percentiles is not very large. Therefore, the use of the 30th percentile of the lowest AOD as the background AOD threshold value is also reasonable for different months. It should be noted that although the BAOD threshold value is the 30th percentile, the monthly mean AOD will vary with seasons, and consequently, the reference BAOD value also has a seasonal variation.

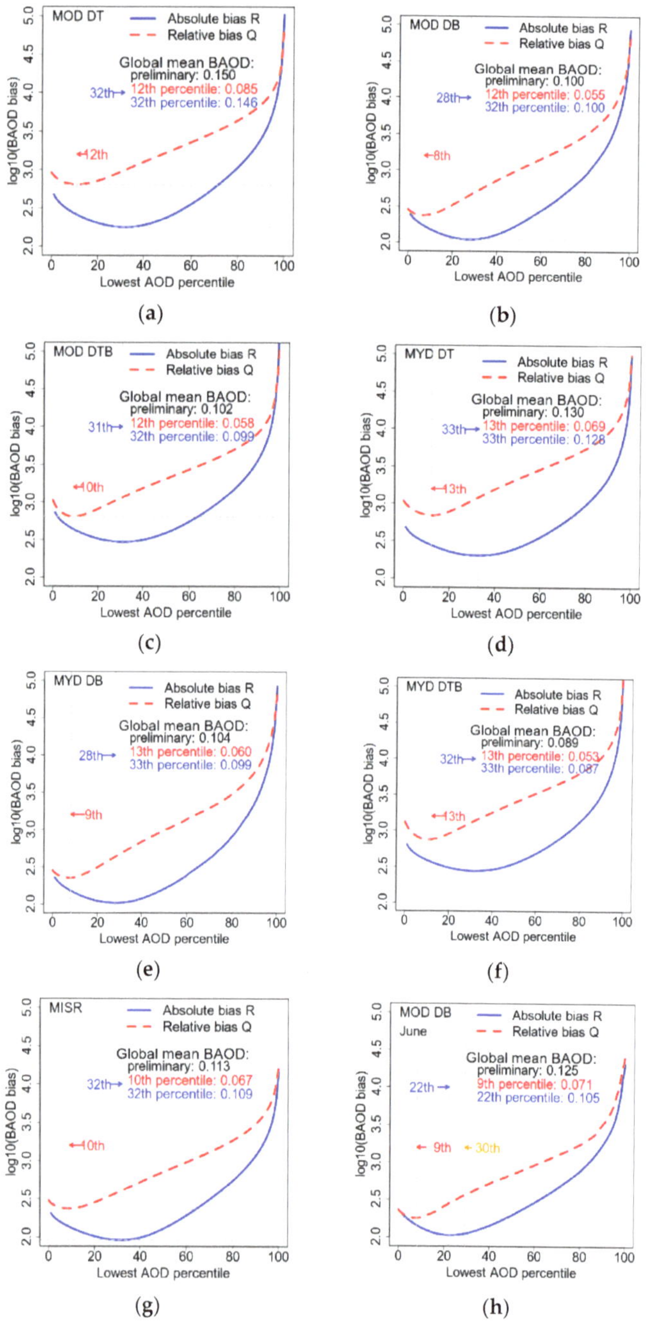

Figure 6. Variation of the BAOD bias with lowest AOD percentiles at 550 nm from (**a**) Terra-MODIS DT, (**b**) Terra-MODIS DB, (**c**) Terra-MODIS DTB, (**d**) Aqua-MODIS DT, (**e**) Aqua-MODIS DB, (**f**) Aqua-MODIS DTB, (**g**) MISR and (**h**) Terra-MODIS DB in June during 2012–2019.

Table 1. Seasonal summary of the best cutoff percentile of the lowest AOD for background aerosol for different AOD products. The seasonal MISR results are not available due to its limited input number (>60).

Dataset	Year 2012–2019	Month												
		January	February	March	April	May	June	July	August	September	October	November	December	Average
VIIRS DB	30%	33%	32%	29%	27%	27%	27%	28%	28%	27%	32%	33%	31%	29.5%
MOD DB	28%	31%	27%	26%	24%	26%	22%	25%	25%	24%	25%	29%	31%	26.3%
MOD DT	32%	31%	28%	27%	28%	30%	27%	27%	27%	30%	28%	32%	31%	28.8%
MOD DTB	31%	27%	27%	27%	26%	27%	25%	26%	27%	38%	27%	28%	28%	26.9%
MYD DB	28%	30%	30%	28%	29%	27%	27%	30%	28%	30%	28%	29%	32%	28.9%
MYD DT	33%	30%	28%	29%	31%	29%	29%	28%	29%	32%	32%	32%	33%	30.2%
MYD DTB	32%	29%	28%	28%	29%	30%	29%	29%	29%	30%	28%	28%	29%	28.8%
MISR	32%	-	-	-	-	-	-	-	-	-	-	-	-	-
Average	30.8%	30.1%	28.6%	27.7%	27.7%	28.0%	26.6%	27.6%	27.6%	28.7%	28.6%	30.0%	30.7%	28.5%

4.2. BAOD Determination from an Aerosol Chemical Perspective

Although we have determined that the 30th percentile of the lowest AOD is the best background AOD threshold through a statistical approach, it is preferred to validate this from a different perspective, such as a chemical view, to prove its rationality. Typical aerosol particles have specific chemical components [5,13], and therefore, the variation of aerosol types could be reflected by the variation of their chemical components [29]. The rationality of the BAOD threshold value is illustrated using the co-located aerosol chemical information from MERRA-2. A spatiotemporal matchup scheme is applied to ensure accurate validation. This is because VIIRS provides a regional snapshot at a single time, and the resolutions of MERRA-2 and VIIRS AOD are different. Specifically, hourly MERRA-2 data are first selected within a time window of $\pm$30 min from the satellite overpass. Then, they are re-scaled at a resolution of $1^\circ \times 1^\circ$.

The global average variation of the normalized standard deviation of column aerosol mass density with the lowest percentile of AOD over land is shown in Figure 7. The increase of the lowest AOD percentile leads to an increase in the normalized standard deviation, indicating a corresponding increase in the degree of chemical component change. When AOD is located in the lowest 30th percentile, aerosol loading is relatively light, and there is little variation in chemical components (as indicated by a low standard deviation of both the average and each chemical).

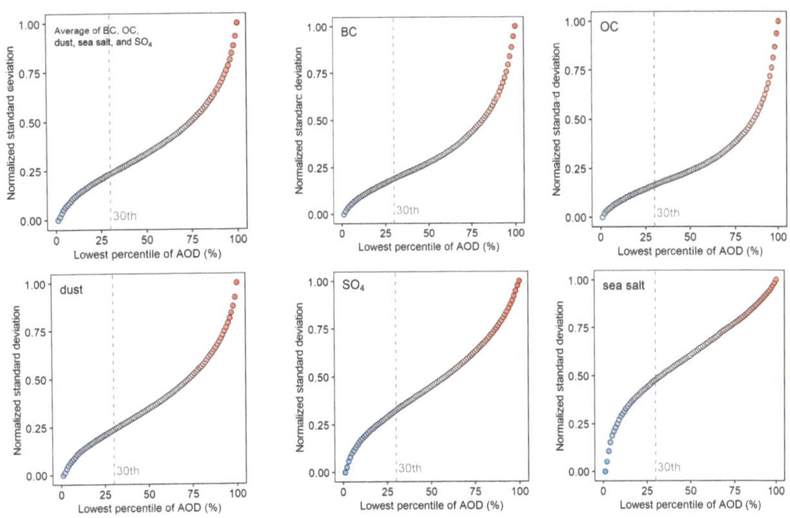

Figure 7. Variations of the normalized standard deviation of column aerosol mass density, blue and red refer to low and high value of normalized standard deviation, respectively.

Figure 8 shows a decreasing trend in slope variation of normalized standard deviation from 0 to the 30th lowest percentile AOD, followed by a stable slope variation from the 30th to around the 75th lowest percentile AOD. This indicates that the chemical component variations are gradually becoming more prominent before the 30th lowest percentile AOD, but they still remain at a low level of change, as indicated by the low standard deviation. After the 30th lowest percentile AOD, a relatively stable rate of change in chemical components is observed, indicating a gradual and significant shift in aerosol type as AOD increases (with a high standard deviation). Thus, from a chemical perspective, the aerosol type is generally stable when the AOD is low but becomes increasingly unstable as the AOD increases. The breaking point of the changing rate of the standard deviation occurs at the 30th lowest percentile AOD, which separates these two aerosol conditions.

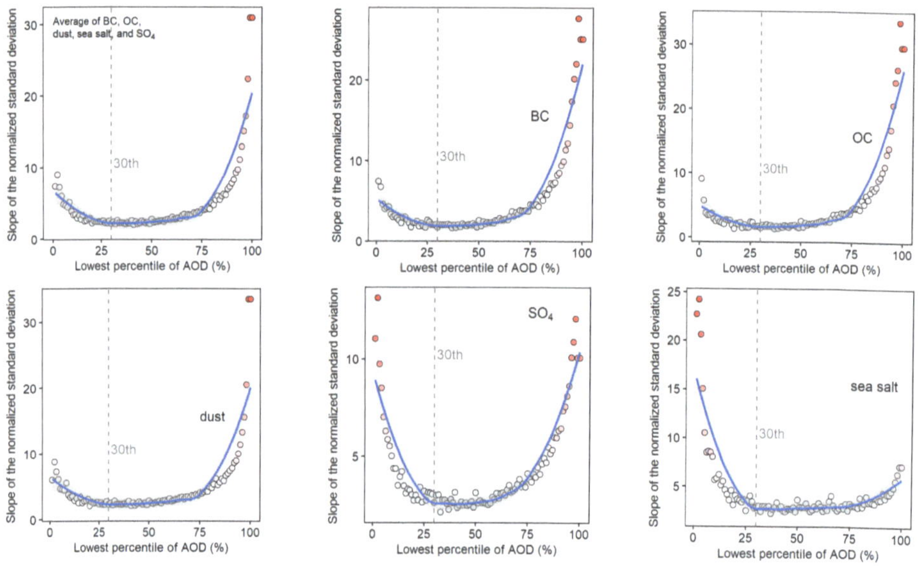

Figure 8. Slope of the normalized standard deviations of the column mass densities. Note the y-coordinate ranges are different, and white and red refer to low and high value of slope of the normalized standard deviation, respectively.

4.3. Case Studies of BAOD Applying on Satellite Aerosol Type Discrimination

For passive observations, particle radius is a key parameter in aerosol type classification, and Angstrom Exponent (AE) is a widely used parameter that describes the bulk aerosol particle size. Previous studies have proposed many well-established aerosol-type classification methods via ground-based datasets [39–41], but they are not applicable in low aerosol loading conditions because the key parameter, SSA, can only be accurately measured or retrieved when AOD is high (e.g., AOD > 0.4 for AERONET inversion product) [28]. Alternatively, the AOD-AE scheme is often used to achieve a rough classification in low aerosol loading conditions [42]. For ground observations, both AOD and AE can be accurately obtained, and aerosol types can be identified as background (continental clean), urban industry/biomass burning, dust, and mixed over land. However, the situation gets worse for satellite observations because AE can not be accurately retrieved in low aerosol loading conditions. Also, due to a lack of proper background AOD threshold value for a different region, a concession is made to simply tag the aerosol as a 'background' type when AOD is less than 0.3 in the current VIIRS dataset, and in fact, the aerosol type is not actually identified.

To illustrate the usefulness of BAOD in aerosol type classification, we update the VIIRS DB aerosol type classification scheme by replacing the original 'background' threshold of

0.3 with BAOD, and two case studies are introduced to compare their classifying results. Figure 9 shows the original and updated VIIRS aerosol-type classification schemes. In the updated scheme, the background AOD threshold value is determined dynamically with consideration of both the reference BAOD and the sensor limitation (AOD < 0.3), choosing the smaller one of BAOD and 0.3. A new aerosol type LP (Low Polluted) is introduced when AOD falls between the reference BAOD (if less than 0.3) and 0.3. This LP type represents a situation when aerosol loading is higher than the background condition with potential influence by weak aerosol sources, such as thin dust layers with AOD less than 0.3. To highlight the differences, we changed the VIIRS background aerosol type to an unknown type since VIIRS does not perform the type identification when AOD < 0.3.

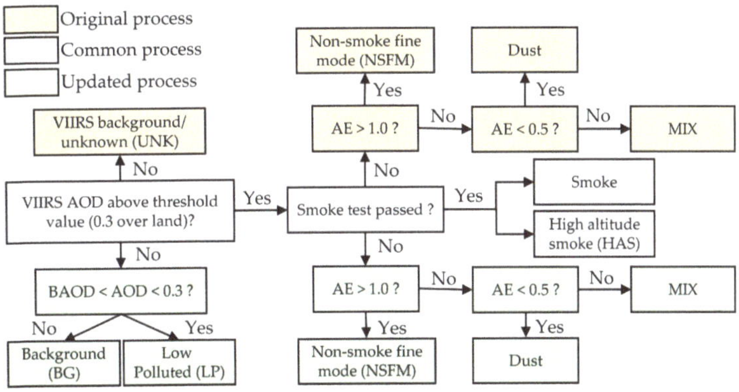

Figure 9. Flowchart illustrating the aerosol type classification scheme over land in the VIIRS DB algorithm. The yellow boxes represent the original scheme, and the blue boxes represent the updated scheme of this study.

The first case is a comparison of detecting thin dust layers over the Mediterranean (Figure 10). A thin dust layer can be observed above the Mediterranean region on 10 April, 26 April, and 9 June 2019 from the MODIS true color images (1st column of Figure 10). The thin yellow layer over the blue Mediterranean region is composed of suspended dust particles blown from the southwest to the northeast, and the white pixels are clouds. AODs over the north part of this region are generally below 0.3. Thus, the original VIIRS typing scheme reports a UNK (VIIRS background) aerosol type over these areas and loses sight of the thin dust layers (2nd column of Figure 10). However, the thin dust layers can generally be detected via the updated typing scheme as LP type, which uses BAOD as the background threshold value (3rd column of Figure 10).

The second case is a comparison of detecting thin smoke layers over South Africa (Figure 11). The thin smoke layer (gray) can be easily distinguished from the cloud (white) on 3 days in May 2019 (1st column of Figure 11). AODs over most of South Africa in the selected days are generally below 0.3. Thus, the original VIIRS typing scheme reports a 'background' aerosol type over this region and fails to identify the thin smoke layers (2nd column of Figure 11). However, using the updated typing scheme developed in this study, the smoke layers can be well detected as LP type (3rd column of Figure 11). From these two dust and smoke case studies, it is clear that the BAOD determined in this study is useful in detecting aerosol-type variations when AOD is lower than 0.3, which cannot be achieved using a unified background threshold of 0.3.

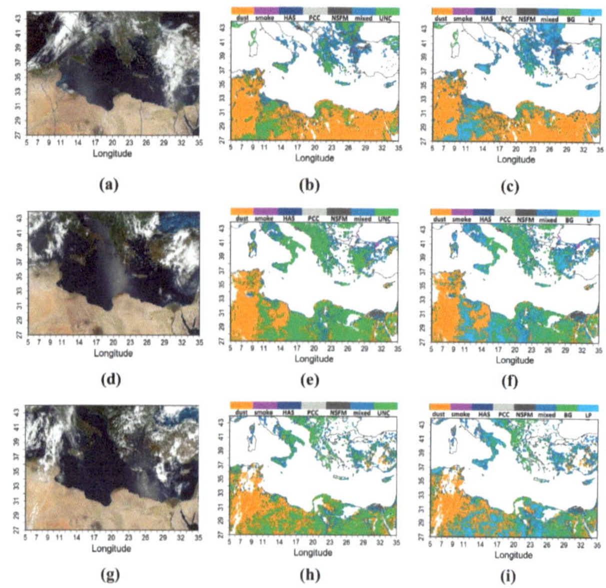

Figure 10. VIIRS RGB images of three dust events over the Mediterranean in (**a**) April 26th, (**d**) 9th June, and (**g**) 10th April 2019. Corresponding aerosol type classification results from VIIRS original AOD-AE scheme are shown in (**b**,**e**,**h**), and the results from the updated scheme are shown in (**c**,**f**,**i**). HAS, NSFM, UNK, BG, and LP represent the high altitude smoke, non-smoke fine mode, unknown, background, and low pollution in low aerosol loading, respectively.

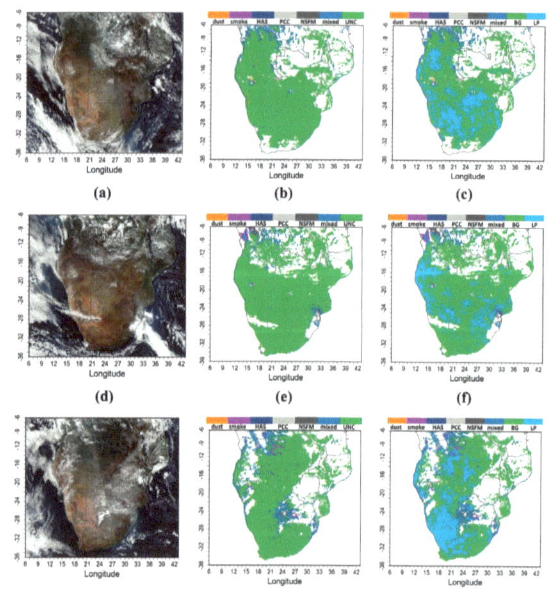

Figure 11. Similar to Figure 10, but for three biomass-burning events over South Africa on 19th May (**a**–**c**), 21st (**d**–**f**), and 24th (**g**–**i**), 2019. HAS, NSFM, UNK, BG, and LP represent the high altitude smoke, non-smoke fine mode, unknown, background, and low pollution in low aerosol loading, respectively.

4.4. Definitions and Identification of Background Aerosols

The definitions of background aerosol in the previous AOD-AE scheme, the current VIIRS DB algorithm, and our study are different. Specifically, the background aerosol used in many previous AOD-AE schemes usually refers to a remote clean continental aerosol type away from anthropogenic influence (hereafter BG-RM), while the background aerosol in the VIIRS DB algorithm is more like an unknown aerosol type. The background aerosol in this study represents a local common aerosol condition without impact from transported aerosols or occasional emission events like wildfires, and it changes with locations. For example, in remote locations, the background aerosol has the same definition as BG-RM; in urban regions, it represents the daily urban/industrial aerosols without influence from dust or wildfire events or pollution episodes; and in the desert region, it refers to the natural dust aerosol.

Many previous studies used a threshold value for background type classification at a single site, and the background aerosol in these studies is similar to BG-RM. However, when classifying aerosol type at a large domain, a unified threshold value may not be proper because background aerosol types for different locations may be different, leading to different background aerosol loadings. For example, Yu, et al. [43] used 0.2 as a background threshold value to classify aerosol types in the urban Beijing area. For a clean region like the Tibetan Plateau, however, a unified threshold of 0.2 will lead to a severe loss of aerosol-type information. Thus, Pokharel, Guang, Liu, Kang, Ma, Holben, Xia, Xin, Ram, Rupakheti, Wan, Wu, Bhattarai, Zhao and Cong [42] chose a much smaller value of 0.05 as a background threshold value to catch the aerosol-type variations. This indicates that background aerosol type and its corresponding aerosol loading should vary with locations. Thus, a location-specified threshold value may be more reasonable to represent local 'clean' conditions (background aerosol). On a global scale, Zhang, Kondragunta, Laszlo, Liu, Remer, Huang, Superczynski and Ciren [6] used the lowest 5th percentile AOD as a background threshold value and Xia [44] used the lowest 25 percentile as a background threshold value, but their selections tend to be personal experience or preference. Based on a statistical basis in this study, we recommend the lowest 30th percentile AOD as the best background threshold value.

Although changing the background threshold value from unified 0.3 to 30th percentiles does not introduce further information regarding the aerosol type (available information for typing aerosol is still AOD and AE, and AE is unreliable when AOD is less than 0.3), it does help to identify possible aerosol type changes in a location with low AOD. The reason is that the BAOD value obtained from the statistical process can be regarded as a priori knowledge of local common aerosol loading. When AOD is lower than BAOD, the aerosol generally represents the local background type. When AOD at a certain location is higher than BAOD, it means that the particular region is influenced by biomass burning, pollution, or dust episodes. Consequently, the local aerosol type may change from background to some other types, e.g., smoke or dust. In other words, the 30th percentile itself contains local background aerosol loading information. Thus, background aerosol can be identified by comparing AOD and BAOD without the use of AE, which becomes unreliable under low AOD conditions.

The multiple models in the AOD histogram can be a representation of aerosol species. A relatively large increase or decrease in AOD is mostly related to the change in local aerosol type. For a specific location, for example, Los Angeles, the background aerosol type should be urban/industrial type without transported aerosols (e.g., smoke or dust), and its AOD histogram tends to be a mono-normal distribution. When wildfire happens, however, AOD increases sharply, and a peak appears toward a larger AOD end. If dust aerosols or pollutants are transported into the area, aerosol loading will also increase, and another peak may appear. Therefore, the first peak could be representative of local background aerosol, but the later peaks are influenced by the transported pollutants or unusual local emissions like wildfires. However, what these later peaks represent or which kind of aerosol they are (e.g., dust, smoke, or some kind of mixture) needs further in-depth analysis because it may vary substantially from location to location.

5. Conclusions

This study develops a statistical method to determine global background AOD threshold values using the VIIRS AOD product. The 30th percentile of the lowest AOD is found to be the best estimation of BAOD for multiple datasets and different seasons. According to the multiple lognormal fitting, quadruple and triple lognormal modes are found to occur most frequently around the world, indicating a distinct aerosol-type variation on a global scale. A high BAOD value of over 0.4 appears in North and Mid Africa, the Arabian Peninsula, North and Mid Africa, and the northern part of South America, while other regions generally have a BAOD value of less than 0.1. The original VIIRS dataset left 80% of observations (AOD < 0.3) unclassified in low aerosol loading regions like America, Europe, and Australia and tagged them as a 'background' type; thus, the aerosol type information for low AOD remains unknown in the original VIIRS aerosol type classification scheme.

Chemical component information from MERRA-2 is involved in examining the selection of the 30th percentile of the lowest AOD as the BAOD threshold value. On a global scale, aerosol chemical components vary slightly when AOD is lower than BAOD, while the degree and rate of change of aerosol chemical components gradually become strong and significant when AOD is greater than BAOD. And the breaking point of the changing trend coincides with our statistical results properly.

Based on the reference BAOD threshold value, we update the AOD-AE aerosol type classification scheme. We redefine the background type and introduce a new aerosol-type cluster, low polluted (LP), into the classification scheme. Through two dust and smoke case studies, it is clear that our updated AOD-AE scheme is able to achieve a pretty good detection of aerosol type changes in low aerosol loading conditions, and it can give us a better view of aerosol type variation for VIIRS-like passive sensors over low aerosol loadings regions like America, Europe, and Australia.

There are also some limitations that should be concluded. VIIRS AOD performs relatively worse over low AOD areas, such as the highly elevated Tibetan Plateau in western China. This can lead to biased BAOD threshold value estimation. Although BAOD provides more details about the variation of aerosol types under low aerosol loading conditions from satellite observation, it is still insufficient to quantitatively identify the specific aerosol particles that cause an increase in aerosol loading above the BAOD threshold when the overall AOD is lower than 0.3. Future efforts will focus on improving the performance of satellite inversion algorithms over rugged mountain regions. Additionally, detailed identification of aerosol types under low aerosol loading conditions may be achieved by including polarized observations or chemical information from real-time model simulations, with possible assistance from machine learning.

Author Contributions: Conceptualization, Q.-X.C.; methodology, Q.-X.C.; software, Q.-X.C.; validation, Q.-X.C. and C.-L.H.; formal analysis, Q.-X.C.; investigation, S.-K.D.; resources, K.-F.L.; data curation, Q.-X.C.; writing—original draft preparation, Q.-X.C.; writing—review and editing, Q.-X.C.; visualization, C.-L.H.; supervision, S.-K.D.; project administration, Q.-X.C.; funding acquisition, Q.-X.C. All authors have read and agreed to the published version of the manuscript.

Funding: This study is supported by the China Postdoctoral Science Foundation, grant number 2022M720943.

Data Availability Statement: VIIRS and MODIS data can be found at NASA's Level-1 and Atmosphere Archive and Distribution System Distributed Active Archive Center (https://ladsweb.modaps.eosdis.nasa.gov/, accessed on 25 March 2024). MISR data can be found at the Atmospheric Science Data Center (https://eosweb.larc.nasa.gov/project/misr/misr_table, accessed on 25 March 2024). The AERONET data used in this study are publicly accessible at the AERONET Team website at NASA Goddard Space Flight Center (http://aeronet.gsfc.nasa.gov/, accessed on 25 March 2024). MERRA-2 data are available at MDISC, managed by the NASA Goddard Earth Sciences (GES) Data and Information Services Center (DISC) (https://disc.gsfc.nasa.gov/datasets?project=MERRA-2, accessed on 25 March 2024).

Acknowledgments: The authors thank the VIIRS, MISR, and MODIS team and also AERONET principal investigators and site managers for providing the data used in this work.

Conflicts of Interest: The authors declare no conflicts of interest.

References

1. Cai, H.; Yang, Y.; Luo, W.; Chen, Q. City-level variations in aerosol optical properties and aerosol type identification derived from long-term MODIS/Aqua observations in the Sichuan Basin, China. *Urban Clim.* **2021**, *38*, 100886. [CrossRef]
2. Zhao, B.; Wang, Y.; Gu, Y.; Liou, K.-N.; Jiang, J.H.; Fan, J.; Liu, X.; Huang, L.; Yung, Y.L. Ice nucleation by aerosols from anthropogenic pollution. *Nat. Geosci.* **2019**, *12*, 602–607. [CrossRef] [PubMed]
3. Hsu, N.C.; Lee, J.; Sayer, A.M.; Kim, W.; Bettenhausen, C.; Tsay, S.C. VIIRS Deep Blue Aerosol Products Over Land: Extending the EOS Long-Term Aerosol Data Records. *J. Geophys. Res. Atmos.* **2019**, *124*, 4026–4053. [CrossRef]
4. Crippa, P.; Sullivan, R.C.; Thota, A.; Pryor, S.C. Sensitivity of Simulated Aerosol Properties Over Eastern North America to WRF-Chem Parameterizations. *J. Geophys. Res. Atmos.* **2019**, *124*, 3365–3383. [CrossRef]
5. Gao, J.; Li, Y.; Xie, Z.; Wang, L.; Hu, B.; Bao, F. Which aerosol type dominate the impact of aerosols on ozone via changing photolysis rates? *Sci. Total Environ.* **2023**, *854*, 158580. [CrossRef] [PubMed]
6. Zhang, H.; Kondragunta, S.; Laszlo, I.; Liu, H.; Remer, L.A.; Huang, J.; Superczynski, S.; Ciren, P. An enhanced VIIRS aerosol optical thickness (AOT) retrieval algorithm over land using a global surface reflectance ratio database. *J. Geophys. Res. Atmos.* **2016**, *121*, 10717–10738. [CrossRef]
7. Bao, F.; Li, Y.; Gao, J. Carbonaceous aerosols remote sensing from geostationary satellite observation, Part I: Algorithm development using critical reflectance. *Remote Sens. Environ.* **2023**, *287*, 113459. [CrossRef]
8. Dubovik, O.; Holben, B.; Eck, T.F.; Smirnov, A.; Kaufman, Y.J.; King, M.D.; Tanré, D.; Slutsker, I. Variability of absorption and optical properties of key aerosol types observed in worldwide locations. *J. Atmos. Sci.* **2002**, *59*, 590–608. [CrossRef]
9. Chen, A.; Zhao, C.; Shen, L.; Fan, T. Influence of aerosol properties and surface albedo on radiative forcing efficiency of key aerosol types using global AERONET data. *Atmos. Res.* **2023**, *282*, 106519. [CrossRef]
10. Eom, S.; Kim, J.; Lee, S.; Holben, B.N.; Eck, T.F.; Park, S.-B.; Park, S.S. Long-term variation of aerosol optical properties associated with aerosol types over East Asia using AERONET and satellite (VIIRS, OMI) data (2012–2019). *Atmos. Res.* **2022**, *280*, 106457. [CrossRef]
11. Xia, X.; Che, H.; Zhu, J.; Chen, H.; Cong, Z.; Deng, X.; Fan, X.; Fu, Y.; Goloub, P.; Jiang, H.; et al. Ground-based remote sensing of aerosol climatology in China: Aerosol optical properties, direct radiative effect and its parameterization. *Atmos. Environ.* **2016**, *124*, 243–251. [CrossRef]
12. Petrenko, M.; Kahn, R.; Chin, M.; Limbacher, J. Refined Use of Satellite Aerosol Optical Depth Snapshots to Constrain Biomass Burning Emissions in the GOCART Model. *J. Geophys. Res. Atmos.* **2017**, *122*, 10983–11004. [CrossRef]
13. Chen, Q.-X.; Shen, W.-X.; Yuan, Y.; Tan, H.-P. Verification of aerosol classification methods through satellite and ground-based measurements over Harbin, Northeast China. *Atmos. Res.* **2019**, *216*, 167–175. [CrossRef]
14. Zhang, L.; Li, J. Variability of Major Aerosol Types in China Classified Using AERONET Measurements. *Remote Sens.* **2019**, *11*, 2334. [CrossRef]
15. Zheng, Y.; Che, H.; Xia, X.; Wang, Y.; Yang, L.; Chen, J.; Wang, H.; Zhao, H.; Li, L.; Zhang, L.; et al. Aerosol optical properties and its type classification based on multiyear joint observation campaign in north China plain megalopolis. *Chemosphere* **2021**, *273*, 128560. [CrossRef] [PubMed]
16. Kalapureddy, M.C.R.; Kaskaoutis, D.G.; Ernest Raj, P.; Devara, P.C.S.; Kambezidis, H.D.; Kosmopoulos, P.G.; Nastos, P.T. Identification of aerosol type over the Arabian Sea in the premonsoon season during the Integrated Campaign for Aerosols, Gases and Radiation Budget (ICARB). *J. Geophys. Res. Atmos.* **2009**, *114*, D17203. [CrossRef]
17. Lee, J.; Kim, J.; Song, C.H.; Kim, S.B.; Chun, Y.; Sohn, B.J.; Holben, B.N. Characteristics of aerosol types from AERONET sunphotometer measurements. *Atmos. Environ.* **2010**, *44*, 3110–3117. [CrossRef]
18. Cappa, C.D.; Kolesar, K.R.; Zhang, X.; Atkinson, D.B.; Pekour, M.S.; Zaveri, R.A.; Zelenyuk, A.; Zhang, Q. Understanding the optical properties of ambient sub- and supermicron particulate matter: Results from the CARES 2010 field study in northern California. *Atmos. Chem. Phys.* **2016**, *16*, 6511–6535. [CrossRef]
19. Cao, C.; Xiong, J.; Blonski, S.; Liu, Q.; Uprety, S.; Shao, X.; Bai, Y.; Weng, F. Suomi NPP VIIRS sensor data record verification, validation, and long-term performance monitoring. *J. Geophys. Res. Atmos.* **2013**, *118*, 11664–11678. [CrossRef]
20. Sayer, A.; Hsu, N.; Bettenhausen, C.; Jeong, M.-J. Validation and uncertainty estimates for MODIS Collection 6 "Deep Blue" aerosol data. *J. Geophys. Res. Atmos.* **2013**, *118*, 7864–7872. [CrossRef]
21. Chen, X.; Ding, H.; Li, J.; Wang, L.; Li, L.; Xi, M.; Zhao, L.; Shi, Z.; Liu, Z. Remote sensing retrieval of aerosol types in China using geostationary satellite. *Atmos. Res.* **2024**, *299*, 107150. [CrossRef]
22. Vadde, S.; Kalluri, R.O.R.; Gugamsetty, B.; Kotalo, R.G.; Kajjer Virupakshappa, U.; Akkiraju, B.; Thotli, L.R.; Lingala, S.S.R.; Rapole, J.K. Classifying aerosol type using in situ and satellite observations over a semi-arid station, Anantapur, from southern peninsular India. *Adv. Space Res.* **2023**, *72*, 1109–1122. [CrossRef]

23. Khademi, F.; Bayat, A. Classification of aerosol types using AERONET version 3 data over Kuwait City. *Atmos. Environ.* **2021**, *265*, 118716. [CrossRef]
24. Zhao, H.; Gui, K.; Ma, Y.; Wang, Y.; Wang, Y.; Wang, H.; Zheng, Y.; Li, L.; Zhang, L.; Che, H.; et al. Climatological variations in aerosol optical depth and aerosol type identification in Liaoning of Northeast China based on MODIS data from 2002 to 2019. *Sci. Total Environ.* **2021**, *781*, 146810. [CrossRef]
25. Zhou, P.; Wang, Y.; Liu, J.; Xu, L.; Chen, X.; Zhang, L. Difference between global and regional aerosol model classifications and associated implications for spaceborne aerosol optical depth retrieval. *Atmos. Environ.* **2023**, *300*, 119674. [CrossRef]
26. Cao, C.; Luccia, F.J.D.; Xiong, X.; Wolfe, R.; Weng, F. Early On-Orbit Performance of the Visible Infrared Imaging Radiometer Suite Onboard the Suomi National Polar-Orbiting Partnership (S-NPP) Satellite. *IEEE Trans. Geosci. Remote Sens.* **2014**, *52*, 1142–1156. [CrossRef]
27. Eck, T.F.; Holben, B.N.; Giles, D.M.; Slutsker, I.; Sinyuk, A.; Schafer, J.S.; Smirnov, A.; Sorokin, M.; Reid, J.S.; Sayer, A.M.; et al. AERONET Remotely Sensed Measurements and Retrievals of Biomass Burning Aerosol Optical Properties During the 2015 Indonesian Burning Season. *J. Geophys. Res. Atmos.* **2019**, *124*, 4722–4740. [CrossRef]
28. Giles, D.M.; Sinyuk, A.; Sorokin, M.G.; Schafer, J.S.; Smirnov, A.; Slutsker, I.; Eck, T.F.; Holben, B.N.; Lewis, J.R.; Campbell, J.R.; et al. Advancements in the Aerosol Robotic Network (AERONET) Version 3 database—Automated near-real-time quality control algorithm with improved cloud screening for Sun photometer aerosol optical depth (AOD) measurements. *Atmos. Meas. Tech.* **2019**, *12*, 169–209. [CrossRef]
29. Qiao, Y.; Ji, D.; Shang, H.; Xu, J.; Xu, R.; Shi, C. The Fusion of ERA5 and MERRA-2 Atmospheric Temperature Profiles with Enhanced Spatial Resolution and Accuracy. *Remote Sens.* **2023**, *15*, 3592. [CrossRef]
30. Liu, C.; Yin, Z.; He, Y.; Wang, L. Climatology of Dust Aerosols over the Jianghan Plain Revealed with Space-Borne Instruments and MERRA-2 Reanalysis Data during 2006–2021. *Remote Sens.* **2022**, *14*, 4414. [CrossRef]
31. Ignatov, A.; Holben, B.; Eck, T. The lognormal distribution as a reference for reporting aerosol optical depth statistics; Empirical tests using multi-year, multi-site AERONET Sunphotometer data. *Geophys. Res. Lett.* **2000**, *27*, 3333–3336. [CrossRef]
32. Povey, A.; Grainger, R. Towards more representative gridded satellite products. *IEEE Geosci. Remote Sens. Lett.* **2018**, *16*, 672–676. [CrossRef]
33. Łukaszyk, S. A new concept of probability metric and its applications in approximation of scattered data sets. *Comput. Mech.* **2004**, *33*, 299–304. [CrossRef]
34. Wylie, D.; Jackson, D.L.; Menzel, W.P.; Bates, J.J. Trends in Global Cloud Cover in Two Decades of HIRS Observations. *J. Clim.* **2005**, *18*, 3021–3031. [CrossRef]
35. Sayer, A.M.; Hsu, N.C.; Lee, J.; Kim, W.V.; Dutcher, S.T. Validation, stability, and consistency of MODIS collection 6.1 and VIIRS version 1 Deep Blue aerosol data over land. *J. Geophys. Res. Atmos.* **2019**, *124*, 4658–4688. [CrossRef]
36. Reid, J.S.; Hyer, E.J.; Johnson, R.S.; Holben, B.N.; Yokelson, R.J.; Zhang, J.; Campbell, J.R.; Christopher, S.A.; Di Girolamo, L.; Giglio, L.; et al. Observing and understanding the Southeast Asian aerosol system by remote sensing: An initial review and analysis for the Seven Southeast Asian Studies (7SEAS) program. *Atmos. Res.* **2013**, *122*, 403–468. [CrossRef]
37. Zhao, B.; Jiang, J.H.; Diner, D.J.; Su, H.; Gu, Y.; Liou, K.-N.; Jiang, Z.; Huang, L.; Takano, Y.; Fan, X.; et al. Intra-annual variations of regional aerosol optical depth, vertical distribution, and particle types from multiple satellite and ground-based observational datasets. *Atmos. Chem. Phys.* **2018**, *18*, 11247–11260. [CrossRef] [PubMed]
38. Gao, M.; Han, Z.; Tao, Z.; Li, J.; Kang, J.E.; Huang, K.; Dong, X.; Zhuang, B.; Li, S.; Ge, B.; et al. Air quality and climate change, Topic 3 of the Model Inter-Comparison Study for Asia Phase III (MICS-Asia III)—Part 2: Aerosol radiative effects and aerosol feedbacks. *Atmos. Chem. Phys.* **2020**, *20*, 1147–1161. [CrossRef]
39. Bibi, H.; Alam, K.; Bibi, S. In-depth discrimination of aerosol types using multiple clustering techniques over four locations in Indo-Gangetic plains. *Atmos. Res.* **2016**, *181*, 106–114. [CrossRef]
40. Schmeisser, L.; Andrews, E.; Ogren, J.A.; Sheridan, P.; Jefferson, A.; Sharma, S.; Kim, J.E.; Sherman, J.P.; Sorribas, M.; Kalapov, I.; et al. Classifying aerosol type using in situ surface spectral aerosol optical properties. *Atmos. Chem. Phys.* **2017**, *17*, 12097–12120. [CrossRef]
41. Shin, S.K.; Tesche, M.; Noh, Y.; Müller, D. Aerosol-type classification based on AERONET version 3 inversion products. *Atmos. Meas. Tech.* **2019**, *12*, 3789–3803. [CrossRef]
42. Pokharel, M.; Guang, J.; Liu, B.; Kang, S.; Ma, Y.; Holben, B.N.; Xia, X.A.; Xin, J.; Ram, K.; Rupakheti, D.; et al. Aerosol Properties Over Tibetan Plateau from a Decade of AERONET Measurements: Baseline, Types, and Influencing Factors. *J. Geophys. Res. Atmos.* **2019**, *124*, 13357–13374. [CrossRef]
43. Yu, X.; Kumar, K.R.; Lü, R.; Ma, J. Changes in column aerosol optical properties during extreme haze-fog episodes in January 2013 over urban Beijing. *Environ. Pollut.* **2016**, *210*, 217–226. [CrossRef] [PubMed]
44. Xia, X. Variability of aerosol optical depth and Angstrom wavelength exponent derived from AERONET observations in recent decades. *Environ. Res. Lett.* **2011**, *6*, 044011. [CrossRef]

remote sensing

MDPI

Article

Analysis of Long-Term Aerosol Optical Properties Combining AERONET Sunphotometer and Satellite-Based Observations in Hong Kong

Xinyu Yu [1], Janet Nichol [2], Kwon Ho Lee [3], Jing Li [1] and Man Sing Wong [1,4,*]

1 Department of Land Surveying and Geo-Informatics, The Hong Kong Polytechnic University, Hong Kong, China
2 Department of Geography, University of Sussex, Brighton BN19RH, UK
3 Department of Atmospheric & Environmental Sciences, Gangneung-Wonju National University, Gangneung 26403, Korea
4 Research Institute for Land and Space, The Hong Kong Polytechnic University, Hong Kong, China
* Correspondence: ls.charles@polyu.edu.hk; Tel.: +852-3400-8959

Abstract: This study analyzes seasonal characteristics and long-term variations in aerosol optical parameters in Hong Kong from 2006 to 2021 using AERONET data and satellite-based observations based on the extreme-point symmetric mode decomposition (ESMD) model. The dominant aerosol types in Hong Kong are mixed aerosols and urban/industrial aerosols with fine-mode sizes, and slightly absorbing or non-absorbing properties. Aerosol optical depth (AOD), Angstrom exponent (AE) and single scattering albedo (SSA) varied seasonally with a lower AOD but higher AE and SSA in summer, and elevated AOD but lower AE and SSA in spring and winter. The long-term variations show the year 2012 to be a turning point, with an upward trend in AOD and AE before 2012 and then downwards after 2012. However, for SSA, a rising trend was exhibited in both pre- and post-2012 periods, but with a larger gradient in the first period. The ESMD analysis shows shorter-term, non-linear fluctuations in aerosol optical parameters, with alternating increasing and declining trends. The examination of the relationships between AOD and meteorological factors based on the extreme gradient boosting (XGBoost) method shows that the effects of weather conditions on AOD are complex and non-monotonic. A lower relative humidity, higher wind speed in southwest directions and lower temperature are beneficial to the abatement of aerosol loads in Hong Kong. In conclusion, the findings of this study enhance the understanding of aerosol properties and the interactions between aerosol loading and meteorological factors.

Keywords: seasonal characteristics; aerosol optical parameters; long-term variations; extreme-point symmetric mode decomposition

Citation: Yu, X.; Nichol, J.; Lee, K.H.; Li, J.; Wong, M.S. Analysis of Long-Term Aerosol Optical Properties Combining AERONET Sunphotometer and Satellite-Based Observations in Hong Kong. *Remote Sens.* **2022**, *14*, 5220. https://doi.org/10.3390/rs14205220

Academic Editors: Xingfa Gu, Jing Wei and Shuaiyi Shi

Received: 25 August 2022
Accepted: 15 October 2022
Published: 18 October 2022

1. Introduction

Aerosols are liquid or solid particles of different sizes suspended in the atmosphere, and are emitted from anthropogenic and natural sources. Extensive studies have been conducted to investigate the aerosol distribution and analyze the related effects on climate, the atmospheric environment and human health [1–6]. They can influence the Earth–atmosphere system and climate both indirectly (changing lifetime, amount and radiative properties of cloud by modifying cloud physical properties) and directly (altering the radiative forcing via scattering and absorbing radiation) [7]. Moreover, aerosol leads to considerable uncertainty in Earth–atmosphere interactions and climate change [8]. Thus, quantifying the long-term trends of aerosol parameters and the underlying causes are beneficial to understanding regional air quality conditions and climate variability.

The optical parameters of aerosols, aerosol optical depth (AOD), Angstrom exponent (AE) and single scattering albedo (SSA) indicate some important characteristics of aerosols, such as column-integrated aerosol loads, particle size and light-absorbing properties. They are also important factors for climate change analysis [9]. AOD is a primary aerosol parameter representing the amount of aerosol loading in the atmosphere and AE is generally used to estimate the aerosol particle size. As suggested by Kaufman [10], coarse-mode aerosols with particle diameters greater than 1 μm are dominant in the atmosphere when the AE value is smaller than 0.7, whereas a higher AE (>1.8) indicates the dominance of fine-mode aerosol with particle diameters smaller than 0.1 μm. An AE value between 0.7 and 1.8 is typical of the accumulation mode and the aerosol diameters vary between 0.1 and 1 μm. SSA indicates the aerosol absorption characteristics, which depend on the chemical composition of the aerosol. Specifically, it measures the ratio of the scattering coefficient to the total extinction coefficient. Currently, the aerosol optical parameters can be acquired from both ground-based and satellite-based measurements. Ground-based observations, such as the Skyradiometer network (SKYNET) [11], Sun-sky radiation observations network (SONET) [12] and Aerosol robotic network (AERONET) [13], provide accurate aerosol measurements with a high temporal resolution. Although some studies have demonstrated that the uneven and sparse distribution limits widespread application (e.g., potential uncertainties in temporal and regional representation) [14,15], the analysis of long-term variations based on the AERONET measurements is still reliable for understanding changes in aerosol characteristics in a specific area.

Numerous studies have been carried out to investigate long-term changes in aerosol properties based on satellite-based and ground-based observations at global or regional scales. For example, Xia [16] suggests that a significant decrease in AOD can be found in Europe and North America by using the data from 79 AERONET stations from 2000 to 2010. By combining multi-source satellite observations, Leeuw et al. [17] observed that AOD trends over mainland China show an initial increase followed by a significant decline between 1995 and 2015. Che et al. [18] analyzed global AOD variations via long-term aerosol datasets and found distinctly different AOD varying trends in different regions. Ramachandran and Rupakheti [19] reveal a rising trend of fine-mode particulate AOD across Asia from 2001 using AERONET data and Moderate resolution Imaging Spectroradiometer (MODIS) observations. Previous studies have mainly focused on exploring variations in AOD, whereas changes in AE and SSA are rarely included, which gives an incomplete view of aerosol variations. Additionally, in the aforementioned studies, variations in aerosol parameters were commonly addressed using the linear regression method. However, the ensemble empirical mode decomposition (EEMD), a non-linear trend analysis method, has been adopted for the estimation of global AOD variations [20]. The results indicated that non-linear varying trends can provide more perceptible changes than linear trends. Therefore, the estimation of long-term variations in aerosol parameters from a non-linear perspective appears a more promising approach.

Basically, changes in aerosol loading are affected by meteorological conditions, and extensive efforts have been made to analyze the correlations [21,22]. Pearson correlation and multiple linear regression (MLR) are commonly used in the above studies to study the associations between AOD and weather conditions. Actually, the effects of meteorological factors on aerosol loading are synergistic and variable. Notably, the aforementioned methods can only derive monotonic relationships. Superior to traditional statistical approaches, machine learning models are capable of estimating the relative importance of driving factors and capturing complex non-linear relationships between predictors and dependent variables. Previous studies have affirmed that when multi-driving variables are involved, it is feasible and reliable to determine the individual impacts of a given factor from synergistic effects on $PM_{2.5}$ pollution using machine learning methods [23,24]. However, to the best of our knowledge, the complex non-monotonic associations of weather conditions with AOD in Hong Kong have not been investigated and interpreted.

Therefore, to fill the aforementioned gaps, and to analyze the impacts of meteorological conditions on AOD, the aerosol parameters measured by the PolyU AERONET station in Hong Kong and the satellite-based observations were integrated and investigated based on ESMD and a data-driven method (XGBoost). This study was conducted to (1) identify the seasonal characteristics in aerosol optical parameters and the dominant aerosol type in Hong Kong; (2) investigate the long-term variations in AOD, AE and SSA from annual and seasonal perspectives between 2006 and 2021; and (3) reveal how AOD responds to the changes in weather conditions.

2. Materials and Methods

2.1. AERONET Data and Sampling Site

AERONET (https://aeronet.gsfc.nasa.gov/ accessed on 22 August 2022) is a global aerosol monitoring network that provides accurate measurements of aerosol optical parameters at multiple wavelengths [13]. There are three quality levels: Levels 1.0, 1.5 and 2.0, which represent raw observations, cloud-screened data with quality controls but not final calibration, and quality-assured data after pre- and post-field calibrations, respectively. Two AERONET stations are deployed in Hong Kong, namely Sheung and PolyU stations. The Sheung station is a suburban site, while the PolyU station shows subtropical urban aerosol characteristics (22.3033°N, 114.1797°E, 30 m) with a longer service period (starting from November 2005), located on the roof of the Hong Kong Polytechnic University (Figure 1). In this study, daily and monthly Version 3 Level 2.0 AE at 440–675 nm, AOD at 500 nm and Level 1.5 SSA at 440 nm from 2006 to 2021 were adopted. The uncertainty in the obtained AOD is up to 0.02, while SSA is an inversion product with an uncertainty of about 0.03–0.05 [25].

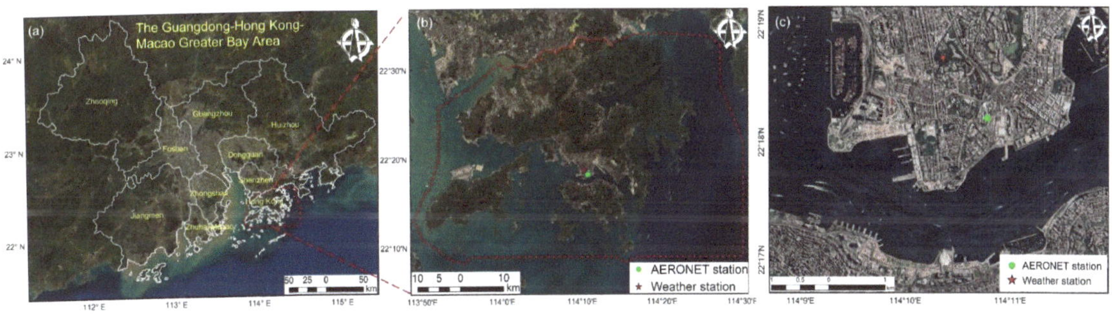

Figure 1. Geographic location of the PolyU AERONET station and King's Park weather station. (**a**) The Guangdong-Hong Kong Macao Greater Bay Area, (**b**) Hong Kong, and (**c**) enlarged view of the PolyU AERONET station and King's Park weather station location.

2.2. Satellite-Based Observations

Due to the maintenance and calibration of sunphotometers, or weather conditions, missing values of AERONET measurements are inevitable. Therefore, satellite-based observations were considered to ensure continuous time-series sequences. MCD19A2 (https://lpdaac.usgs.gov/ accessed on 22 August 2022), an AOD product at 470 nm and 550 nm wavelengths provided by MODIS with daily intervals and a 1 km spatial resolution via the multi-angle inversion atmospheric correction (MAIAC) approach, was adopted, which is highly consistent with AERONET ground observations [26,27]. A comparison between AERONET AOD measurements and the MODIS MAIAC AOD dataset in different seasons in Hong Kong is displayed in Figure 2. Daily mean satellite-based AOD at 470 nm and 550 nm was extracted by defining with the PolyU AERONET station in the center of an area of 5 × 5 pixels to interpolate the AE at 440–675 nm and AOD at 500 nm according to the equations presented in this section [28]. The results show that the satellite-based observations are in good agreement with the ground-based measurements, with a correlation coefficient from 0.72 to 0.77.

$$\tau(\lambda_i) = \beta \lambda_i^{-\alpha} \tag{1}$$

$$\alpha = -\frac{\ln(\tau_1/\tau_2)}{\ln(\lambda_1/\lambda_2)} \quad (2)$$

where $\tau(\lambda_i)$ represents AOD at λ_i wavelength, α denotes the Angstrom exponent and β is the turbidity coefficient. Subsequently, monthly mean AOD and AE were calculated based on valid daily measurements to complement the missing AERONET data. The percentage of the satellite-based observations used in this study to substitute the AERONET missing data accounts for about 38%.

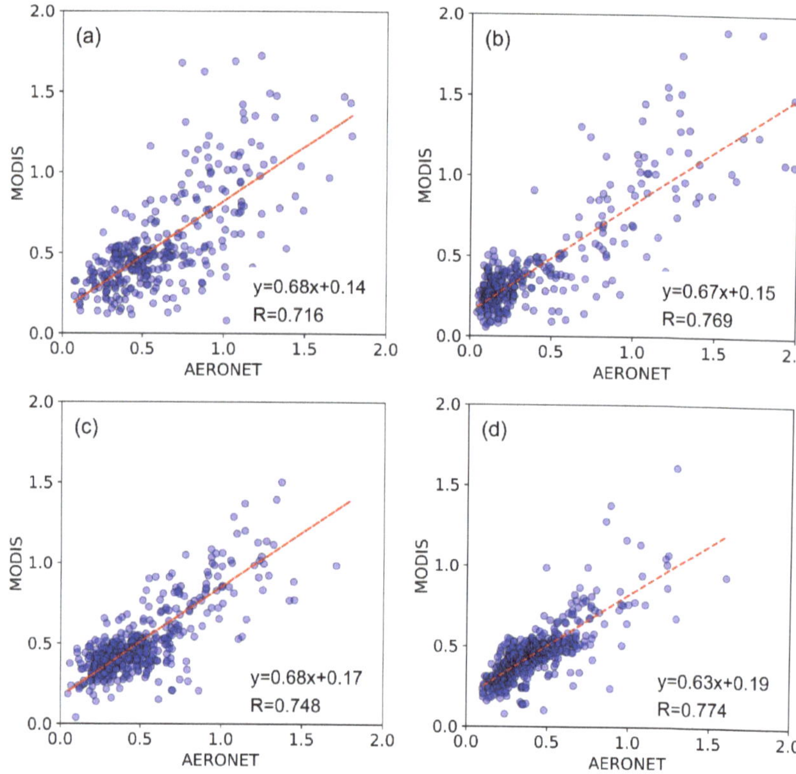

Figure 2. Scatter plots of multi-seasonal MODIS AOD data versus AERONET AOD measurements. (**a**) Spring, (**b**) summer, (**c**) autumn and (**d**) winter. The red dashed line represents the linear regression line.

For the missing values of SSA, the Multi-angle Imaging SpectroRadiometer (MISR) measurements were used, which can provide SSA inversion results with a spatial resolution of 4.4 km and a repeat coverage of about seven to nine days for the sampling site (https://misr.jpl.nasa.gov/ accessed on 22 August 2022). Daily mean SSA at 446 nm was extracted by defining the PolyU AERONET station in the center of an area of 3 × 3 pixels. Some studies have stated that the retrieval SSA of MISR was overestimated as compared to the AERONET, indicating that the effects of some highly absorbing particles were ignored [29–31]. Thus, the retrieved SSA of MISR was corrected by using the SSA inversion results of AERONET using the linear regression. After this, the monthly average SSA was calculated to replace the missing values of AERONET.

2.3. Meteorological Data

In this study, the measurements of the King's Park weather station (22.3119°N, 114.1728°E, 40 m) were used, which is the closest to the PolyU AERONET station (about

1.2 km, red star in Figure 1b). The daily ground-level mean measurements can be acquired from the Hong Kong Observatory, including wind direction, wind speed, relative humidity, temperature, pressure and precipitation (https://www.hko.gov.hk/en/index.html accessed on 22 August 2022). Precipitation was not involved since the ground-level measurements of aerosol optical parameters are not available on rainy days. Additionally, to ensure accuracy and reduce uncertainties, the weather station records consistent with valid measurements of the PolyU AERONET site were extracted and used to explore the impacts of meteorological variables on AOD.

2.4. Extreme-Point Symmetric Mode Decomposition (ESMD)

ESMD was developed based on ensemble empirical mode decomposition (EEMD) [32], which can be used to detect the time-dependent non-linear tendency without antecedent assumption. This is a newly popular approach to probing non-linear varying trends of climate-related issues [33,34]. The original signals can be decomposed into several intrinsic mode functions (IMFs) and a residual curve (R).

$$Y(m) = \sum_{i=1}^{n} M_i(m) + R(m) \tag{3}$$

where $M_i(m)$ is the ith IMF, $R(m)$ is the tendency curve and n denotes the total number of IMFs. More detailed descriptions can be found in the Supplementary Materials. In our study, R denotes the variation tendency of aerosol optical parameters from 2006 to 2021.

2.5. Extreme Gradient Boosting (XGBoost)

XGBoost is an ensemble decision tree model of machine learning with high accuracy for classification and prediction issues. It has been widely applied to explore the relationships between dependent and independent variables. More detailed descriptions can be found in the literature [35]. In this study, XGBoost was adopted to identify the relative importance of meteorological factors and investigate their interactions with AOD, which can be generally demonstrated as follows:

$$y_{AOD} = f(x_i) \tag{4}$$

where y_{AOD} means the daily averaged measurements of the PolyU AERONET station, and x_i represents meteorological factors, including pressure, temperature, wind direction, wind speed and relative humidity.

The relationships between AOD and meteorological factors were estimated by the partial dependence plot results based on the XGBoost model, which can visualize how AOD responds to changes in a specific variable by marginalizing other impact factors [36]. In other words, the variable sensitivity can be illustrated by how the dependent variable reacts when a given predictor changes while other predictors remain constant. To avoid overfitting and to examine reliability, the 10-fold cross-validation (CV) was adopted, whereby the dataset was divided into ten equal-sized subgroups. One subgroup was used as the test dataset and the other samples are for training. The training process is repeated ten times to ensure that every subgroup is tested once. Mean absolute error (MAE), root-mean-square error (RMSE) and the coefficient of determination R-squared (R^2) were used to assess the performance of XGBoost.

3. Results

3.1. Aerosol Optical Characteristics and Aerosol Type Classification

3.1.1. Statistical Description of Aerosol Optical Parameters

Figure 3 displays the multi-seasonal frequency histogram and corresponding statistical descriptions (i.e., minimum, mean, maximum values and standard deviation) based on daily mean measurements. The AOD has a distinct seasonal variation with a unimodal structure (Figure 3a). The lowest and highest daily average AODs occur in summer with values of 0.04 and 1.98, respectively, resulting in the largest standard deviation (SD) of 0.43. Hygroscopic growth and air stagnation may be responsible for the highest AOD in summer [37]. A much higher AOD was observed in spring, with about 58% of AOD levels above 0.5. Summer, autumn and winter have lower AOD values, especially in summer, when almost 50% of AOD is below 0.3. Thus, spring has the largest mean AOD (0.58), followed by autumn and winter. This may be attributed to long-distance dust and biomass-burning sources, as well as rapid production mechanisms by secondary aerosols [38]. Summer has the lowest mean AOD, as the prevailing wind is southerly in summer, bringing clean air from the South China Sea [39]. Additionally, there is a large amount of rainfall in summer, which can scavenge the pollutants and cleanse the atmosphere [7].

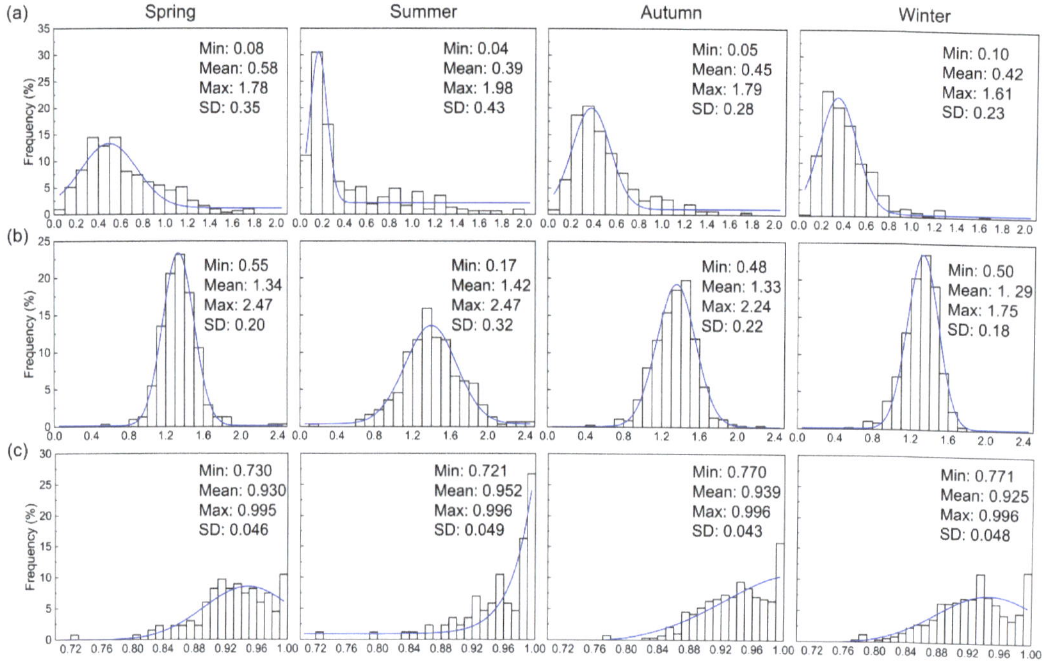

Figure 3. Multi-seasonal frequency histograms of (**a**) AOD, (**b**) AE and (**c**) SSA. The blue line is the fitting curve of the frequency histogram.

Analogously, AE also exhibited a unimodal structure centered at 1.3–1.4 in spring, summer and winter, while centered at 1.4–1.5 in autumn, suggesting the dominance of fine-mode aerosol (Figure 3b). Significant seasonal disparities of AE can be found in summer, with a wide range from 0.17 to 2.47 and a high SD of 0.32. This indicates variations in aerosol particle types in the atmosphere (i.e., fine-mode pollutants and coarse-mode dust). However, Figure 3b suggests that anthropogenic fine-mode particles make more contributions than coarse-mode pollutants over the study period in Hong Kong.

Figure 3c indicates that SSA has a similar distribution frequency histogram in all seasons except summer. The wide range of SSA, 0.721 to 0.996, indicates different aerosol-absorbing characteristics from strongly scattering to strongly absorbing aerosols. A large percentage of high SSA above 0.9 can be observed in all seasons, suggesting that strongly scattering aerosols dominated in Hong Kong during the study period. The highest mean SSA was found in summer, indicating strong scattering properties. This is partly ascribed to the aerosol particles tending to capture atmospheric water vapor under high relative humidity in summer, making a larger particle radius and further increasing the aerosol scattering [40]. Due to the effects of the winter monsoon, the transport of pollutants from China's mainland was facilitated by the prevailing northerly wind [39], and hence absorbing particles (i.e., black carbon) are increased in winter in Hong Kong.

3.1.2. Aerosol Types and Absorbing Aerosol Types

Two different classification criteria suggested by Salinas et al. [41] and Zheng et al. [42] were used to define the aerosol types and absorbing aerosol types in Hong Kong during the study period (see Table 1). For the first one, a daily averaged AOD at 500 nm and AE at 440–675 nm were used to classify the aerosol types, as suggested by Salinas et al. [41]. As depicted in Figure 4a, mixed aerosol (MIXA1 and MIXA2) is the representative type in Hong Kong, accounting for nearly 50%. Instead of pure aerosol, it is composed of different types of aerosol (i.e., dust, urban and biomass burning aerosols). The second dominant type is urban/industrial aerosol (UIA), with a percentage of 33%, followed by biomass-burning aerosol (BBA). Marine aerosol (MA) and dust aerosol (DA) only make up a small fraction (6%). For the second classification criteria, to further explore the dominant aerosol size and properties of absorbing and scattering aerosols, a daily mean Level 2.0 AE at 440–675 nm, fine-mode fraction (FMF) at 500nm and Level 1.5 SSA at 440 nm were used as the aerosol optical parameters, as proposed by Zheng et al. [42]. FMF represents the proportion of fine-mode aerosols in the atmosphere as compared to total aerosols [43,44]. The results are displayed in Figure 4b. Consistent with the above statements, anthropogenic fine-mode aerosols with non-absorbing (FNA) or slight absorbing (FSA) characteristics are the prevailing type in Hong Kong, accounting for over 66% overall. Meanwhile, there are some fine-mode aerosols exhibiting medium (FMA) and highly absorbing (FHA), with percentages of 16.3% and 4.9%, respectively. Marine absorbing (M-A) and non-absorbing aerosols (M-NA) have comparable proportions, 6.5% and 4.8%. The percentage of coarse-mode aerosols is only small (0.7%), and most of them show absorption properties.

Table 1. Classification criteria based on aerosol optical parameters.

	Types	Aerosol Optical Parameters Thresholds
Criterion 1 (AOD, AE; Salinas et al. [41])	Marine aerosol (MA)	AOD < 0.2, AE < 1.0
	Dust aerosol (DA)	AOD > 0.2, AE < 1.0
	Mixed aerosol (MIXA1)	AOD < 0.2, AE > 1.0
	Urban/industrial aerosol (UIA)	0.2 < AOD < 0.4, AE > 1.0
	Mixed aerosol (MIXA2)	0.4 < AOD < 0.8, AE > 1.0
	Biomass burning aerosol (BBA)	AOD > 0.8, AE > 1.0
Criterion 2 (AE, SSA, FMF; Zheng et al. [42])	Coarse absorbing (C-A)	SSA < 0.95, AE < 0.6, FMF < 0.4
	Coarse non-absorbing (C-NA)	SSA > 0.95, AE < 0.6, FMF < 0.4
	Mixed absorbing (M-A)	SSA < 0.95, 0.6 < AE < 1.2, 0.4 < FMF < 0.6
	Mixed non-absorbing (M-NA)	SSA > 0.95, 0.6 < AE < 1.2, 0.4 < FMF < 0.6
	Fine highly absorbing (FHA)	SSA < 0.85, AE > 1.2, FMF > 0.6
	Fine medium absorbing (FMA)	0.85 < SSA < 0.9, AE > 1.2, FMF > 0.6
	Fine slightly absorbing (FSA)	0.9 < SSA < 0.95, AE > 1.2, FMF > 0.6
	Fine non-absorbing (FNA)	SSA > 0.95, AE > 1.2, FMF > 0.6

Additionally, seasonal aerosol types and absorbing aerosol types were also investigated to identify the discrepancies between aerosol types in different seasons. As shown in Figure S1, MIXA and UIA are the predominant types in all seasons, and the percentage of UIA in winter is the highest, reaching up to 42.6%. This may be because of the long-range transport of increased urban aerosols during the cold winter heating season in the Chinese mainland [37]. For BBA, it has a higher proportion in spring, possibly caused by frequent forest fires and the increasing burden of biomass burning in spring. The percentages of MA and DA remain relatively constant in different seasons, except in spring, during which MA particles were not found. The seasonal absorbing aerosol types are defined and displayed in Figure S2. Discernible differences occur between summer and winter, with larger quantities of FNA in summer and decreasing frequencies in winter, implying strong scattering properties in summer but more absorption in winter. This is consistent with the statistics of SSA given in Figure 3c. Additionally, there are no significant fluctuations in the percentages of other types (i.e., M-A, M-NA, C-A and C-NA) across different seasons.

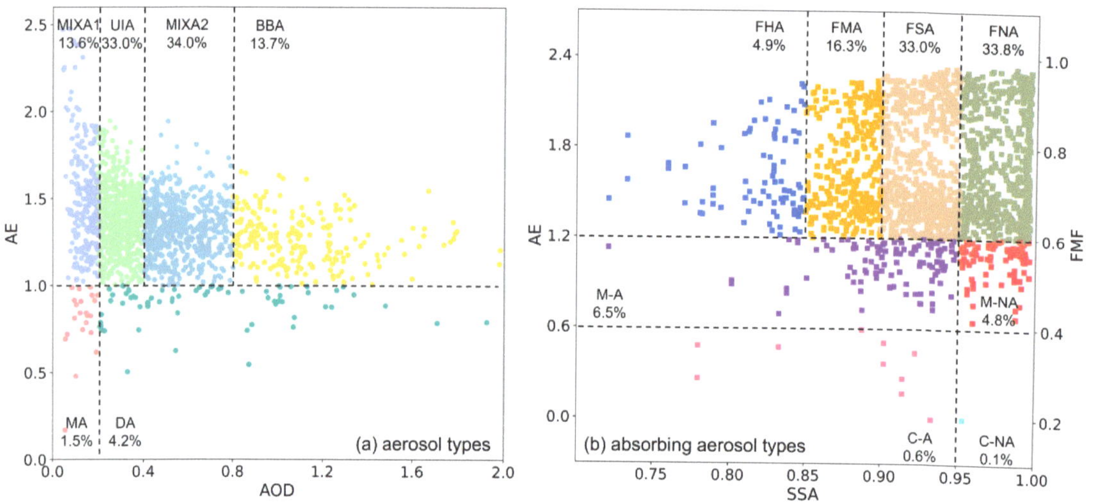

Figure 4. Two aerosol type classification schemes based on the daily measurements of PolyU AERONET station. (**a**) Aerosol types (MIXA: mixed aerosol; BBA: biomass-burning aerosol; MA: marine aerosol; DA: dust aerosol; UIA: urban/industrial aerosol); (**b**) absorbing aerosol types (C-A: coarse absorbing; C-NA: coarse non-absorbing; M-A: mixed absorbing; M-NA: mixed non-absorbing; FSA: fine slightly absorbing; FMA: fine medium absorbing; FHA: fine highly absorbing; FNA: fine non-absorbing).

3.2. Trend Analysis of Aerosol Optical Characteristics

3.2.1. Long-Term Variations of Aerosol Optical Parameters

Figure 5 shows inter-annual varying trends in aerosol optical parameters from 2006 to 2021. The linear and non-linear tendencies were estimated by using monthly averaged measurements of the PolyU AERONET station. The year 2012 can be deemed a turning point with the highest annual mean AOD (0.57). AOD in Hong Kong shows an upward trend before 2012 and a declining trend after 2012 (see Figure S3); additionally, the annual mean AOD decreased gradually and a significant decline trend can be seen for 2012 to 2021, which is consistent with AOD trends in the PRD region suggested by previous studies [45]. Thus, the study period was divided into two subperiods, namely pre-2012 and post-2012. The linear regression shows an increasing tendency in AOD and AE during the pre-2012 period with a significant decreasing trend after 2012 (Figure 5a,b). Contrary to the decrease in AOD and AE, a slight increase in SSA was found, at a rate of 0.0001 during the post-2012 period (Figure 5c).

A time series of aerosol optical parameters were decomposed into five IMFs and one variation trend curve based on the ESMD approach. The variance contribution rate (VCR) and the significance test at 95% confidence level of each IMF and the residual curve were calculated and are displayed in Table 2. It can be seen that the IMFs and trend curves are above the 95% confidence level and significantly correlated with the original time series, except the IMF5 of AOD. As suggested by Zheng et al. [21], the IMFs contain periodic oscillations and irregular non-periodic changes. Take the decomposed results of AOD time series as an example: for the irregular fluctuations (i.e., IMF1 and IMF2 in Figure S4a), the unstable amplitudes of IMFs may be caused by various unexpected factors, including weather conditions, short-term pollution control policies and human activity intensity. Periodic oscillations generally include seasonal (IMF3 in Figure S4a) and inter-annual periodic changes (IMF4 and IMF5 in Figure S4a) related to geographical phenomena. The residual curve can be used to denote the variation trend, which is the main focus of our study. The varying trends analyzed by ESMD are not completely consistent with the linear trends, presenting more fluctuating and detailed variations in aerosol optical parameters. Using AOD as an illustration, the ESMD-based AOD trend (blue dashed line in Figure 4a) shows that there is a decline from 2006 to 2008 with the value changing from above 0.5

to around 0.4, and the values increased subsequently in 2009. The descending trend seen up to 2008 may be caused by emissions control [46] and the reduction in industrial air pollution during the global economic crisis, and an increasing trend can be seen after 2009 since the emissions from economic production and transport returned to pre-recession levels, as suggested by Buchholz et al. [47]. After this, a remarkable descending trend can be found, with the value decreasing to around 0.3 in 2021, which has been attributed to the encouragement of renewable sources of energy and relative air pollution control policies [48,49]. For AE, an increasing tendency similar to the linear trend can be observed before 2012, whereas noticeable fluctuating variations (decreasing–increasing–decreasing) are present in the second period. Especially, there is a steep declining trend after 2020, possibly caused by the reduction in human activities due to the effects of COVID-19 lockdown policies [50]. Nonetheless, most of the monthly averaged AE values are larger than 1.0, suggesting the dominance of anthropogenic aerosols. Figure 5c shows more oscillations in SSA, alternating between upward and downward tendencies. SSA increased substantially after 2017, implying a reduction in absorbing aerosols (long-range transport biomass burning and local vehicles) and more aerosols with strong scattering properties. The underlying cause appears to be the formulation of air pollution control policies and the promotion of new energy sources.

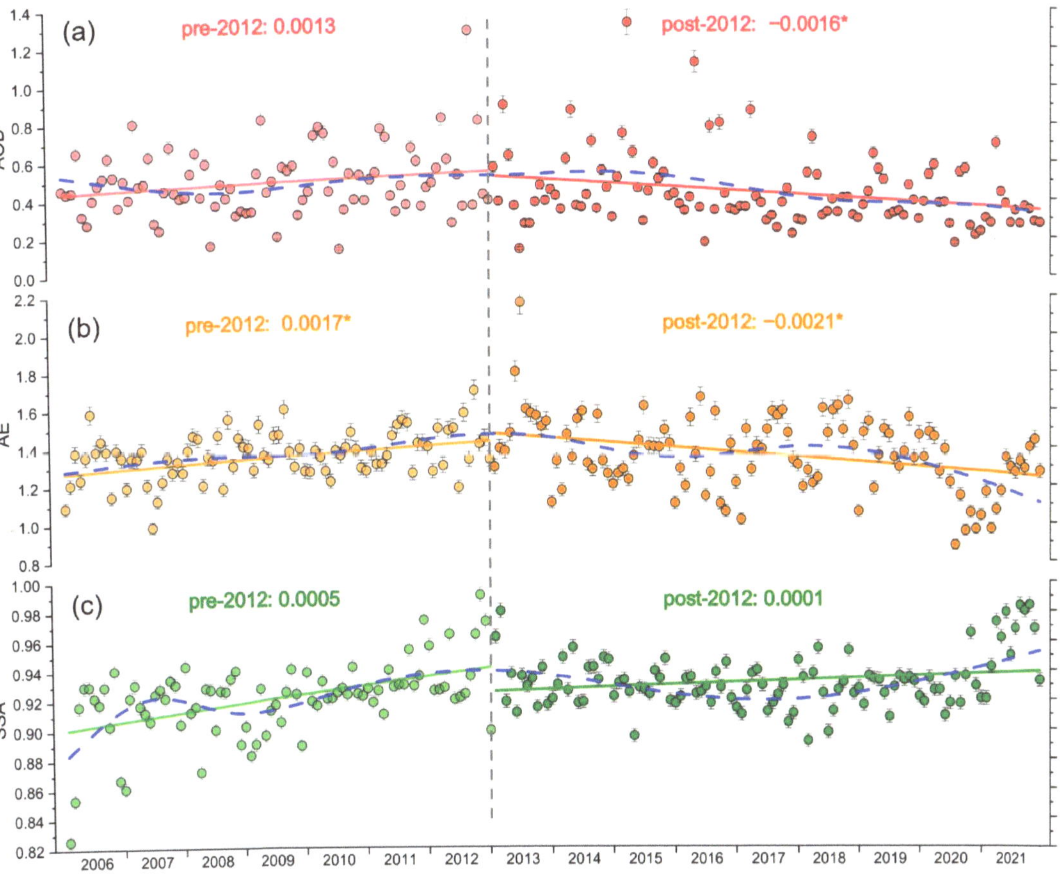

Figure 5. Linear (colored solid lines) and ESMD-based non-linear (blue dashed lines) annual varying trends of (**a**) AOD, (**b**) AE, and (**c**) SSA from 2006 to 2021 based on monthly averaged measurements. (* denotes the linear trends passing the 95% confidence level).

Table 2. Variance contribution rate (VCR) and correlation coefficient of the decomposed results.

Parameters		IMF1	IMF2	IMF3	IMF4	IMF5	R
AOD	VCR (%)	39.36	29.03	14.60	5.04	2.42	9.55
	Correlation coefficient	0.55 *	0.54 *	0.33 *	0.18 *	0.14	0.30 *
AE	VCR (%)	32.52	14.42	27.18	4.47	4.54	16.87
	Correlation coefficient	0.54 *	0.37 *	0.47 *	0.15 *	0.23 *	0.39 *
SSA	VCR (%)	37.97	13.16	9.44	6.53	3.01	29.89
	Correlation coefficient	0.55 *	0.35 *	0.38 *	0.41 *	0.14 *	0.57 *

* represents the significance test at the 95% confidence level.

In conclusion, the linear trend is useful for summarizing the overall variable trends during the study period, whereas the ESMD-based non-linear trends can accurately capture more detailed fluctuations in data, and provide more detailed and plausible variations of aerosol optical properties. The overall results indicate that AOD and AE levels, as well as the proportion of absorbing aerosols in Hong Kong, were reduced during the study period, suggesting that the aforementioned air quality control countermeasures were effective.

3.2.2. Seasonal Variations of Aerosol Optical Parameters

To reveal a detailed pattern of seasonal variations in aerosol optical parameters, linear and ESMD-based non-linear trends were derived using monthly averaged measurements directly. Figure 6a shows that the overall linear seasonal trends in AOD are consistent with the long-term variation directions, but the magnitudes of trends vary in different seasons. For AE, the linear seasonal trends are similar to the long-term variations, except in winter. There is an increase from 2013 to 2021 in winter, attenuating the substantial decline presented in other three seasons and leading to a moderate decrease in the annual trend in the post-2012 period. Similarly, descending variations in linear seasonal trends of SSA are observed in winter, which is different from the increasing pattern in other seasons.

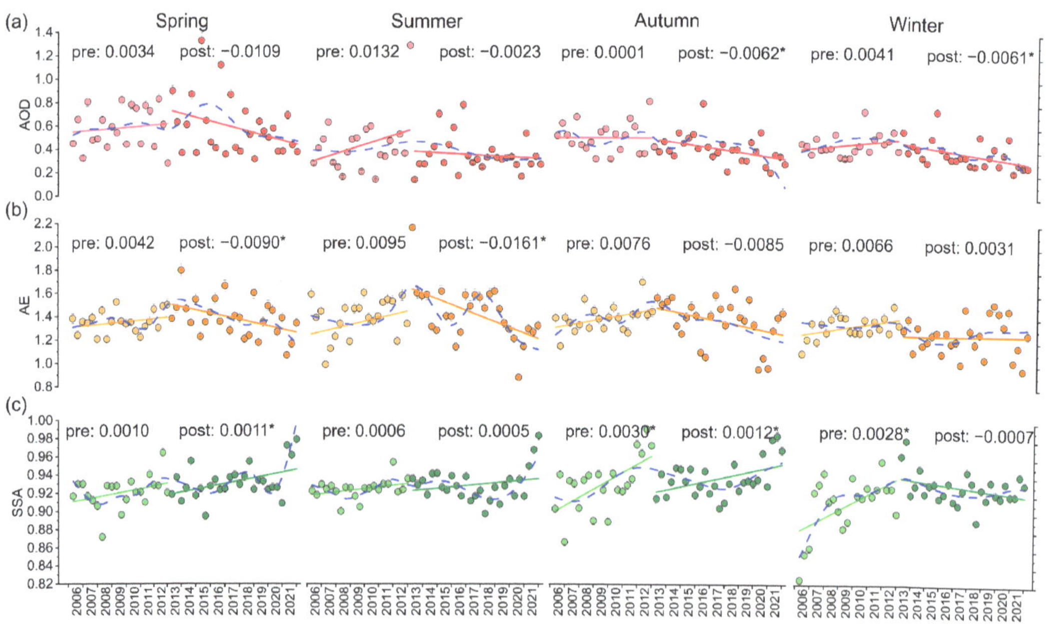

Figure 6. Linear (colored solid lines) and ESMD-based non-linear (blue dashed lines) seasonal varying trends of (**a**) AOD, (**b**) AE, and (**c**) SSA from 2006 to 2021 based on monthly averaged measurements. (* denotes the linear trends passing the 95% confidence level).

A time series of seasonal aerosol optical parameters was also decomposed into several IMFs and the trend curve was based on the ESMD approach. The variance contribution rate (VCR) and the significance test at a 95% confidence level of the varying trends were calculated and are displayed in Table 3, and the estimation results of each IMF are shown in Table S1. The results show that the varying trends are above the 95% confidence level and have a stronger correlation with the original time series, except for the varying trend of summer AOD. For the ESMD-based non-linear trends, these are not entirely aligned with the long-term tendencies shown by the linear analysis. Specifically, in the pre-2012 period, a decreasing–increasing trend of AOD was exhibited in summer and autumn, while a steady upward trend was seen in spring and winter. In terms of the post-2012 period, significant declining trends dominated in all seasons. Strikingly, during the COVID-19 lockdown period (2020–2021), AOD variations leveled off in spring and summer but pronounced decreasing trends occurred in autumn and winter. Apart from the effects of restricted human activities during COVID-19, the reduction in background pollution in Guangdong province is beneficial as well, the upwind area of Hong Kong under winter monsoon [51,52]. Fan et al. [53] also suggested that a considerable mitigation of air pollution can be found across the Pearl River Delta (PRD) area due to the COVID-19 lockdown. As depicted in Figure 6b, spring and autumn have similar changing patterns, with fluctuating variations (increasing–decreasing–increasing) in the pre-2012 period and a continuous downward trend from 2013 to 2021. Although a slightly rising trend in winter AE was presented after 2017, it was offset by the rapid decline in summer AE and a sustained descent in autumn and spring, leading to an overall continuous downward trend in AE after 2017. With regard to the seasonal variations of SSA, the downward trend in spring SSA occurred more frequently, presumably caused by the considerable biomass-burning activities across Indochina Peninsula in spring [54,55]. In addition, during the pre-2012 period, the noticeable increase in winter compensated for the decreasing trend in other seasons, making a similar varying trend to the long-term changes. Although there is a slight descending linear trend in winter SSA during the post-2012 period, a significant rising trend dominated after 2017, indicating a reduction in absorbing aerosols (i.e., black carbon) in recent years. The striking increase in SSA can be found in all seasons from 2020 to 2021, which may be ascribed to the reduction in absorbing aerosols emanating from long-range biomass burning, as well as the restrictions on vehicle transport and human activities during COVID-19 lockdown period [56,57]. Overall, both linear and ESMD-based non-linear seasonal variations of aerosol optical parameters are not completely consistent with the long-term trends. This is because a higher AOD can be observed in spring than in other seasons and spring AOD has been exhibiting a stable trend in recent years. The results imply that attempting to mitigate spring AOD is beneficial to achieving a remarkable drop in overall AOD in Hong Kong.

Table 3. Variance contribution rate (VCR) and correlation coefficient of the residual curve.

		R-Spring	R-Summer	R-Autumn	R-Winter
AOD	VCR (%)	13.03	5.85	35.46	29.60
	Correlation coefficient	0.32 *	0.26	0.47 *	0.49 *
AE	VCR (%)	31.09	36.89	25.01	14.77
	Correlation coefficient	0.50 *	0.58 *	0.44 *	0.29 *
SSA	VCR (%)	37.87	50.98	28.38	48.60
	Correlation coefficient	0.69 *	0.76 *	0.58 *	0.71 *

* represents the significance test at the 95% confidence level.

3.3. *Quantitative Impacts of Meteorological Factors*

Meteorological conditions are vital influencing factors for AOD, which have non-negligible impacts on the composition and transmission path of aerosols [9]. Pressure (hpa), wind speed (km/h), wind direction (degrees), relative humidity (%), and temperature (°C) are used to quantify the relative contributions of each factor and complex relationships between explanatory variables and AOD based on the Xgboost model. In our study, after the 10-fold CV test, the XGBoost method achieves an R^2 of 0.80, MAE of 0.08 and RMSE of 0.11. The estimated relative importance of meteorological factors is displayed in Table 4, which is calculated by evaluating the increase in the model prediction error when randomly shuffling a single feature, with the range from 0 to 1 [58]. The results suggest that temperature exerts the greatest influence on AOD, while relative humidity is the second primary influencing factor, followed by wind direction, with an importance rate of about 0.2. Wind speed and pressure make lower contributions to AOD, accounting for 0.18 and 0.16, respectively. Overall, the meteorological variables used in this study have comparable influences in estimating AOD, with a relative importance of around 0.2.

Table 4. Relative importance rate of meteorological factors on AOD.

	Wind Speed	Wind Direction	Relative Humidity	Temperature	Pressure
importance rate	0.18	0.20	0.21	0.25	0.16

Instead of Pearson correlation analysis and multi-linear regression, partial dependence derived by XGBoost was employed to investigate the non-monotonic relationship between a given variable and AOD (Figure 7). The results indicate that AOD increases significantly with relative humidity, increasing from 50% to 70% due to the hygroscopic growth of aerosols [59], whereas there is a noticeable decline in AOD when the relative humidity exceeds 70%. This may be because a higher relative humidity tends to induce frequent precipitation events and the aerosol particles may be scoured, resulting in lower AOD levels [22]. Notably, AOD remains almost unchanged when relative humidity is lower than 50%, which may be responsible for the sparse records of lower relative humidity (see Figure S5), making the correlation insignificant [60]. In contrast, a strong negative correlation can be found between wind speed and AOD, as a higher wind speed promotes aerosol diffusion, dilution and circulation, leading to better air quality [61]. Nonetheless, there are some slight positive trends between wind speed and AOD. This can be attributed to the surface soil and dust raised by high wind speeds, thus increasing the aerosol particles [62]. As depicted in Figure 7c, there is a strong relationship between wind direction and AOD. Specifically, significant upward trends in AOD occur when wind directions change from north to northeast (0–45 degrees) and southwest to west (225–270 degrees). Higher AOD levels can be observed in the northeast (45 degrees) and west (270 degrees) directions. This is more likely caused by the contributions of the long-distance transport of industrial emissions from the mainland (i.e., Henan, Shandong and Anhui provinces) [9], as well as dust and agricultural pollution from Southeast Asia [53]. With the wind directions changing from northeast to east (45–90 degrees) and from west to northwest (270–315 degrees), the neighboring regions (i.e., Guangdong, Fujian and Guangxi provinces) of Hong Kong have cleaner background pollution, leading to a decrease in AOD. The southern part of Hong Kong is surrounded by sea (see Figure 1), which is most likely the reason why AOD remains stable at a wind direction of 90–225 degrees. As can be seen in Figure 7d, AOD is sensitive to variations in temperature, and AOD increases with increased temperature when the temperature is below 25 °C. However, when the temperature is in the range of 28–29 °C, a negative association with AOD is presented. According to the partial dependence results of pressure, AOD is less susceptible to changes in pressure (Figure 7e). AOD decreases gradually with an increasing pressure from 1005 to 1010 hpa, whereas a positive correlation can be observed when the pressure is in the range of 1010–1015 hpa. Subsequently, AOD stabilizes between 0.45 and 0.5 when the pressure exceeds 1015 hpa. In summary, unlike for typical linear correlations, partial dependence analysis implemented by the XGBoost can derive complex and non-linear relationships between AOD and meteorological factors. How AOD responds to the changes in driving factors can be estimated and visualized, thus recommending the use of black-box models for facilitating the understanding of interactions between AOD and meteorological conditions.

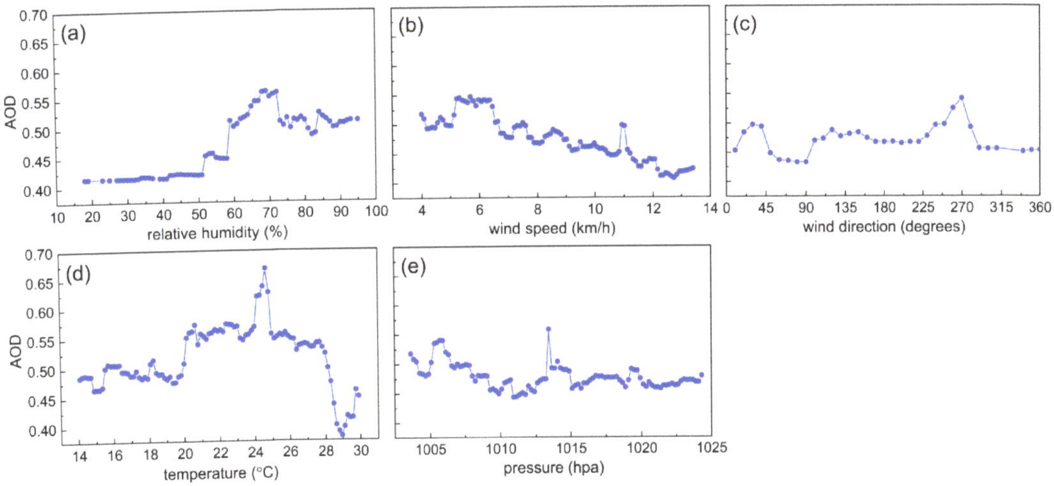

Figure 7. Interactions between AOD and meteorological factors derived by XGBoost.

4. Discussion

Using ground-based AERONET measurements taken over 16 years and combined with satellite-based observations, our study explores the dominant aerosol types, long-term varying trends and the interactions with meteorological factors. The main contributions of our study can be summarized as follows: First, in addition to classifying the aerosol types based on AOD and AE values in previous studies [63–66], the aerosol-absorbing properties in Hong Kong were also categorized using AE, SSA and FMF. Moreover, the multi-seasonal AOD, AE and SSA distribution patterns in Hong Kong were discussed, which provides a basis for understanding the seasonal aerosol optical characteristics in subtropical urban areas such as Hong Kong. Notably, the thresholds of the aerosol optical parameters are generally used for dominant aerosol type classification, while the proportions of specific aerosol types depend on the selected aerosol parameters and the thresholds used for the classification [63]. Additionally, using aerosol optical parameters for aerosol type identification is suitable for areas with a stable aerosol composition, but may not be well-suited for locations with complex topography, Arctic regions and remote marine areas, as aerosols in these regions are difficult to represent using median optical parameters [67]. Second, variations in AOD were mainly analyzed in most previous studies in the literature [16–19], whereas changes in AE and SSA are rarely included, leading to an incomplete view of aerosol variations. Apart from the linear trend, the non-linear fluctuations of aerosol optical parameters were also discussed based on the ESMD method, which can reflect more detailed varying trends. Finally, different from previous studies [21,22], non-monotonic interactions between AOD and meteorological factors were analyzed using a data-driven model, XGBoost, which contributes to the knowledge of how AOD responds to the changes in weather conditions.

However, there are still some limitations in our study. First, the minimum threshold for the number of AERONET AOD observations per day was not considered. To enrich the sample size of the AOD observations, the daily mean AOD was not filtered, although there was only one observation per day. In the future, multi-source AOD datasets (i.e., reanalysis data and satellite-based observations) or more accurately retrieved AOD should be included for the substitute of AERONET AOD to generate accurate and continuous AOD time series. Second, due to the availability of meteorological factors, the interactions between AOD and other meteorological factors (i.e., planetary boundary layer, PBL) were ignored, which has been suggested to have a significant impact on aerosol vertical distributions and the dispersion of air pollutants in previous studies [68,69].

5. Conclusions

In this study, combining aerosol optical parameters measured by the PolyU AERONET station with satellite observations (i.e., MODIS and MISR aerosol products) from 2006 to 2021 was adopted to analyze the seasonal distribution patterns, varying trends and the association with meteorological factors in Hong Kong. Four primary conclusions can be summarized from the results. Firstly, aerosol optical parameters present significant seasonal characteristics. The seasonal mean AOD is at its highest in spring, while summer has the lowest mean AOD but the largest averaged AE and SSA. Secondly, mixed aerosol and urban/industrial aerosol, which have a fine-mode size with slight or non-absorbing properties, are predominant types in Hong Kong during the study period. Thirdly, the year 2012 can be deemed as the turning point. Both AOD and AE exhibit an increasing trend in the pre-2012 period and a downward tendency after 2012, whereas a consistent rising trend of SSA can be observed over both periods. The use of ESMD to investigate the non-linear annual and seasonal varying trends, was able to provide more detailed insights than linear trends. AOD, AE and SSA present "decreasing (2006–2008)–increasing (2009–2013)–decreasing (2014–2021)", "increasing (2006–2012)–decreasing (2013–2015)-increasing (2016–2018)–decreasing (2019–2021)" and "increasing (2006–2007)–fluctuating (2008–2017)–increasing (2018–2021)" variations during the study period, respectively. Additionally, both linear and ESMD-based non-linear seasonal variations are not completely consistent with the annual trends. The results suggest that the sharp AOD decrease in autumn and winter in recent years contributed to the decline in overall AOD in Hong Kong. Finally, instead of a linear correlation, XGBoost was employed to quantify the non-monotonic relationships between meteorological factors and AOD, as well as rank the relative importance of each variable. It can be concluded that meteorological factors have comparable contributions to AOD and that interactions are complex. The results imply that meteorological conditions with lower relative humidity, higher wind speed, winds from east and southwest directions, lower temperature and higher pressure tend to lower AOD levels. In conclusion, our study can enhance the understanding of aerosol characteristics and long-term variations in Hong Kong. Furthermore, the interactions between meteorological factors and AOD can serve as a basis for developing air quality promotion strategies.

Supplementary Materials: The following supporting information can be downloaded at: https://www.mdpi.com/article/10.3390/rs14205220/s1. Figure S1: Seasonal aerosol types based on daily measurements of the PolyU AERONET station. (a) spring; (b) summer; (c) autumn and (d) winter; Figure S2: Seasonal absorbing aerosol types based on daily measurements of the PolyU AERONET station. (a) spring; (b) summer; (c) autumn and (d) winter; Figure S3: Variations of annual mean AOD with the linear trend (yr-1) in two different periods, namely pre-2012 (red) and post-2012 (blue). * represents the trends passing the 95% confidence level; Figure S4: The decomposed results of aerosol optical parameters based on ESMD. (a) AOD, (b) AE and (c) SSA; Figure S5: Distribution frequency of meteorological factors; Figure S6: Distribution frequency of multi-seasonal meteorological factors; Table S1: Variance contribution rate (VCR, %) and correlation coefficient (CC) of the IMFs.

Author Contributions: Conceptualization, X.Y. and M.S.W.; methodology, X.Y. and M.S.W.; formal analysis, X.Y., M.S.W., J.N., K.H.L. and J.L.; data curation, X.Y.; writing—original draft preparation, X.Y.; writing—review and editing, M.S.W., J.N., K.H.L. and J.L.; supervision, M.S.W., J.N. and K.H.L.; funding acquisition, M.S.W. All authors have read and agreed to the published version of the manuscript.

Funding: This research was funded by the Research Institute for Land and Space (Grant No. 1-CD81), The Hong Kong Polytechnic University; and General Research Fund (Grant No. 15603920 and 15609421), and Collaborative Research Fund (Grant No. C5062-21GF, C7064-18GF), from the Hong Kong Research Grants Council, Hong Kong, China.

Data Availability Statement: All datasets used in this study are available to download by the web links given in the article.

Conflicts of Interest: The authors declare no conflict of interest.

References

1. Kim, M.; Kim, J.; Wong, M.S.; Yoon, J.; Lee, J.; Wu, D.; Chan, P.; Nichol, J.E.; Chung, C.-Y.; Ou, M.L. Improvement of aerosol optical depth retrieval over Hong Kong from a geostationary meteorological satellite using critical reflectance with background optical depth correction. *Remote Sens. Environ.* **2014**, *142*, 176–187. [CrossRef]
2. Wong, M.S.; Lee, K.H.; Nichol, J.E.; Li, Z. Retrieval of Aerosol Optical Thickness Using MODIS 500 × 500 m^2, a Study in Hong Kong and the Pearl River Delta Region. *IEEE Trans. Geosci. Remote Sens.* **2010**, *48*, 3318–3327. [CrossRef]
3. Wong, M.S.; Nichol, J.E.; Lee, K.H. An operational MODIS aerosol retrieval algorithm at high spatial resolution, and its application over a complex urban region. *Atmos. Res.* **2011**, *99*, 579–589. [CrossRef]
4. Huebert, B.J.; Bates, T.; Russell, P.B.; Shi, G.; Kim, Y.J.; Kawamura, K.; Carmichael, G.; Nakajima, T. An overview of ACE-Asia: Strategies for quantifying the relationships between Asian aerosols and their climatic impacts. *J. Geophys. Res. Atmos.* **2003**, *108*, 8633. [CrossRef]
5. Sheesley, R.J.; Schauer, J.J.; Chowdhury, Z.; Cass, G.R.; Simoneit, B.R. Characterization of organic aerosols emitted from the combustion of biomass indigenous to South Asia. *J. Geophys. Res. Atmos.* **2003**, *108*, 4285. [CrossRef]
6. Pye, H.O.; Ward-Caviness, C.K.; Murphy, B.N.; Appel, K.W.; Seltzer, K.M. Secondary organic aerosol association with cardiorespiratory disease mortality in the United States. *Nat. Commun.* **2021**, *12*, 5721. [CrossRef] [PubMed]
7. Wang, Z.; Liu, D.; Wang, Z.; Wang, Y.; Khatri, P.; Zhou, J.; Takamura, T.; Shi, G. Seasonal characteristics of aerosol optical properties at the SKYNET Hefei site (31.90 N, 117.17 E) from 2007 to 2013. *J. Geophys. Res. Atmos.* **2014**, *119*, 6128–6139. [CrossRef]
8. IPCC. Climate change 2013: The physical science basis. In *Contribution of Working Group I to the Fifth Assessment*; Stocker, T.F., Qin, D., Plattner, G.K., Tignor, M.M.H.L., Allen, S.K., Boschung, J., Nauels, A., Xia, Y., Midgley, P.M., Eds.; Cambridge University Press: Cambridge, UK; New York, NY, USA, 2013; p. 1535.
9. Deng, X.; Shi, C.; Wu, B.; Chen, Z.; Nie, S.; He, D.; Zhang, H. Analysis of aerosol characteristics and their relationships with meteorological parameters over Anhui province in China. *Atmos. Res.* **2012**, *109*, 52–63. [CrossRef]
10. Kaufman, Y.J. Remote sensing of direct and indirect aerosol forcing. In *Aerosol Forcing of Climate*; Charlson, R.J., Heintzenberg, J., Eds.; John Wiley: New York, NY, USA, 1995; pp. 297–332.
11. Nakajima, T.; Campanelli, M.; Che, H.; Estellés, V.; Irie, H.; Kim, S.W.; Kim, J.; Liu, D.; Nishizawa, T.; Pandithurai, G.; et al. An overview of and issues with sky radiometer technology and SKYNET. *Atmos. Meas. Tech.* **2020**, *13*, 4195–4218. [CrossRef]
12. Li, Z.Q.; Xu, H.; Li, K.T.; Li, D.H.; Xie, Y.S.; Li, L.; Zhang, Y.; Gu, X.F.; Zhao, W.; Tian, Q.J.; et al. Comprehensive study of optical, physical, chemical, and radiative properties of total columnar atmospheric aerosols over China: An overview of Sun–Sky Radiometer Observation Network (SONET) measurements. *Bull. Am. Meteorol. Soc.* **2018**, *99*, 739–755. [CrossRef]
13. Holben, B.N.; Eck, T.F.; Slutsker, I.A.; Tanre, D.; Buis, J.P.; Setzer, A.; Vermote, E.; Reagan, J.A.; Kaufman, Y.J.; Nakajima, T.; et al. AERONET—A federated instrument network and data archive for aerosol characterization. *Remote Sens. Environ.* **1998**, *66*, 1–16. [CrossRef]
14. Kinne, S.; Lohmann, U.; Feichter, J.; Schulz, M.; Timmreck, C.; Ghan, S.; Easter, R.; Chin, M.; Ginoux, P.; Takemura, T.; et al. Monthly averages of aerosol properties: A global comparison among models, satellite data, and AERONET ground data. *J. Geophys. Res. Atmos.* **2003**, *108*, 4634. [CrossRef]
15. Li, J.; Liu, L.; Lacis, A.A.; Carlson, B.E. An optimal fitting approach to improve the GISS ModelE aerosol optical property parameterization using AERONET data. *J. Geophys. Res. Atmos.* **2010**, *115*, D16211. [CrossRef]
16. Xia, X. Variability of aerosol optical depth and Angstrom wavelength exponent derived from AERONET observations in recent decades. *Environ. Res. Lett.* **2011**, *6*, 044011. [CrossRef]
17. De Leeuw, G.; Sogacheva, L.; Rodriguez, E.; Kourtidis, K.; Georgoulias, A.K.; Alexandri, G.; Amiridis, V.; Proestakis, E.; Marinou, E.; Xue, Y.; et al. Two decades of satellite observations of AOD over mainland China using ATSR-2, AATSR and MODIS/Terra: Data set evaluation and large-scale patterns. *Atmos. Chem. Phys.* **2018**, *18*, 1573–1592. [CrossRef]
18. Che, H.; Gui, K.; Xia, X.; Wang, Y.; Holben, B.N.; Goloub, P.; Cuevas-Agulló, E.; Wang, H.; Zheng, Y.; Zhao, H.; et al. Large contribution of meteorological factors to inter-decadal changes in regional aerosol optical depth. *Atmos. Chem. Phys.* **2019**, *19*, 10497–10523. [CrossRef]
19. Ramachandran, S.; Rupakheti, M. Trends in the types and absorption characteristics of ambient aerosols over the Indo-Gangetic Plain and North China Plain in last two decades. *Sci. Total Environ.* **2022**, *831*, 154867. [CrossRef]
20. Zhang, Z.Y.; Wong, M.S.; Nichol, J. Global trends of aerosol optical thickness using the ensemble empirical mode decomposition method. *Int. J. Climatol.* **2016**, *36*, 4358–4372. [CrossRef]
21. Zheng, Y.; Wang, X.; Zhang, X.; Hu, G. Multi-spatiotemporal patterns of aerosol optical depth and influencing factors during 2000–2020 from two spatial perspectives: The entire Yellow River Basin region and its urban agglomerations. *Int. J. Appl. Earth Obs. Geoinf.* **2022**, *106*, 102643. [CrossRef]
22. He, L.; Wang, L.; Huang, B.; Wei, J.; Zhou, Z.; Zhong, Y. Anthropogenic and meteorological drivers of 1980–2016 trend in aerosol optical and radiative properties over the Yangtze River Basin. *Atmos. Environ.* **2020**, *223*, 117188. [CrossRef]

23. Zhang, Z.; Xu, B.; Xu, W.; Wang, F.; Gao, J.; Li, Y.; Li, M.; Feng, Y.; Shi, G. Machine learning combined with the PMF model reveal the synergistic effects of sources and meteorological factors on $PM_{2.5}$ pollution. *Environ. Res.* **2022**, *212*, 113322. [CrossRef] [PubMed]
24. Yang, N.; Shi, H.; Tang, H.; Yang, X. Geographical and temporal encoding for improving the estimation of $PM_{2.5}$ concentrations in China using end-to-end gradient boosting. *Remote Sens. Environ.* **2022**, *269*, 112828. [CrossRef]
25. Dubovik, O.; Smirnov, A.; Holben, B.N.; King, M.D.; Kaufman, Y.J.; Eck, T.F.; Slutsker, I. Accuracy assessments of aerosol optical properties retrieved from Aerosol Robotic Network (AERONET) Sun and sky radiance measurements. *J. Geophys. Res. Atmos.* **2000**, *105*, 9791–9806. [CrossRef]
26. Lyapustin, A.; Martonchik, J.; Wang, Y.; Laszlo, I.; Korkin, S. Multiangle implementation of atmospheric correction (MAIAC): 1. Radiative transfer basis and look-up tables. *J. Geophys. Res. Atmos.* **2011**, *116*, D03210. [CrossRef]
27. Tao, M.; Wang, J.; Li, R.; Wang, L.; Wang, L.; Wang, Z.; Tao, J.; Che, H.; Chen, L. Performance of MODIS high-resolution MAIAC aerosol algorithm in China: Characterization and limitation. *Atmos. Environ.* **2019**, *213*, 159–169. [CrossRef]
28. Ångström, A. On the atmospheric transmission of sun radiation and on dust in the air. *Geogr. Ann.* **1929**, *11*, 156–166. [CrossRef]
29. Kahn, R.A.; Gaitley, B.J. An analysis of global aerosol type as retrieved by MISR. *J. Geophys. Res. Atmos.* **2015**, *120*, 4248–4281. [CrossRef]
30. Si, Y.; Li, S.; Chen, L.; Shang, H.; Wang, L.; Letu, H. Assessment and improvement of MISR Angstrom exponent and single-scattering albedo products using AERONET data in China. *Remote Sens.* **2017**, *9*, 693. [CrossRef]
31. Kahn, R.A.; Gaitley, B.J.; Garay, M.J.; Diner, D.J.; Eck, T.F.; Smirnov, A.; Holben, B.N. Multiangle Imaging SpectroRadiometer global aerosol product assessment by comparison with the Aerosol Robotic Network. *J. Geophys. Res. Atmos.* **2010**, *115*, D23209. [CrossRef]
32. Wang, J.L.; Li, Z.J. Extreme-point symmetric mode decomposition method for data analysis. *Adv. Adapt. Data Anal.* **2013**, *5*, 1350015. [CrossRef]
33. Qin, Y.; Li, B.; Chen, Z.; Chen, Y.; Lian, L. Spatio-temporal variations of nonlinear trends of precipitation over an arid region of northwest China according to the extreme-point symmetric mode decomposition method. *Int. J. Climatol.* **2018**, *38*, 2239–2249. [CrossRef]
34. Wang, X.; Chen, Y.; Li, Z.; Fang, G.; Wang, F.; Liu, H. The impact of climate change and human activities on the Aral Sea Basin over the past 50 years. *Atmos. Res.* **2020**, *245*, 105125. [CrossRef]
35. Chen, T.; Guestrin, C. XGBoost: A Scalable Tree Boosting System. In Proceedings of the 22nd ACM SIGKDD International Conference on Knowledge Discovery and Data Mining, San Francisco, CA, USA, 13–17 August 2016; pp. 785–794. [CrossRef]
36. Friedman, J.H. Greedy function approximation: A gradient boosting machine. *Ann. Stat.* **2001**, *29*, 1189–1232. [CrossRef]
37. Yu, X.; Lü, R.; Liu, C.; Yuan, L.; Shao, Y.; Zhu, B.; Lei, L. Seasonal variation of columnar aerosol optical properties and radiative forcing over Beijing, China. *Atmos. Environ.* **2017**, *166*, 340–350. [CrossRef]
38. Kumar, K.R.; Kang, N.; Sivakumar, V.; Griffith, D. Temporal characteristics of columnar aerosol optical properties and radiative forcing (2011–2015) measured at AERONET's Pretoria_CSIR_DPSS site in South Africa. *Atmos. Environ.* **2017**, *165*, 274–289. [CrossRef]
39. Liu, Y.; He, J.; Lai, X.; Zhang, C.; Zhang, L.; Gong, S.; Che, H. Influence of atmospheric circulation on aerosol and its optical characteristics in the pearl river delta region. *Atmosphere* **2020**, *11*, 288. [CrossRef]
40. Xia, X.; Che, H.; Zhu, J.; Chen, H.; Cong, Z.; Deng, X.; Fan, X.; Fu, Y.; Goloub, P.; Jiang, H.; et al. Ground-based remote sensing of aerosol climatology in China: Aerosol optical properties, direct radiative effect and its parameterization. *Atmos. Environ.* **2016**, *124*, 243–251. [CrossRef]
41. Salinas, S.V.; Chew, B.N.; Liew, S.C. Retrievals of aerosol optical depth and Ångström exponent from ground-based Sun-photometer data of Singapore. *Appl. Opt.* **2009**, *48*, 1473–1484. [CrossRef]
42. Zheng, C.; Zhao, C.; Zhu, Y.; Wang, Y.; Shi, X.; Wu, X.; Chen, T.; Wu, F.; Qiu, Y. Analysis of influential factors for the relationship between $PM_{2.5}$ and AOD in Beijing. *Atmos. Chem. Phys.* **2017**, *17*, 13473–13489. [CrossRef]
43. Bellouin, N.; Boucher, O.; Haywood, J.; Reddy, M.S. Global estimate of aerosol direct radiative forcing from satellite measurements. *Nature* **2005**, *438*, 1138–1141. [CrossRef] [PubMed]
44. Yan, X.; Zang, Z.; Li, Z.; Luo, N.; Zuo, C.; Jiang, Y.; Li, D.; Guo, Y.; Zhao, W.; Shi, W.; et al. A global land aerosol fine-mode fraction dataset (2001–2020) retrieved from MODIS using hybrid physical and deep learning approaches. *Earth Syst. Sci. Data* **2022**, *14*, 1193–1213. [CrossRef]
45. Zhang, L.; Zhang, M.; Yao, Y. Multi-time scale analysis of regional aerosol optical depth changes in national-level urban agglomerations in China using modis collection 6.1 datasets from 2001 to 2017. *Remote Sens.* **2019**, *11*, 201. [CrossRef]
46. Cai, S.; Wang, Y.; Zhao, B.; Wang, S.; Chang, X.; Hao, J. The impact of the "air pollution prevention and control action plan" on $PM_{2.5}$ concentrations in Jing-Jin-Ji region during 2012–2020. *Sci. Total Environ.* **2017**, *580*, 197–209. [CrossRef] [PubMed]
47. Buchholz, R.R.; Worden, H.M.; Park, M.; Francis, G.; Deeter, M.N.; Edwards, D.P.; Emmons, L.K.; Gaubert, B.; Gille, J.; Martínez-Alonso, S.; et al. Air pollution trends measured from Terra: CO and AOD over industrial, fire-prone, and background regions. *Remote Sens. Environ.* **2021**, *256*, 112275. [CrossRef]
48. Zhang, Q.; He, K.; Huo, H. Cleaning China's air. *Nature* **2012**, *484*, 161–162. [CrossRef]
49. Hong Kong Environmental Protection Department (HKEPD). Air Pollution Control Strategies. 2019. Available online: https://www.epd.gov.hk/epd/english/environmentinhk/air/prob_solutions/strategies_apc.html (accessed on 22 August 2022).

50. Wang, S.; Zhang, Y.; Ma, J.; Zhu, S.; Shen, J.; Wang, P.; Zhang, H. Responses of decline in air pollution and recovery associated with COVID-19 lockdown in the Pearl River Delta. *Sci. Total Environ.* **2021**, *756*, 143868. [CrossRef]
51. Zeren, Y.; Guo, H.; Lyu, X.; Zhou, B.; Liu, X.; Yang, L.; Yuan, Z.; Wang, Y. Remarkable spring increase overwhelmed hard-earned autumn decrease in ozone pollution from 2005 to 2017 at a suburban site in Hong Kong, South China. *Sci. Total Environ.* **2022**, *831*, 154788. [CrossRef]
52. Jiang, F.; Guo, H.; Wang, T.J.; Cheng, H.R.; Wang, X.M.; Simpson, I.J.; Ding, A.; Saunders, S.M.; Lam, S.H.M.; Blake, D.R. An ozone episode in the Pearl River Delta: Field observation and model simulation. *J. Geophys. Res. Atmos.* **2010**, *115*, D22305. [CrossRef]
53. Fan, L.; Fu, S.; Wang, X.; Fu, Q.; Jia, H.; Xu, H.; Qin, G.; Hu, X.; Cheng, J. Spatiotemporal variations of ambient air pollutants and meteorological influences over typical urban agglomerations in China during the COVID-19 lockdown. *J. Environ. Sci.* **2021**, *106*, 26–38. [CrossRef]
54. Liao, Z.; Ling, Z.; Gao, M.; Sun, J.; Zhao, W.; Ma, P.; Quan, J.; Fan, S.; Liao, Z.; Ling, Z.; et al. Tropospheric ozone variability over Hong Kong based on recent 20 years (2000–2019) ozonesonde observation. *J. Geophys. Res. Atmos.* **2021**, *126*, e2020JD033054. [CrossRef]
55. Xue, L.; Ding, A.; Cooper, O.; Huang, X.; Wang, W.; Zhou, D.; Wu, Z.; McClure-Begley, A.; Petropavlovskikh, I.; Andreae, M.O.; et al. ENSO and Southeast Asian biomass burning modulate subtropical trans-Pacific ozone transport. *Natl. Sci. Rev.* **2021**, *8*, nwaa132. [CrossRef]
56. Jia, M.; Evangeliou, N.; Eckhardt, S.; Huang, X.; Gao, J.; Ding, A.; Stohl, A. Black carbon emission reduction due to COVID-19 lockdown in China. *Geophys. Res. Lett.* **2021**, *48*, e2021GL093243. [CrossRef]
57. Lin, Y.C.; Zhang, Y.L.; Xie, F.; Fan, M.Y.; Liu, X. Substantial decreases of light absorption, concentrations and relative contributions of fossil fuel to light-absorbing carbonaceous aerosols attributed to the COVID-19 lockdown in east China. *Environ. Pollut.* **2021**, *275*, 116615. [CrossRef] [PubMed]
58. Casalicchio, G.; Molnar, C.; Bischl, B. Visualizing the feature importance for black box models. In Proceedings of the Joint European Conference on Machine Learning and Knowledge Discovery in Databases, Würzburg, Germany, 16–20 September 2018; Springer: Cham, Switzerland; pp. 655–670.
59. Malm, W.C.; Day, D.E. Estimates of aerosol species scattering characteristics as a function of relative humidity. *Atmos. Environ.* **2001**, *35*, 2845–2860. [CrossRef]
60. Zhu, X.; Li, Y.; Wang, X. Machine learning prediction of biochar yield and carbon contents in biochar based on biomass characteristics and pyrolysis conditions. *Bioresour. Technol.* **2019**, *288*, 121527. [CrossRef] [PubMed]
61. Tai, A.P.; Mickley, L.J.; Jacob, D.J. Correlations between fine particulate matter ($PM_{2.5}$) and meteorological variables in the United States: Implications for the sensitivity of $PM_{2.5}$ to climate change. *Atmos. Environ.* **2010**, *44*, 3976–3984. [CrossRef]
62. Csavina, J.; Field, J.; Félix, O.; Corral-Avitia, A.Y.; Sáez, A.E.; Betterton, E.A. Effect of wind speed and relative humidity on atmospheric dust concentrations in semi-arid climates. *Sci. Total Environ.* **2014**, *487*, 82–90. [CrossRef]
63. Liu, Y.; Yi, B. Aerosols over East and South Asia: Type identification, optical properties, and implications for radiative forcing. *Remote Sens.* **2022**, *14*, 2058. [CrossRef]
64. Cúneo, L.; Ulke, A.G.; Cerne, B. Advances in the characterization of aerosol optical properties using long-term data from AERONET in Buenos Aires. *Atmos. Pollut. Res.* **2022**, *13*, 101360. [CrossRef]
65. Stefan, S.; Voinea, S.; Iorga, G. Study of the aerosol optical characteristics over the Romanian Black Sea Coast using AERONET data. *Atmos. Pollut. Res.* **2020**, *11*, 1165–1178. [CrossRef]
66. Ozdemir, E.; Tuygun, G.T.; Elbir, T. Application of aerosol classification methods based on AERONET version 3 product over eastern Mediterranean and Black Sea. *Atmos. Pollut. Res.* **2020**, *11*, 2226–2243. [CrossRef]
67. Schmeisser, L.; Andrews, E.; Ogren, J.A.; Sheridan, P.; Jefferson, A.; Sharma, S.; Kim, J.E.; Sherman, J.P.; Sorribas, M.; Kalapov, I.; et al. Classifying aerosol type using in situ surface spectral aerosol optical properties. *Atmos. Chem. Phys.* **2017**, *17*, 12097–12120. [CrossRef]
68. Su, T.; Li, Z.; Zheng, Y.; Wu, T.; Wu, H.; Guo, J. Aerosol-boundary layer interaction modulated entrainment process. *NPJ Clim. Atmos. Sci.* **2022**, *5*, 64. [CrossRef]
69. Su, T.; Li, Z.; Li, C.; Li, J.; Han, W.; Shen, C.; Tan, W.; Wei, J.; Guo, J. The significant impact of aerosol vertical structure on lower atmosphere stability and its critical role in aerosol–planetary boundary layer (PBL) interactions. *Atmos. Chem. Phys.* **2020**, *20*, 3713–3724. [CrossRef]

remote sensing

MDPI

Article

The Identification and Analysis of Long-Range Aerosol Transport Pathways with Layered Cloud-Aerosol Lidar with Orthogonal Polarization Datasets from 2006 to 2016

Lingyu Wang [1,2,3], Wensheng Wang [1,2], Baolei Lyu [4,5,6,*], Jinghua Zhang [1,2], Yilun Han [3], Yuqi Bai [3] and Zhi Guo [1,2]

1 Key Laboratory of Network Information System Technology (NIST), Institute of Electronics, Chinese Academy of Sciences, Beijing 100190, China; wanglingyu@aircas.ac.cn (L.W.); wangws@aircas.ac.cn (W.W.); guozhi@aircas.ac.cn (Z.G.)
2 The Aerospace Information Research Institute, Chinese Academy of Sciences, Beijing 100094, China
3 Department of Earth System Science, Ministry of Education Key Laboratory for Earth System Modeling, Institute for Global Change Studies, Tsinghua University, Beijing 100084, China; hanyilun@mail.tsinghua.edu.cn (Y.H.)
4 Huayun Sounding Meteorology Technology Corporation, Beijing 100081, China
5 Jiangsu Collaborative Innovation Center of Atmospheric Environment and Equipment Technology (CICAEET), Nanjing University of Information Science & Technology, Nanjing 210044, China
6 China Meteorological Administration Xiong'an Atmospheric Boundary Layer Key Laboratory, Xiong'an 071000, China
* Correspondence: lvbaolei@hysdqx.com

Abstract: In this study, we used Cloud-Aerosol Lidar with Orthogonal Polarization (CALIOP) aerosol products acquired from 2006 to 2016 to identify global long-range aerosol transport pathways, including the trans-Atlantic, the trans-Pacific, and the trans-Arabian Sea pathways. Deep analyses were subsequently conducted focusing on two significant paths within the range of the trans-Pacific transport pathway, from which we generated a three-stage conceptual model mainly identifying aerosols from the Taklimakan Desert and aerosols from the North China Plain. The results show that in the first stage of the model, the dust or polluted-dust aerosols were emitted, raised, and mixed within the planetary boundary layer (PBL), characterized by high percentages (>70%) of aerosols in the PBL (AOD_{PBL}), while in the second stage, some aerosols were further raised into the free troposphere where the AOD_{PBL} percentages decreased to less than 40%, driven by vertical movements and turbulences; in the last stage, the aerosols gradually settled back to the surface layer due to gravity and wet deposition, inferred by increasing AOD_{PBL} percentages. We demonstrated that the proposed model is capable of characterizing different aerosol types and climate conditions on spatiotemporal scales, providing a straightforward and evident approach to exploring long-range aerosol transport pathways.

Keywords: CALIOP; aerosols; three-stage model; long-range transport

Citation: Wang, L.; Wang, W.; Lyu, B.; Zhang, J.; Han, Y.; Bai, Y.; Guo, Z. The Identification and Analysis of Long-Range Aerosol Transport Pathways with Layered Cloud-Aerosol Lidar with Orthogonal Polarization Datasets from 2006 to 2016. *Remote Sens.* **2023**, *15*, 4537. https://doi.org/10.3390/rs15184537

Academic Editor: Jing Wei

Received: 5 July 2023
Revised: 11 September 2023
Accepted: 12 September 2023
Published: 15 September 2023

1. Introduction

Atmospheric aerosols have important influences on human health, air quality, and climate change, mainly from natural and anthropogenic sources [1–3]. Natural sources include dust, sea salt, oceanic dimethyl sulfide (DMS), volcanoes, and secondary organic aerosols (SOAs) from volatile organic compounds (VOCs) [4–6]. Anthropogenic emissions are closely associated with industrialization and agricultural production [7] and significantly impact human health [2,8,9]. Most aerosols have short lifetimes (about a week) and have limited transport distances. Some aerosols, such as dust and persistent pollutants, that have longer lifetimes (probably a few weeks) make long-distance aerosol transport possible [10,11].

On the global scale, three long-range aerosol transport pathways have been widely explored and discussed, namely trans-Atlantic transport, trans-Pacific transport, and trans-Arabian transport [12–16]. These three pathways drive local aerosol emissions to have global influences. For example, the westward transport of Sahara Desert dust can reach the Atlantic Ocean and even Central America [17–19]. Dust emissions and transport in the Arabian Peninsula are vital to the climate of the Arabian Peninsula and the Arabian Sea [20]. Dust from the Taklimakan and Gobi deserts potentially influences Northern China, Mongolia, the Korean Peninsula, Japan, the northeastern Pacific Ocean, and the western coast of North America through activation and transport [11]. Among all these pathways, the trans-Pacific transport pathways are some of the most complicated due to their distinct sources, large spatial impacts, and dense populations along the pathways [21–23]. Therefore, we selected the trans-Pacific pathways for deep aerosol long-distance transport analysis in this study.

In East Asia, aerosol sources have enormous diversity and wide spatial distribution, mainly including dust emissions from the Taklimakan Desert and the Gobi Desert, pollutant emissions in Eastern China, and other large Asian cities [24–26]. Due to aerosol transport, these emissions would not just influence local or regional air quality but also have impacts in a much larger area [14]. Moreover, some transcontinental dust and pollutant aerosols have been found from East Asia to the western part of North America [27]. Compared with the trans-Atlantic and trans-Arabian Sea dust transport pathways, a mixture of dust and pollutants is the primary characteristic of trans-Pacific aerosol transport. They also have more profound impacts considering the large population density in this area. Therefore, studying and comparing the characteristics of dust and pollutant transport would be worthwhile. Previous works have seldom focused on multiple-source observation data [28,29].

In previous studies, atmospheric models were the primary approach for long-distance aerosol transport analyses. Models like the Weather Research and Forecasting coupled with the Community Multiscale Air Quality (WRF/CMAQ) model, the Weather Research and Forecasting model with chemistry (WRF-Chem) [30–32], the Spectral Radiation-Transport Model for Aerosol Species (SPRINTARS) [26,33,34], and the Goddard Chemistry Aerosol Radiation and Transport (GOCART) model were widely used in aerosol regional and long-distance transport analyses. However, model computation is time-consuming, especially as the simulation areas are usually large in these studies. Meanwhile, satellite observations provide direct descriptions for studying the vertical structure of aerosols and the characteristics of aerosol transport [25,35,36]. For example, trans-Pacific aerosol transport has been analyzed using CALIOP aerosol profile datasets [27]. The global trends in layered aerosol properties from CALIOP level-2 (L2) datasets were systematically analyzed using the PBLH values as vertical thresholds [35]. However, systematic and comprehensive analyses of the long-distance aerosol transport characteristics from direct 3-dimensional (3D) aerosol observations are rare [37]. In particular, discussions from the perspective of aerosol distributions within/out PBL along long-range transport pathways are subject to ongoing research.

This paper used CALIOP aerosol profiles to calculate layered data within the PBL and in the free troposphere. Globally, the characteristics of long-distance aerosol transport were determined from the layered aerosol data. Furthermore, using multiple datasets, a three-stage conceptual model was proposed and assessed with long-range transport pathways of Taklimakan Desert dust and North China Plain pollutants.

This paper is organized as follows. The experimental data sources and preprocessing methods are given in Section 2. The global long-range aerosol transport characteristics are presented in Section 3. Finally, Section 4 provides the conclusions and discussions.

2. Materials and Methods

2.1. CALIOP L3 Aerosol Observation Dataset

CALIOP, a nadir-view, two-wavelength (532 and 1064 nm), polarization-sensitive (at 532 nm) lidar instrument, is mounted on a sun-synchronous satellite, the Cloud-Aerosol Lidar and Infrared Pathfinder Satellite Observation (CALIPSO). The CALIOP L3 aerosol profile product of version 3.00 is used, which is aggregated from quality-assured daily L2 aerosol profiles at 532 nm. The CALIOP L3 dataset has global coverage (85°S to 85°N, 180°W to 180°E). Its horizontal resolution is 2° (latitude) × 5° (longitude), and the vertical resolution is 60 m, with an upper altitude limit of 12 km [38]. The CALIOP L3 data contain products under four atmosphere conditions, which are all-sky, cloud-free, cloudy-sky transparent, and cloudy-sky opaque. This study used the cloud-free product from June 2006 (initial observation time) to November 2016 for analysis. The CALIOP total-atmosphere AOD (AOD_{TA}) is calculated by integrating the vertical aerosol extinction profile,

$$AOD = \sum (Exc_i \times \Delta h_i), \tag{1}$$

where Exc_i is the extinction coefficient of the ith layer, and Δh_i is the corresponding height difference. The AOD values in the PBL (AOD_{PBL}) and the free troposphere (AOD_{FTL}) are calculated using the same method. Furthermore, the ratios of the AOD_{PBL} values (denoted as R_AOD_{PBL}) can be calculated as in the equation below,

$$R\_AOD_{PBL} = \frac{AOD_{PBL}}{AOD_{TA}}. \tag{2}$$

At the same time, the ratios of the AOD_{FTL} values (denoted as R_AOD_{FTL}) can also be calculated. This study used the daytime CALIOP aerosol profile observations to represent aerosol vertical structures. In addition, aerosol types were used to understand 3-D aerosol distributions and emissions better. The CALIOP L3 product contains aerosol extinction profile data for different aerosol types (dust, smoke, and polluted dust). In particular, dust, smoke, and polluted-dust products were used to calculate global layered AOD values. Dust aerosol products were also used to calculate aerosol profiles along aerosol transport pathways from the source of the Taklimakan Desert, while polluted-dust aerosol products were used to calculate aerosol profiles along aerosol transport pathways from the source of the North China Plain. In this study, AOD_{PBL}, AOD_{FTL}, and AOD_{TA} values and corresponding ratios for dust and polluted dust were also processed using the same method.

2.2. PBLH Reanalysis Data

In this study, we used the Re-Analysis-Interim (ERA-Interim) reanalysis dataset provided by the European Centre for Medium-Range Weather Forecasts (ECMWF) [39]. Specifically, the ERA-Interim PBLH variable divides the bottom well-mixed planetary boundary layer from the top free troposphere layer. This product is available every three hours at 3:00, 6:00, 9:00, 12:00, 15:00, 18:00, 21:00, and 24:00 in Coordinated Universal Time (UTC). As the CALIPSO satellite crosses the equator at around 13:30 (local time) in the daytime, the ERA-Interim data closest to the CALIOP L3 aerosol product are used to ensure temporal consistency. The seasonal mean values of the global PBLH distributions are shown in Figure S1 in the Supplementary Information (SI). In order to match the horizontal resolution of the CALIOP L3 product, the ERA-Interim PBLH data were processed to the same horizontal resolution of 2° × 5° by calculating the arithmetic mean values of 2° × 5° grids.

2.3. Air Mass Forward Trajectories

The Hybrid Single-Particle Lagrangian Integrated Trajectory (HYSPLIT-4) version 4 model provided by the Air Resources Laboratory (ARL) of the National Oceanic and Atmospheric Administration (NOAA) was utilized herein to calculate the air mass forward trajectories [26,40]. Furthermore, this study used $1^{\circ} \times 1^{\circ}$ archived meteorological data from the Global Data Assimilation System (GDAS) of the National Center for Atmospheric Research (NCAR) as the driving meteorological input data. This study tested five configurations of 10, 11, 12, 13, and 14 days for air mass forward trajectory calculation, with 3987 trajectories for each condition from 2006 to 2016. Considering the representativeness of the air mass forward trajectories for guiding aerosol long-distance transport characteristics, 11-day air mass forward trajectories starting at 4:00 a.m. (UTC) were calculated to help better understand the aerosol long-distance transmission pathways [41,42]. With an adequate consideration of the spatial distribution of aerosols, the air mass forward trajectories of the Taklimakan Desert and North China Plain were calculated at (41°N, 88°E) and (38.85°N, 115.49°E) starting at 100 m AGL, 500 m AGL, and 1000 m AGL (Figure 1, Figures S2 and S3). A seasonal analysis of trajectory clusters starting at the heights of different proportions to PBLH was also conducted (Figures S4 and S5). The seasonal variations were not significant in terms of trajectory length and direction, and also the corresponding AOD_{PBL} along the trajectory clusters. Therefore, an annual analysis was performed in this study to investigate the long-term patterns of AOD_{PBL} distributions in different general transport paths.

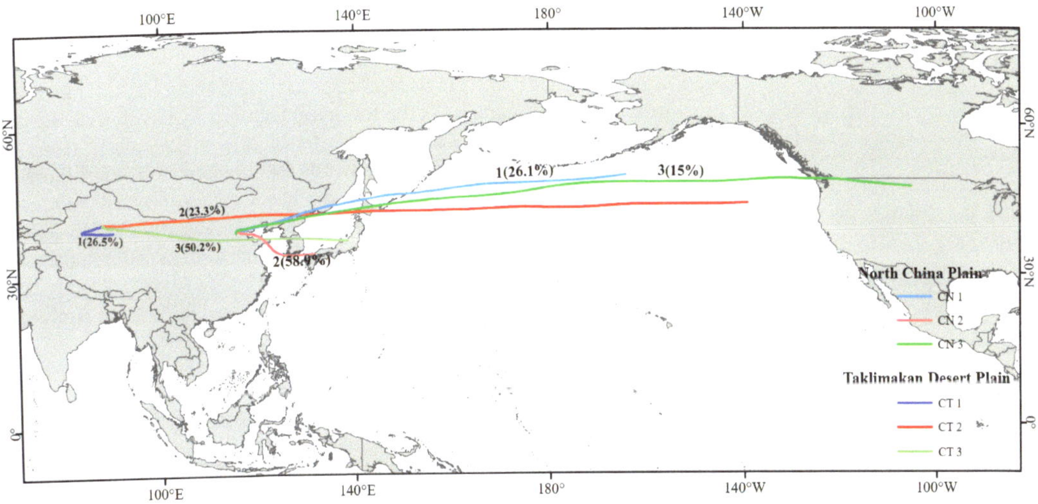

Figure 1. Eleven-day air mass forward trajectory clusters originating from the North China Plain and the Taklimakan Desert (starting at 100 m AGL).

The air mass forward trajectories were clustered together based on the wind direction and speed of the driving meteorological input data. Based on prior knowledge of the dominant transport regimes in the Taklimakan Desert and the North China Plain, the air mass forward trajectories were clustered into 3 clusters using a three-dimensional K-means method and the TrajStat software v1.2.2.6 [43]. After that, each trajectory was regarded as a string of points (X, Y, Z). The K-means method was performed by calculating and minimizing the distances among these trajectories using the following equation:

$$d_{12} = \sqrt{\sum_i^N \left((X_1(i) - X_2(i))^2 + (Y_1(i) - Y_2(i))^2 + (Z_1(i) - Z_2(i))^2\right)}, \tag{3}$$

where d_{12} denotes the Euclidean distance between trajectories (X_1, Y_1, Z_1) and (X_2, Y_2, Z_2). The number N is the number of points along the trajectory, which is 265 in this study. In the Taklimakan Desert, three representative air mass forward trajectories (CT1, CT2, and CT3) were used to help understand aerosol transport. The ratios of the air mass forward trajectories clustered into CT1, CT2, and CT3 were 26.5%, 23.3%, and 50.2% (starting at 100 m AGL), respectively. Similarly, the North China Plain's representative air mass forward trajectories were CN1, CN2, and CN3, and the ratios of the air mass forward trajectories were 26.1%, 58.9%, and 15% (starting at 100 m AGL), respectively. The clustering results starting at 100, 500, and 1000 m AGL tended to be similar for air masses originating from the North China Plain and the Taklimakan Desert, and the transport distances increased with the increased in the starting level. Considering the limited differences in clustering results, we chose the data starting at 100 m AGL as the main result data. This study also counted the seasons to which the air mass trajectories of each cluster belonged, and the statistical results are shown in Tables S1 and S2. Furthermore, aerosol profiles along six air mass forward trajectory clusters from CALIOP L3 aerosol profiles were used to reflect the aerosol distribution patterns, since long-range pathways longer than 100 degrees were explored in this study.

3. Results

3.1. Aerosol Transport Pathway Identification

3.1.1. Generation of Global Distributions of High AOD Values

This study used CALIOP aerosol extinction datasets to help better understand the vertical aerosol structures. Based on Equation (1), the column-integrated AOD values in the different layers, i.e., in the PBL, the FTL, and the total atmosphere, were calculated from the raw CALIOP L3 data using the reanalysis PBLH values as the threshold altitudes. Furthermore, the ratios of the AOD_{PBL} and AOD_{FTL} values (denoted as R_AOD_{PBL} and R_AOD_{FTL}) were also calculated from the CALIOP L3 aerosol extinction datasets (Figure 2). AOD_{PBL}, AOD_{FTL}, AOD_{TA}, and the corresponding ratios were extracted along three North China pathways (CN1, CN2, and CN3) and three Taklimakan pathways (CT1, CT2, and CT3).

In general, high-AOD_{TA} regions occurred in Western Africa (WAF), the Arabian Peninsula, the Taklimakan Desert, Central Africa, India, and East Asia. However, AOD_{PBL} and AOD_{FTL} values exhibited different spatial patterns compared to those of AOD_{TA}. For example, due to intensive biomass burning, AOD_{PBL} values were higher in Central Africa and northern South America [36]. In East Asia, the AOD_{TA} values were high in the Taklimakan Desert and East Asia. The AOD_{FTL} values were also high in the North China Plain and the eastern coast of Asia (Figures S6 and S7). Aerosol transport may have occurred in these three regions, as reflected by these features. There were also apparent seasonal variations in the AOD_{PBL} values. For example, the AOD_{PBL} values were high in June–July–August (JJA) and September–October–November (SON) in Central Africa (Figure S8). In India, the AOD_{PBL} values were high all year round, especially in March–April–May (MAM), SON, and December–January–February (DJF).

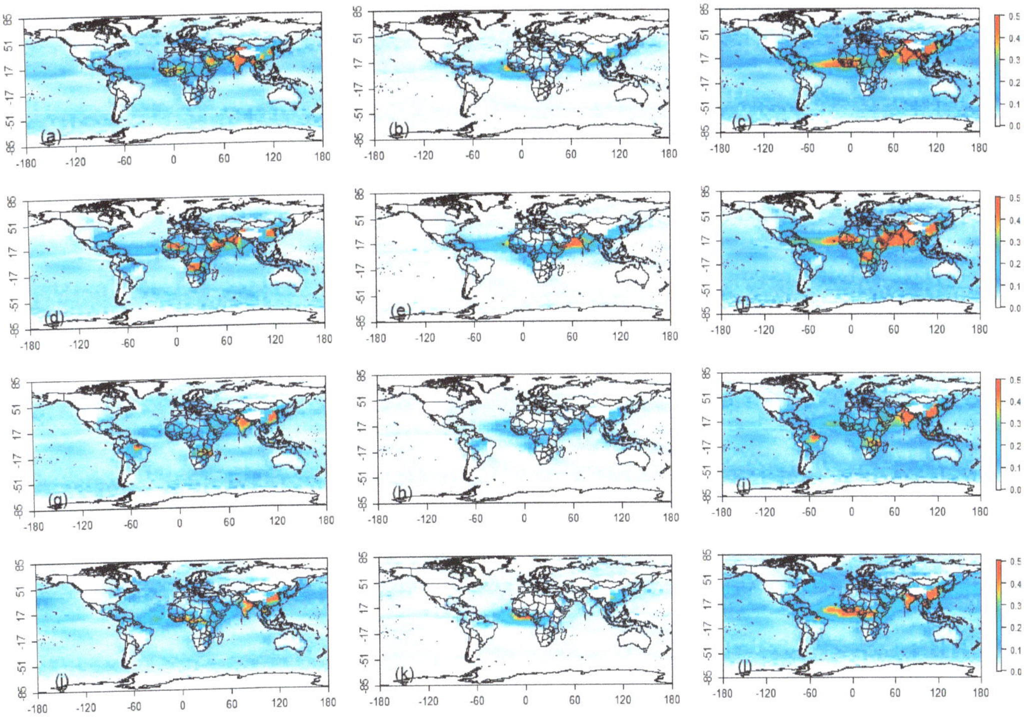

Figure 2. Seasonal layered AOD values for all aerosols. The four rows represent the four seasons: MAM, JJA, SON, and DJF. The three columns are for AOD_{PBL}, AOD_{FTL}, and AOD_{TA}. (**a**,**d**,**g**,**j**) mean the AOD_{PBL} values in MAM, JJA, SON, and DJF, respectively. (**b**,**e**,**h**,**k**) mean the AOD_{FTL} values in MAM, JJA, SON, and DJF, respectively. (**c**,**f**,**i**,**l**) represent the AOD_{TA} values in MAM, JJA, SON, and DJF, respectively.

3.1.2. Identification of Global Aerosol Transport Pathways

Over the land, the R_AOD_{PBL} values were ~70%, with seasonal mean R_AOD_{PBL} values of 74.1% (MAM), 71.6% (JJA), 71.6% (SON), and 65.4% (DJF), respectively (Figure S9 and Table S3). The R_AOD_{PBL} values were significantly high in the plateau regions, such as the Rocky Mountains and the Tibetan areas, where the ratios were close to 100%. Comparatively, over the oceans, the mean R_AOD_{PBL} values were 71%, 71.3%, 73.1%, and 76.1% in MAM, JJA, SON, and DJF, respectively (Table S4). The R_AOD_{PBL} and R_AOD_{FTL} values can help better understand the transport patterns of aerosols.

It should be noted that the R_AOD_{FTL} values were significantly improved in regions such as the west coast of Central Africa, the west coast of Central America, and the Pacific Ocean near Northeast Asia. Considering that aerosol emissions in the ocean are low [4], elevated AOD_{FTL} values in these regions should be mainly caused by emissions in the nearby continental regions [44–46]. In the far oceanic areas, the R_AOD_{FTL} values gradually decreased, indicating that the aerosols settled back into the surface air. Based on the analysis of the R_AOD_{PBL} values, the WAF, the Arabian Peninsula, and East Asia could be the places where aerosol transport happens intensively and frequently. In addition, the high R_AOD_{FTL} values over Northeast Siberia and Canadian Archipelago and nearest parts of Arctic Ocean are likely to originate from Asian dust transport [11].

It is reasonable to assume that areas with greater R_AOD_{FTL} values could be pathways of higher possibilities of long-distance aerosol transport (Figure 3) [17,47]. Accordingly, three regions of the trans-Atlantic, trans-Arabian Sea, and trans-Pacific would be potential transport pathways. The aerosols in these three long-distance aerosol transport regions mainly come from the Sahara Desert, the Arabian Peninsula, and East Asia. Considering the complex emission sources and significant impacts in East Asia, deep analyses of aerosol transport processes were further conducted with comprehensive explanatory datasets.

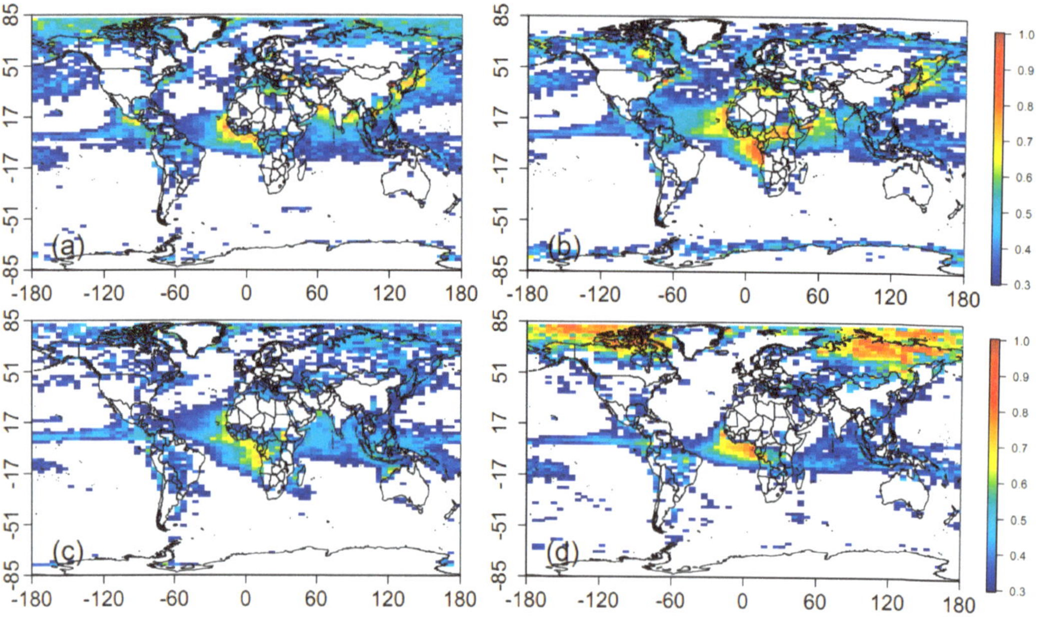

Figure 3. Regions with R_AOD_{FTL} values >30% on the seasonal scale: (**a–d**) R_AOD_{FTL} values in MAM, JJA, SON, and DJF.

3.2. Three-Stage Conceptual Model

3.2.1. Polluted Dust Aerosol Transport Pathway

Polluted-dust aerosol profiles along the air mass trajectory clusters CN1, CN2, and CN3 were used to explore the polluted-dust aerosol transport characteristics from the North China Plain. Long-range aerosol transport sourced from the North China Plain has transcontinental influences on aerosol distributions and air qualities along the pathways [12]. In the source areas of the North China Plain, the seasonal mean AOD_{TA} values were 0.396, 0.658, 0.618, and 0.487, respectively, in MAM, JJA, SON, and DJF, which are generally larger than those in other regions along aerosol transport pathways (Figure S10). Meanwhile, the R_AOD_{PBL} values were also large at 74.4% (MAM), 68.5% (JJA), 79.6% (SON), and 77.2% (DJF) (Figure 4), indicating strong local emission intensities as discussed previously. Along the three representative transport pathways of CN1, CN2, and CN3 (shown in Figure 1), CN2 was the shortest pathway representing regional transport relatively close to source regions. The aerosols from the North China Plain were brought to areas as far as Southern Japan. There were ~30% of CN2 air mass trajectories from JJA. The summary of the ratios of air mass trajectories from SON and DJF was 50.25%, which means over half of the trajectories were from these two seasons (Table S1). As for the transport characteristics, along the pathway CN2, most aerosols were concentrated in the PBLH in the source area, as the R_AOD_{PBL} values were almost 70% in this stage. Then, there wsa an apparent low-level interval with R_AOD_{PBL} values at ~120°E, especially in warm seasons of MAM and JJA (the lowest value was <20%). As for the reason, it means that in a coastal zone, aerosols starting in daytime PBL over land are highly likely to be transported further in the free atmosphere over the nearby sea. Also, the decrease can also be relative to the PBLH values over the Yellow Sea, which were lower than over land in warm seasons and vice versa in cold seasons. This means more air layers are involved in AOD_{PBL} over the Yellow

Sea in cold seasons than in warm ones. However, the R_AOD$_{FTL}$ values in this interval were even higher than the R_AOD$_{PBL}$ values, while aerosol emissions from the coast zone were rare [7]. This strongly indicated transport effects within this area. As the pathway CN2 extended, the R_AOD$_{PBL}$ and the AOD$_{PBL}$ values exhibited increasing trends, and the pollutant aerosols settled back to the ground. At the same time, the changes in the PBLH values influenced by sea–land changes also had an impact on the R_AOD$_{PBL}$ and the AOD$_{PBL}$ values.

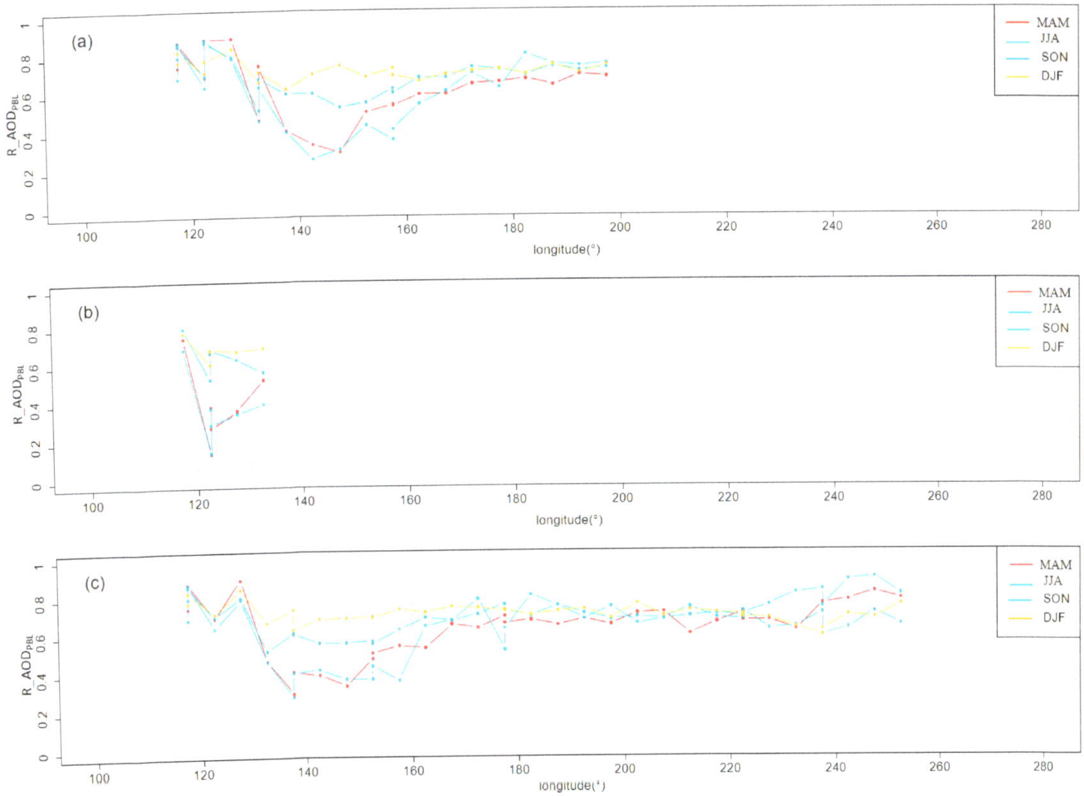

Figure 4. R_AOD$_{PBL}$ values along three aerosol transport pathways (**a**) CN1, (**b**) CN2, and (**c**) CN3 on a seasonal scale (from the North China Plain).

The CN1 and CN3 are both long-range across-Pacific pathways, and there were 53.4% and 52.3% of air mass trajectories of CN1 and CN3 from MAM and JJA, respectively (Table S1). The CN1 trajectories have long lengths of destination to the Eastern Pacific Ocean. The CN3 trajectories are even longer from the North China Plain and arriving at the Northwest United States. In both pathways, AOD$_{PBL}$ ratios were mostly >70% in the first stage (Figure 5). In this stage, the simulated air particle heights were lower than the PBLH values in the source regions, and the air particle heights gradually rose above PBLH along with the trajectory (Figure 6). The AOD$_{PBL}$ and AOD$_{TA}$ values rapidly decreased as the air masses moved northeast (Figure S10). As the trajectory extended to the east, the aerosols gradually rose into the FTL by vertical air movements (Figure 6), leading to decreased R_AOD$_{PBL}$ values from the North China Plain to around 140°E (Figure 4a,c). The lowest R_AOD$_{PBL}$ values along CN1 were located in the northern part of Hokkaido, while the lowest R_AOD$_{PBL}$ values along CN3 were in the southeastern part of South Korea, with ratios <40% in MAM and JJA. Since the emissions over the sea areas were quite low, high R_AOD$_{FTL}$ values indicated there should be transport pathways in these regions.

Specifically, the R_AOD$_{FTL}$ values could reach 60% in MAM and JJA, while the values were around 30% in SON and DJF.

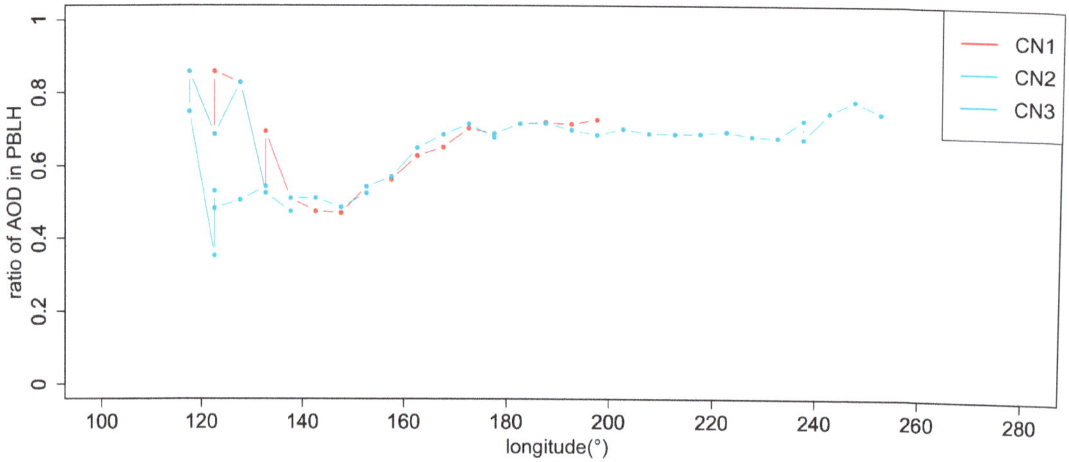

Figure 5. R_AOD$_{PBL}$ values along three aerosol transport pathways on a multiannual scale (from the North China Plain).

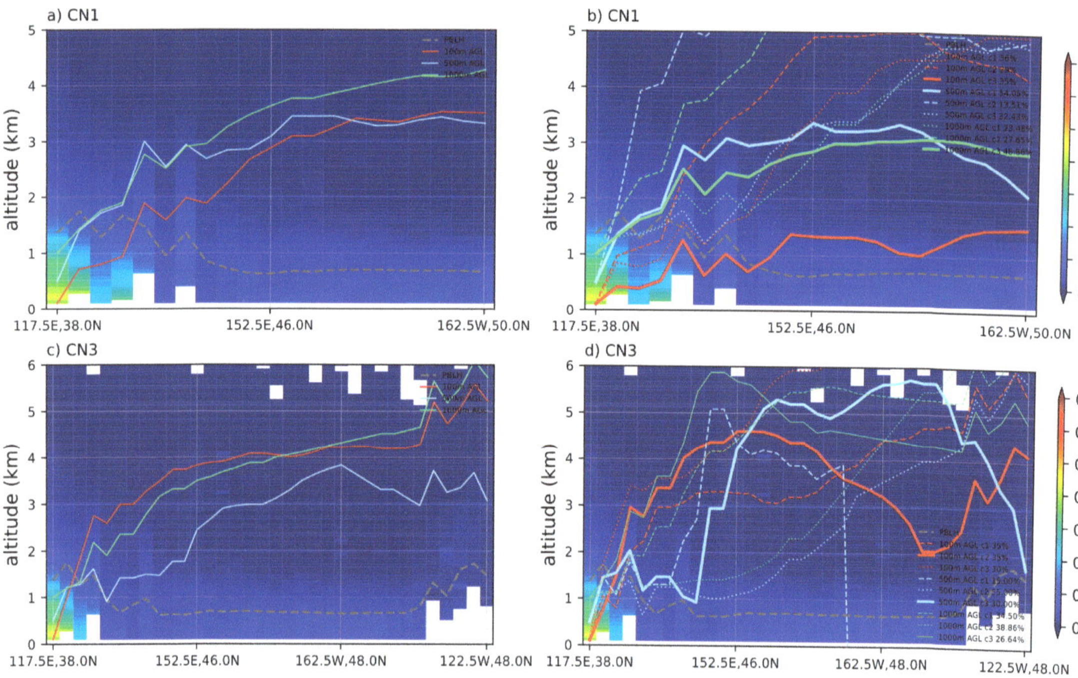

Figure 6. Comparison of air particle heights (100 m AGL, 500 m AGL, and 1000 m AGL) and PBLH values in (**a**) CN1 and (**b**) CN3. (**a**,**c**) The air particle heights of CN1 and CN3. (**b**,**d**) are the clusters of original air mass trajectories clustering to CN1 and CN3.

As the air masses moved further east, the aerosols began settling back into the surface boundary layer. At this stage, approximately 35% of the trajectories for both CN1 and CN3 were responsible for transporting the aerosols to lower layers (Figure 6b,d). The R_AOD$_{PBL}$ values gradually reached up to around 70% in the areas over the Central and Eastern Pacific Oceans in both pathways of CN1 and CN3. The deposition would cause changes in ground-level air quality in remote regions, such as the northwestern part of the United States [13]. In this stage, seasonal differences in R_AOD$_{PBL}$ values were small.

Based on the aforementioned analysis, we propose a conceptual model to describe a long-range transport pathway, which includes three stages based on layered aerosol distributions (Figure S11). In the first stage, the local aerosol emissions are first well mixed within the boundary layer with the improved AOD$_{PBL}$ values over the source areas. Then, in the second stage, the aerosols are further elevated into the free troposphere, characterized by growing R_AOD$_{FTL}$ values. This stage is crucial for a transport pathway as it determines the scale and areas that aerosols would be brought to. In the last stage, most dust and pollutant aerosols settle to the surface layer caused by gravity settling and wet deposition [48], characterized by elevated R_AOD$_{PBL}$ values. As discussed before, the spatial scale of the three stages would be influenced by aerosol particle sizes, air mass movement speed, and terrain features. For example, the stages would have shorter spatial distances for larger-particle aerosols than that for finer-particle aerosols (Figures S6 and S7).

3.2.2. Dust Aerosol Transport Pathway

Similarly, duct aerosol profiles along the air mass trajectory clusters CN1, CN2, and CN3 were used to explore the duct aerosol transport characteristics from the Taklimakan Desert. Dust aerosol transport from the Taklimakan Desert Plain was also analyzed, considering its significant impact on air quality over very large areas. In the Taklimakan Desert Plain, seasonal mean R_AOD$_{PBL}$ values were 87.0% (MAM), 95.7% (JJA), 77.5% (SON), and 55.2% (DJF) (Figure 7), and the R_AOD$_{PBL}$ values were higher compared to those in the North China Plain on the seasonal scale, especially in MAM and JJA. The seasonal mean AOD$_{TA}$ values were highest in MAM (0.573) and lowest in DJF (0.262), which is consistent with frequent dust storms occurring in East Asia [49,50].

Aerosol transport along CN1 mainly influences local air quality. There was 52.3% of air mass trajectories from SON and DJF (Table S2). In detail, the CT1 trajectories are short, indicating its localities for aerosol transport. The R_AOD$_{PBL}$ values were almost >75% in MAM, JJA, and SON (Figure 7a). The aerosols rapidly settled to the surface, as indicated by the decreasing AOD$_{PBL}$ values of CT1 (Figure S12).

The aerosol types in the Taklimakan Desert Plain and the North China Plain were dust- and pollutant-dominated, respectively, which would influence regional aerosol transport patterns. Along the long-distance transport pathways of CT2 and CT3, the air masses passed over East Asia and the Pacific Ocean. In total, 76% of the air mass trajectories of CT2 were from MAM and SON (38.9% from MAM and 37.1% from SON). At the same time, the contributions of air mass movement trajectories to CT3 in each season were relatively uniform (Table S2). The seasonal mean R_AOD$_{PBL}$ values were significantly higher in MAM and JJA than those in SON and DJF, and dust activation in the former two seasons was frequent. In stage 1, aerosol transport pathways CT2 and CT3 extended from Northwestern China to Eastern China, whereas the aerosols in CN1 and CN3 were transported over East Asia. In this stage, the aerosols in CT2 were mostly distributed in the PBLH, with stable AOD$_{PBL}$ ratios of over 75% in MAM, JJA, and SON. While the air mass advanced along CT2, the air particle heights exhibited a consistent linear increase, progressively surpassing the PBLH values, as depicted in Figure 8. This upward trend in air particle heights signifies that the aerosols were being transported into the free troposphere by the advancing air masses. The R_AOD$_{PBL}$ values slowly increased in DJF along CT2 and CT3. In contrast to that, the AOD$_{PBL}$ values along CN1 and CN3 decreased rapidly in stage 1 (Figure S10), and there was a significant second peak in the AOD$_{PBL}$ values for CT2 and CT3 before the R_AOD$_{PBL}$ values began to decrease (Figure S12). The second peak may be associated with a mixture of local aerosol emissions in Eastern China [43]. Lower peak values of CT2 than those of CT3 were because the aerosol emissions in the bypassing areas of CT3 were more severe than those of CT2. The CT2 trajectories passed over Northeastern China, while CT3 trajectories passed over the North China Plain. When the aerosols moved from stage 1 to stage 2, there were significant decreases in the R_AOD$_{PBL}$ values along CT2 and CT3 (Figure 9).

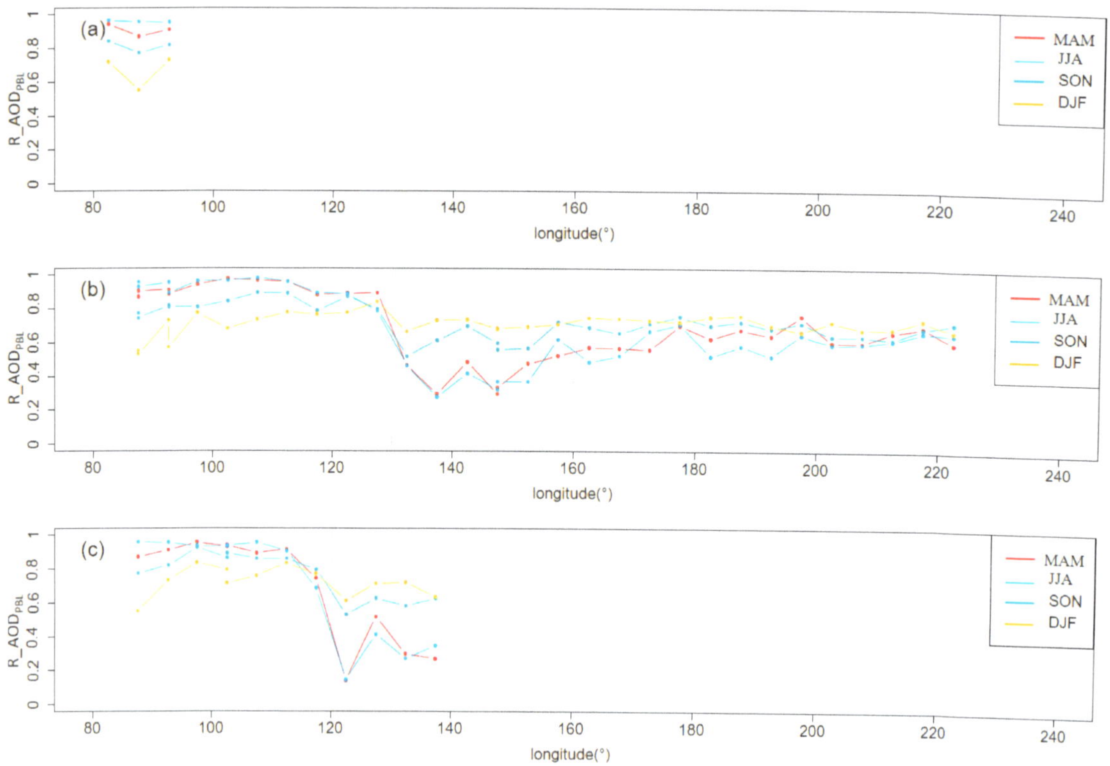

Figure 7. R_AOD$_{PBL}$ values along three aerosol transport pathways (**a**) CT1, (**b**) CT2, and (**c**) CT3 on the seasonal scale (from the Taklimakan Desert).

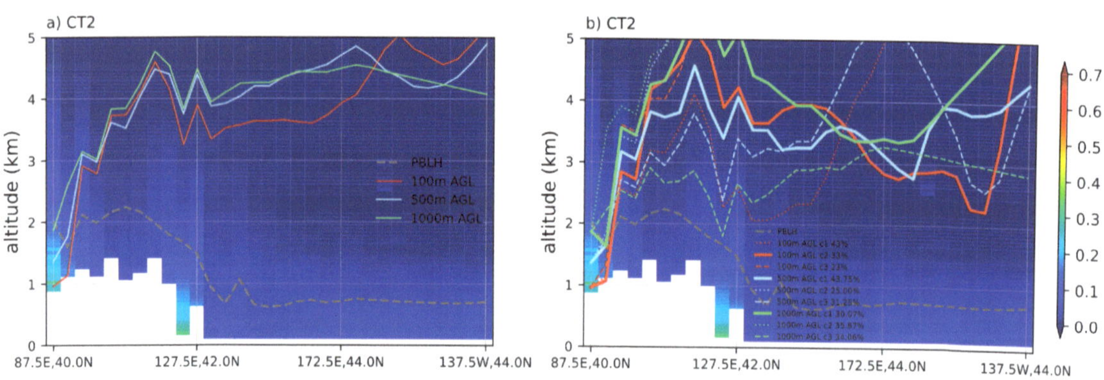

Figure 8. Comparison of air particle heights (100 m AGL, 500 m AGL, and 1000 m AGL) and PBLH values in CT2. (**a**) The air particle heights of CN1 and CN3. (**b**) The clusters of original air mass trajectories clustering to CN1 and CN3.

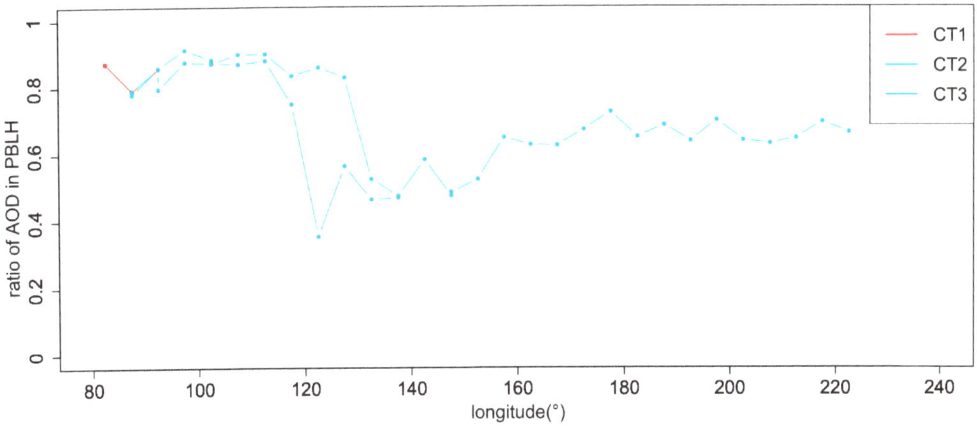

Figure 9. R_AOD_{PBL} values along three aerosol transport pathways on a multiannual scale (from the Taklimakan Desert).

In stage 2, the R_AOD_{PBL} values maintained stable values. Specifically, the R_AOD_{PBL} values were ~60% in SON and DJF, while the R_AOD_{PBL} values decreased to ~50% and even 20% in MAM and JJA. Thus, the R_AOD_{FTL} values of CT3 were significantly higher than the R_AOD_{FTL} values of CN1, CN3, and CT2, and higher ratios of aerosols of CT3 were transported into the free troposphere than those of CN1, CN3, and CT2. Furthermore, there was a small peak in the AOD_{FTL} values for CT2 after the second peak in the AOD_{PBL} values, probably due to rising aerosols into the FTL (Figure S12e,f).

In stage 3, the R_AOD_{PBL} values of CT2 gradually became similar along with the air mass movement on the seasonal scale. At this stage, as the air mass moved downward, the aerosols were also transported to lower atmospheric layers (Figure 8b). The R_AOD_{PBL} values were >60% over the Eastern Pacific Ocean, which may mean the aerosols were settled back into the PBL (Figure 9). Compared with CT2, the transport distance of CT3 was shorter. At the end of CT3, the R_AOD_{PBL} values were still less than 40% in MAM and JJA, and more than 60% of the aerosols were in the FTL. Therefore, the aerosols in the FTL at the end of CT3 continued to be transported farther distances.

4. Conclusions and Discussions

This study used 2006–2016 CALIOP layered aerosol data to explore long-distance aerosol transport pathways. In particular, three apparent long-distance aerosol transport pathways were identified on the global scale: the trans-Pacific transport pathway, the trans-Atlantic transport pathway, and the trans-Arabian Sea transport pathway. Different from aerosol transport from the Sahara Desert and the Arabian Peninsula, complex aerosols influence aerosol trans-Pacific transport. To better understand the long-distance aerosol transport characteristics, we selected two representative aerosol trans-Pacific transport pathways from the North China Plain and the Taklimakan Desert, and meanwhile, used layered AOD values within/out of the PBL and specific aerosol profiles along air mass trajectories to analyze long-distance aerosol transport from East Asia. The results show that the globally covered CALIOP L3 aerosol product and trajectory analyses on a daily timescale adequately identified the characteristics of long-distance aerosol transport from East Asia, eliminating the need for time-consuming model calculations. Additionally, the trajectory model, such as HYSPLIT, simulating air mass movement trajectories provided valuable guidance for aerosol long-distance transport pathways.

A three-stage conceptual aerosol transport model was proposed to explore long-distance aerosol transport characteristics. The aerosols were emitted, raised, and mainly distributed in the PBLH in the first stage. For polluted-dust aerosols, the R_AOD_{PBL} values were almost >70%. Then, the aerosols were further elevated into the free troposphere, and the R_AOD$_{PBL}$ values decreased to <40%. In this stage, there was a significant decrease in the R_AOD$_{PBL}$ values at around 120°–140°E. Finally, in the third stage, the aerosols settled back to the ground due to gravity. For dust aerosols, the R_AOD$_{PBL}$ values were almost >75%, meaning that there was a higher ratio of aerosols concentrated in the PBLH compared with the polluted-dust aerosols in the first stage. In the second stage, the R_AOD$_{PBL}$ values decreased to <50% and even 20%. The decrease was at around 130°–160°E. Last, the R_AOD$_{PBL}$ values were >60% over the Eastern Pacific Ocean. This three-stage conceptual model can be used as a common tool for aerosol long-range transport analysis by describing the dominant features in an aerosol transport process. The method highlighted the factor of R_AOD$_{PBL}$ as an important indicator in the long-range transport feature.

In terms of future work, quantitative studies will require improved spatial resolutions in observational and atmospheric data, such as aerosol optical depth (AOD) observations from the Aerosol Robotic Network (AERONET). Expanding on that, the trajectory model will be cross-validated and enhanced through integration with atmospheric models or deep-learning-based remote analysis. That is to say, a more comprehensive understanding of the physical and statistical aspects is essential for analyzing aerosol transport. At the same time, we can further evaluate the influences of aerosol emissions from the ocean for aerosol long-distance transport. Daily aerosol profiles can be brought in the future to precisely explore aerosol vertical structures along aerosol transport pathways and used as an important data source for quantitative analysis to make up for missing details in general aerosol transport patterns from monthly data. Moreover, the performance of the proposed three-stage aerosol conceptual model deserves further validation regarding the trans-Atlantic and trans-Arabian Sea pathways in future studies.

Supplementary Materials: The following supporting information can be downloaded at: https://www.mdpi.com/article/10.3390/rs15184537/s1. The following supporting information includes 12 figures and 4 tables. Figure S1: PBLH on the global scale; Figure S2: Eleven-day air mass forward trajectory clusters originating from the North China Plain and the Taklimakan Desert (starting at 500 m AGL); Figure S3: Eleven-day air mass forward trajectory clusters originating from the North China Plain and the Taklimakan Desert (starting at 1000 m AGL); Figure S4: Eleven-day air mass forward trajectory clusters originating from the North China Plain (a–d) and the Taklimakan Desert (e–h); Figure S5: AOD$_{PBL}$ values of total aerosols along air mass forward trajectory clusters originating from the North China Plain (a–d), and AOD$_{PBL}$ values of dust aerosols along air mass forward trajectory clusters originating from the Taklimakan Desert (e–h); Figure S6: Seasonal layered AOD values for dust; Figure S7: Seasonal layered AOD values for polluted dust; Figure S8: Seasonal layered AOD values for smoke aerosols; Figure S9: (a–d) R_AOD$_{PBL}$ values and (e–h) R_AOD$_{FTL}$ values in four seasons; Figure S10: (a,d,g) AOD$_{TA}$, (b,e,h) AOD$_{FTL}$, and (c,f,i) AOD$_{PBL}$ along three long-distance aerosol transport pathways (CN1, CN2, and CN3) on a seasonal scale (from the North China Plain); Figure S11. Three-stage conceptual model depicting the changes in R_AOD$_{PBL}$ in long-range aerosol transportation; Figure S12: (a,d,g) AOD$_{TA}$, (b,e,h) AOD$_{FTL}$, and (c,f,i) AOD$_{PBL}$ along three long-distance aerosol transport pathways on a seasonal scale (from the Taklimakan Desert); Table S1: The ratios of air mass forward trajectories in each cluster on a seasonal scale (the North China Plain); Table S2: The ratios of air mass forward trajectories in each cluster on a seasonal scale (the Taklimakan Desert); Table S3: Summary of the R_AOD$_{PBL}$ values over the land; Table S4: Summary of the R_AOD$_{PBL}$ values over the ocean.

Author Contributions: Conceptualization, B.L. and L.W.; methodology, B.L. and L.W.; validation, L.W., W.W. and Y.H.; investigation, L.W. and J.Z.; data curation, J.Z.; writing—original draft preparation, L.W.; review and editing, B.L., W.W. and Y.H.; supervision, Z.G. and Y.B.; funding acquisition, W.W. All authors have read and agreed to the published version of the manuscript.

Funding: This research presented herein was performed in the framework of the China Postdoctoral Science Foundation, grant number 325025, and the Youth Innovation Promotion Association of Chinese Academy of Sciences, grant number 2023137. Parts of this paper were funded by the Science and Technology Key Project of China Huayun Group, grant number HYKJXMZ-202002.

Data Availability Statement: Publicly available datasets were analyzed in this study. CALIOP L3 aerosol data can be found here: [https://www.earthdata.nasa.gov/ (accessed on 11 September 2023)]. PBLH data can be found here: [https://www.ecmwf.int/ (accessed on 11 September 2023)]. Archived meteorological data can be found here: [https://ncar.ucar.edu/ (accessed on 11 September 2023)].

Acknowledgments: The authors are grateful to Kun Fu from the Aerospace Information Research Institute, Chinese Academy of Sciences (AIRCAS), who helped in coordinating raw CALIOP data.

Conflicts of Interest: The authors declare no conflict of interest.

References

1. Chang, D.; Song, Y.; Liu, B. Visibility trends in six megacities in China 1973–2007. *Atmos. Res.* **2009**, *94*, 161–167. [CrossRef]
2. Gao, M.; Saide, P.E.; Xin, J.; Wang, Y.; Liu, Z.; Wang, Y.; Wang, Z.; Pagowski, M.; Guttikunda, S.K.; Carmichael, G.R. Estimates of health impacts and radiative forcing in winter haze in eastern China through constraints of surface $PM_{2.5}$ predictions. *Environ. Sci. Technol.* **2017**, *51*, 2178–2185. [CrossRef] [PubMed]
3. Ding, X.; Kong, L.; Du, C.; Zhanzakova, A.; Wang, L.; Fu, H.; Chen, J.; Yang, X.; Cheng, T. Long-range and regional transported size-resolved atmospheric aerosols during summertime in urban Shanghai. *Sci. Total Environ.* **2017**, *583*, 334–343. [CrossRef]
4. Stier, P.; Feichter, J.; Kinne, S.; Kloster, S.; Vignati, E.; Wilson, J.; Ganzeveld, L.; Tegen, I.; Werner, M.; Balkanski, Y.; et al. The aerosol-climate model ECHAM5-HAM. *Atmos. Chem. Phys.* **2005**, *5*, 1125–1156. [CrossRef]
5. Carslaw, K.S.; Boucher, O.; Spracklen, D.V.; Mann, G.W.; Rae, J.G.L.; Woodward, S.; Kulmala, M. A review of natural aerosol interactions and feedbacks within the Earth system. *Atmos. Chem. Phys.* **2010**, *10*, 1701–1737. [CrossRef]
6. Rap, A.; Scott, C.E.; Spracklen, D.V.; Bellouin, N.; Forster, P.M.; Carslaw, K.S.; Schmidt, A.; Mann, G. Natural aerosol direct and indirect radiative effects. *Geophys. Res. Lett.* **2013**, *40*, 3297–3301. [CrossRef]
7. Dentener, F.; Kinne, S.; Bond, T.; Boucher, O.; Cofala, J.; Generoso, S.; Ginoux, P.; Gong, S.; Hoelzemann, J.J.; Ito, A.; et al. Emissions of primary aerosol and precursor gases in the years 2000 and 1750 prescribed data-sets for AeroCom. *Atmos. Chem. Phys.* **2006**, *6*, 4321–4344. [CrossRef]
8. Fang, D.; Wang, Q.; Li, H.; Yu, Y.; Lu, Y.; Qian, X. Mortality effects assessment of ambient $PM_{2.5}$ pollution in the 74 leading cities of China. *Sci. Total Environ.* **2016**, *569–570*, 1545–1552. [CrossRef]
9. Pui, D.Y.H.; Chen, S.-C.; Zuo, Z. $PM_{2.5}$ in China: Measurements, sources, visibility and health effects, and mitigation. *Particuology* **2014**, *13*, 1–26. [CrossRef]
10. Yu, H.; Chin, M.; Bian, H.; Yuan, T.; Prospero, J.M.; Omar, A.H.; Remer, L.A.; Winker, D.M.; Yang, Y.; Zhang, Y.; et al. Quantification of trans-Atlantic dust transport from seven-year (2007–2013) record of CALIPSO lidar measurements. *Remote Sens. Environ.* **2015**, *159*, 232–249. [CrossRef]
11. Yu, Y.; Kalashnikova, O.V.; Garay, M.J.; Notaro, M. Climatology in Asian dust activation and transport based on MISR satellite observations and trajectory analysis. *Atmos. Chem. Phys.* **2019**, *1*, 363–378. [CrossRef]
12. Hadley, O.L.; Ramanathan, V., Carmichael, G.R.; Tang, Y.; Corrigan, C.E.; Roberts, G.C.; Mauger, G.S. Trans-Pacific transport of black carbon and fine aerosols (D < 2.5 μm) into North America. *J. Geophys. Res.* **2007**, *112*, D05309.
13. Hu, Z.; Zhao, C.; Huang, J.; Ruby, L.L.; Qian, Y.; Yu, H.; Huang, L.; Kalashnikova, O.V. Trans-Pacific transport and evolution of aerosols: Evaluation of quasi-global WRF-Chem simulation with multiple observations. *Geosci Model Dev.* **2016**, *9*, 1725–1746. [CrossRef]
14. Kim, D.; Chin, M.; Yu, H.; Pan, X.; Bian, H.; Tan, Q.; Kahn, R.A.; Tsigaridis, K.; Bauer, S.E.; Takemura, T. Asian and trans-Pacific Dust: A Multimodel and Multiremote Sensing Observation Analysis. *J. Geophys. Res. Atmos.* **2019**, *124*, 13534–13559. [CrossRef]
15. Ali, P.; Joseph, P.; Arash, S. Geochemical Fingerprinting of Trans-Atlantic African Dust Based on Radiogenic Sr-Nd-Hf Isotopes and Rare Earth Element Anomalies. *Geophys. Res. Abstr.* **2015**, *17*, EGU2015-1328.
16. Yu, H.; Remer, L.A.; Chin, M.; Bian, H.; Kleidman, R.G.; Diehl, T. A satellite-based assessment of transpacific transport of pollution aerosol. *J. Geophys. Res.* **2008**, *113*, D14S12. [CrossRef]
17. Yu, H.; Tan, Q.; Chin, M.; Bian, H.; Kim, D.; Winker, D.M.; Levy, R.C.; Zhang, Z.; Remer, L.A.; Kahn, R.A.; et al. Estimates of African Dust Deposition Along the Trans-Atlantic Transit Using the Decadelong Record ofAerosol Measurements from CALIOP, MODIS, MISR, and IASI. *J. Geophys. Res. Atmos.* **2019**, *124*, 7975–7996. [CrossRef]
18. Morris, V.; Clemente-Colón, P.; Nalli, N.R.; Joseph, E.; Armstrong, R.A.; Detrés, Y.; Goldberg, M.D.; Minnett, P.J.; Lumpkin, R. Measuring Trans-Atlantic aerosol transport from Africa. *Eos Trans. Am. Geophys. Union* **2011**, *87*, 565–571. [CrossRef]
19. Chen, S. P.; Lu, C.-H.; McQueen, J.; Lee, P. Application of satellite observations in con-junction with aerosol reanalysis to characterize long-range transport of African and Asian dust on air quality in the contiguous U.S. *Atmos. Environ.* **2018**, *187*, 174–195. [CrossRef]
20. Shalaby, A.; Rappenglueck, B.; Eltahir, E.A.B. The climatology of dust aerosol over the arabian peninsula. *Atmos. Chem. Phys.* **2015**, *15*, 1523–1571.
21. Rahn, K.A.; Borys, R.D.; Shaw, G.E. The Asian source of Arctic haze bands. *Nature* **1977**, *268*, 713–715. [CrossRef]
22. Singh, P.; Dey, S. Crop burning and forest fires: Long-term effect on adolescent height in India. *Atmos. Res.* **2021**, *65*, 101244. [CrossRef]

23. Sohail, M.T.; Ullah, S.; Majeed, M.T.; Usman, A.; Andlib, Z. The shadow economy in South Asia: Dynamic effects on clean energy consumption and environmental pollution. *Environ. Sci. Pollut. Res.* **2021**, *28*, 29265–29275. [CrossRef] [PubMed]
24. Lyu, B.; Hu, Y.; Zhang, W.; Du, Y.; Luo, B.; Sun, X.; Sun, Z.; Deng, Z.; Wang, X.; Liu, J.; et al. Fusion Method Combining Ground-Level Observations with Chemical Transport Model Predictions Using an Ensemble Deep Learning Framework: Application in China to Estimate Spatiotemporally-Resolved $PM_{2.5}$ Exposure Fields in 2014–2017. *Environ. Sci. Technol.* **2019**, *53*, 7306–7315. [CrossRef]
25. Mehta, M.; Singh, N. Global trends of columnar and vertically distributed properties of aerosols with emphasis on dust, polluted dust and smoke—Inferences from 10-year long CALIOP observations. *Remote Sens. Environ.* **2018**, *208*, 120–132. [CrossRef]
26. Uno, I.; Eguchi, K.; Yumimoto, K.; Takemura, T.; Shimizu, A.; Uematsu, M.; Liu, Z.; Wang, Z.; Hara, Y.; Sugimoto, N. Asian dust transported one full circuit around the globe. *Nat. Geosci.* **2009**, *2*, 557–560. [CrossRef]
27. Yu, H.; Remer, L.A.; Chin, M.; Bian, H.; Tan, Q.; Yuan, T.; Zhang, Y. Aerosols from Overseas Rival Domestic Emissions over North America. *Science* **2012**, *337*, 566–569. [CrossRef]
28. Zhong, H.; Huang, R.-J.; Lin, C.; Xu, W.; Duan, J.; Gu, Y.; Huang, W.; Ni, H.; Zhu, C.; You, Y.; et al. Measurement report: On the contribution of long-distance transport to the secondary aerosol formation and aging. *Atmos. Chem. Phys.* **2022**, *22*, 9513–9524. [CrossRef]
29. Zhao, T.L.; Gong, S.L.; Zhang, X.Y.; Blanchet, J.-P.; McKendry, I.G.; Zhou, Z.J. A Simulated Climatology of Asian Dust Aerosol and Its Trans-Pacific Transport. Part I: Mean Climate and Validation. *J. Clim.* **2006**, *19*, 88–103. [CrossRef]
30. Kanaya, Y.; Matsui, H.; Taketani, F.; Pan, X.; Komazaki, Y.; Wang, Z.; Chang, L.; Kang, D.; Choi, M.; Kim, S.-Y.; et al. Observed and Modeled Mass Concentrations of Organic Aerosols and $PM_{2.5}$ at Three Remote Sites around the East China Sea: Roles of Chemical Aging. *Aerosol Air Qual. Res.* **2017**, *17*, 3091–3105. [CrossRef]
31. Kanaya, Y.; Miyazaki, K.; Taketani, F.; Miyakawa, T.; Takashima, H.; Komazaki, Y.; Pan, X.; Kato, S.; Sudo, K.; Sekiya, T.; et al. Ozone and carbon monoxide observations over open oceans on R/V Mirai from 67°S to 75°N during 2012 to 2017: Testing global chemical reanalysis in terms of Arctic processes, low ozone levels at low latitudes, and pollution transport. *Atmos. Chem. Phys.* **2019**, *19*, 7233–7254. [CrossRef]
32. Kanaya, Y.; Yamaji, K.; Miyakawa, T.; Taketani, F.; Zhu, C.; Choi, Y.; Komazaki, Y.; Ikeda, K.; Kondo, Y.; Klimont, Z. Rapid reduction in black carbon emissions from China: Evidence from 2009–2019 observations on Fukue Island, Japan. *Atmos. Chem. Phys.* **2020**, *20*, 6339–6356. [CrossRef]
33. Eguchi, K.; Uno, I.; Yumimoto, K.; Takemura, T.; Shimizu, A.; Sugimoto, N.; Liu, Z. Trans-pacific dust transport: Integrated analysis of NASA/CALIPSO and a global aerosol transport model. *Atmos. Chem. Phys.* **2009**, *9*, 3137–3145. [CrossRef]
34. Meng, L.; Yang, X.; Zhao, T.; He, Q.; Mamtimin, A.; Wang, M.; Huo, W.; Yang, F.; Zhou, C.; Pan, H. Simulated regional transport structures and budgets of dust aerosols during a typical springtime dust storm in the Tarim Basin, Northwest China. *Atmos. Res.* **2020**, *238*, 104892. [CrossRef]
35. Bourgeois, Q.; Ekman, A.M.L.; Renard, J.-B.; Krejci, R.; Devasthale, A.; Bender, F.A.-M.; Riipinen, I.; Berthet, G.; Tackett, J.L. How much of the global aerosol optical depth is found in the boundary layer and free troposphere? *Atmos. Chem. Phys.* **2018**, *18*, 7709–7720. [CrossRef]
36. Wang, L.; Lyu, B.; Bai, Y. Global aerosol vertical structure analysis by clustering gridded CALIOP aerosol profiles with fuzzy k-means. *Sci. Total Environ.* **2021**, *761*, 144076. [CrossRef] [PubMed]
37. Shukurov, K.A.; Simonenkov, D.V.; Nevzorov, A.V.; Rashki, A.; Hamzeh, N.H.; Abdullaev, S.F.; Shukurova, L.M.; Chkhetiani, O.G. CALIOP-Based Evaluation of Dust Emissions and Long-Range Transport of the Dust from the Aral−Caspian Arid Region by 3D-Source Potential Impact (3D-SPI) Method. *Remote Sens.* **2023**, *15*, 2819. [CrossRef]
38. Tackett, J.L.; Winker, D.M.; Getzewich, B.J.; Vaughan, M.A.; Young, S.A.; Kar, J. CALIPSO lidar level 3 aerosol profile product: Version 3 algorithm design. *Atmos. Meas. Tech.* **2018**, *11*, 4129–4152. [CrossRef]
39. Dee, D.P.; Uppala, S.M.; Simmons, A.J.; Berrisford, P.; Poli, P.; Kobayashi, S.; Andrae, U.; Balmaseda, M.A.; Balsamo, G.; Bauer, P.; et al. The ERA-Interim reanalysis: Configuration and performance of the data assimilation system. *Q. J. R. Meteorol. Soc.* **2011**, *137*, 553–597. [CrossRef]
40. Draxler, R.R.; Hess, G.D. An overview of the HYSPLIT 4 modeling system for trajectories, dispersion, and deposition. *Aust. Meteorol. Mag.* **1998**, *47*, 295–308.
41. Lee, Y.H.; Chen, K.; Adams, P.J. Development of a global model of mineral dust aerosol microphysics. *Atmos. Chem. Phys.* **2009**, *9*, 2441–2458. [CrossRef]
42. Miller, R.L.; Tegen, I.; Perlwitz, J. Surface radiative forcing by soil dust aerosols and the hydrologic cycle. *J. Geophys. Res.* **2004**, *109*, D04203. [CrossRef]
43. Wang, L.; Lyu, B.; Bai, Y. Aerosol vertical profile variations with seasons, air mass movements and local $PM_{2.5}$ levels in three large China cities. *Atmos. Environ.* **2020**, *224*, 117329. [CrossRef]
44. Han, Y.; Fang, X.; Zhao, T.; Kang, S. Long range trans-Pacific transport and deposition of Asian dust aerosols. *J. Environ. Sci.* **2008**, *20*, 424–428. [CrossRef]
45. Pease, P.P.; Tchakerian, V.P.; Tindale, N.W. Aerosols over the Arabian Sea: Geochemistry and source areas for aeolian desert dust. *J. Arid. Environ.* **1998**, *39*, 477–496. [CrossRef]

46. Velasco-Merino, C.; Mateos, D.; Toledano, C.; Prospero, J.M.; Molinie, J.; Euphrasie-Clotilde, L.; González, R.; Cachorro, V.E.; Calle, A.; Frutos, A.M.d. Impact of long-range transport over the Atlantic Ocean on Saharan dust optical and microphysical properties based on AERONET data. *Atmos. Chem. Phys.* **2018**, *18*, 9411–9424. [CrossRef]
47. Guo, J.; Lou, M.; Miao, Y.; Wang, Y.; Zeng, Z.; Liu, H.; He, J.; Xu, H.; Wang, F.; Min, M.; et al. Trans-Pacific transport of dust aerosols from East Asia: Insights gained from multiple observations and modeling. *Environ. Pollut.* **2017**, *230*, 1030–1039. [CrossRef]
48. Guelle, W.; Balkanski, Y.J.; Schulz, M.; Dulac, F.; Monfray, P. Wet deposition in a global size-dependent aerosol transport model: 1. Comparison of a 1-year 210Pb simulation with ground measurements. *J. Geophys. Res. Atmos.* **1998**, *103*, 11429–11445. [CrossRef]
49. Wang, Y.Q.; Zhang, X.Y.; Gong, S.L.; Zhou, C.H.; Hu, X.Q.; Liu, H.L.; Niu, T.; Yang, Y.Q. Surface observation of sand and dust storm in East Asia and its application in CUACE/Dust. *Atmos. Chem. Phys.* **2008**, *8*, 545–553. [CrossRef]
50. Zhang, X.-X.; Sharratt, B.; Liu, L.-Y.; Wang, Z.-F.; Pan, X.-L.; Lei, J.-Q.; Wu, S.-X.; Huang, S.-Y.; Guo, Y.-H.; Li, J.; et al. East Asian dust storm in May 2017: Observations, modeling, and its influence on the Asia-Pacific region. *Atmos. Chem. Phys.* **2018**, *18*, 8353–8371. [CrossRef]

remote sensing

MDPI

Article

Vertical Profiles of Aerosols Induced by Dust, Smoke, and Fireworks in the Cold Region of Northeast China

Lingjian Duanmu [1,2], Weiwei Chen [2,3], Li Guo [1,*], Yuan Yuan [4], Hongwu Yang [5], Jing Fu [2], Guoqing Song [1,2] and Zixuan Xia [1,2]

[1] College of Biological and Agricultural Engineering, Jilin University, Changchun 130022, China; dmlj20@mails.jlu.edu.cn (L.D.); songgq21@mails.jlu.edu.cn (G.S.); zxxia21@mails.jlu.edu.cn (Z.X.)
[2] State Key Laboratory of Black Soils Conservation and Utilization, Northeast Institute of Geography and Agroecology, Chinese Academy of Sciences, Changchun 130102, China; chenweiwei@iga.ac.cn (W.C.); fujing@iga.ac.cn (J.F.)
[3] College of New Energy and Environment, Jilin University, Changchun 130021, China
[4] School of Energy and Science Engineering, Harbin Institute of Technology, Harbin 150001, China; yuanyuan83@hit.edu.cn
[5] Hongke Photonics Company, Liaoyuan 136200, China; hwyang168@126.com
* Correspondence: liguo2012@jlu.edu.cn; Tel.: +86-431-13404306633

Abstract: Despite the long-term implementation of air pollution control policies in northeast China, severe haze pollution continues to occur frequently. With the adoption of a megacity (Changchun) in northeast China, we analysed the vertical characteristics of aerosols and the causes of aerosol pollution throughout the year using multisource data for providing recommendations for controlling pollution events (i.e., straw burning and fireworks). Based on a ground-based LiDAR, it was found that the extinction coefficient (EC) of aerosols at a height of 300 m in Changchun was highest in winter (0.44 km^{-1}), followed by summer (0.28 km^{-1}), with significant differences from those in warmer regions, such as the Yangtze River Delta. Therefore, it is recommended that air pollution control policies be differentiated between winter and summer. On Chinese New Year's Eve in Changchun, the ignition of firecrackers during the day and night caused increases in the EC at a height of 500 m to 0.37 and 0.88 km^{-1}, respectively. It is suggested that the regulation of firecracker ignition should be reduced during the day and strengthened at night. Based on the CALIPSO and backward trajectory analysis results, two events of dust–biomass-burning composite pollution were observed in March and April. In March, the primary aerosol component was dust from western Changchun, whereas in April, the main aerosol component was biomass-burning aerosols originating from northern and eastern Changchun. Hence, reducing the intensity of spring biomass burning can mitigate the occurrence of dust–biomass-burning composite pollution. These findings can provide emission policy suggestions for areas facing similar issues regarding biomass-burning transmission pollution and firework emissions.

Keywords: LiDAR; $PM_{2.5}$; extinction coefficient; planetary boundary layer; biomass burning

Citation: Duanmu, L.; Chen, W.; Guo, L.; Yuan, Y.; Yang, H.; Fu, J.; Song, G.; Xia, Z. Vertical Profiles of Aerosols Induced by Dust, Smoke, and Fireworks in the Cold Region of Northeast China. *Remote Sens.* **2024**, *16*, 1098. https://doi.org/10.3390/rs16061098

Academic Editor: Manuel Antón

Received: 17 January 2024
Revised: 12 March 2024
Accepted: 15 March 2024
Published: 20 March 2024

1. Introduction

Aerosols are significant constituents of Earth's atmosphere; they comprise suspended solid and liquid particles with varying chemical properties and sizes [1]. Much attention has been devoted to the consequential impacts of aerosols on the air quality (e.g., atmospheric haze, acid precipitation, and particulate matter), climate change (e.g., global warming and extreme weather events), the ecological environment (e.g., carbon emissions and biodiversity), and human health (e.g., physical ailments and mental disorders) [2,3].

Acquiring the long-term baseline characteristics of aerosols and tracking their short-term variations are essential for gauging the extent of aerosol pollution and assessing the downwind effects. Furthermore, understanding the dynamic evolution of aerosol properties lays the foundation for formulating strategies to manage anthropogenic emissions, thereby contributing to improvement in air pollution, which poses risks to human health [4]. Based on remote sensing satellites (e.g., the moderate-resolution imaging spectroradiometer, MODIS; medium-resolution imaging spectrometer, MERIS; and cloud–aerosol LiDAR with orthogonal polarization, CALIOP), multiple datasets, such as the aerosol optical depth (AOD), radiance, and water cloud properties, can be retrieved [5–8]. Notably, cloud–aerosol LiDAR pathfinder satellite observations (CALIPSO) have provided openly accessible observation data on a global scale since June 2006 [9]. Subsequently, the analysis of aerosol vertical profiles in Europe, North America, and East Asia using satellite remote sensing data has achieved significant progress [10,11]. However, its accuracy gradually decreases with decreasing altitude (distance from the surface) of the measurements.

During the last two decades, northeast China has faced fluctuating levels of air pollution from natural emissions, intense anthropogenic activities, and cross-regional transport of aerosols from nearby regions [12–15]. Studies have focused on proposing control measures to reduce the intensity of anthropogenic activities, including energy combustion, illegal emissions, and biomass burning, as well as leveraging favourable meteorological conditions to mitigate the formation of intense haze over megacities in northeast China [16–21]. However, further development is needed to establish efficient air pollution control measures. Understanding the vertical distribution patterns of air pollution is also essential for revealing the mechanisms underlying extreme air pollution events and air-quality regulation [22]. Research was conducted in northeast China using a ground-based LiDAR to investigate the vertical characteristics and transport mechanisms of dust- and biomass-burning aerosol pollution [23,24]. Satellites such as MODIS/Terra, MODIS/Aqua, and CALIOP were employed to analyse the long-term (over ten years) vertical structure of aerosols in this region [25]. Furthermore, studies have focused on exploring the vertical distribution of particulate matter and its correlation with the atmospheric boundary layer structure in northeast China [26,27]. However, most research on the vertical characteristics of aerosols in northeast China has focused on either the temporal variation in the vertical structure or the generation or transmission mechanisms of severe pollution events, with limited exploration of potential policy applications.

In this study, using the ground-based LiDAR, the seasonal aerosol vertical structures and characteristics of severe haze pollution events were studied in Changchun, a megacity in northeast China, from October 2020 to September 2021, and recommendations for air-quality control were determined based on the obtained vertical aerosol characteristics. The results and policy suggestions could serve as a reference for other similar hotspots facing severe pollution.

2. Materials and Methods

2.1. Study Area

In this study, atmospheric monitoring was conducted in the cold region of northeast China (CRC), specifically in the provincial city of Changchun (43°5′–45°15′N, 124°18′–127°5′E) (Figure 1). The region experiences winter from November to March of the following year, characterized by long-term coal-fired heating. June, July, and August are classified as summer, while September and October are considered autumn, and April and May are categorized as spring. Additionally, Changchun is located in a large, cultivated area spanning over 1.32 million hectares and is renowned for its automobile industry, optoelectronic information, and applied chemistry [28]. The city of Changchun falls within the UTC+8 time zone. Hence, all the times referred to in this paper are in UTC+8 (Beijing time), unless otherwise specified.

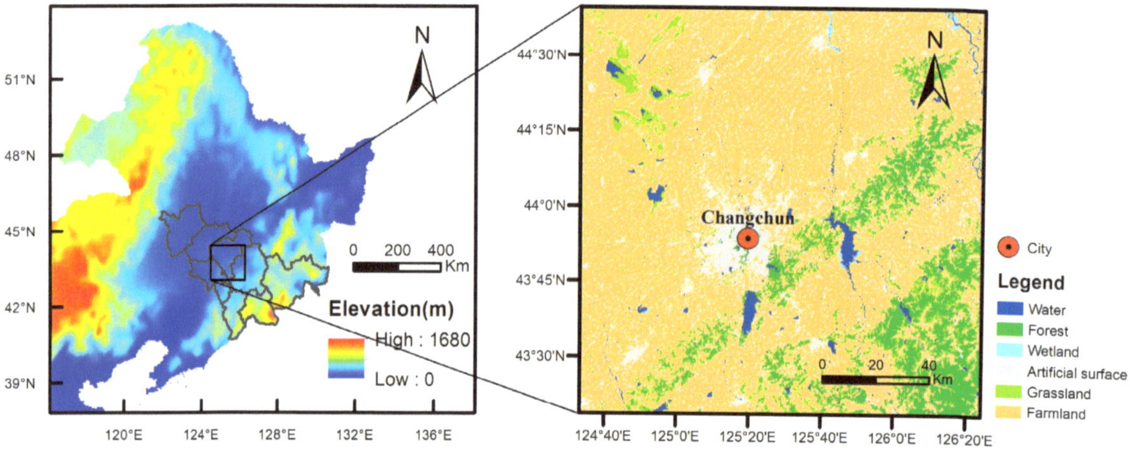

Figure 1. The elevation and land-use types surrounding Changchun City in the cold region of China. The grey outline shows the administrative boundary of Jilin Province; the red circle represents the LiDAR monitoring site in Changchun.

2.2. Data Source

The daily concentration variations of six air pollutants ($PM_{2.5}$, PM_{10}, NO_2, SO_2, CO, and O_3) in Changchun were analysed using hourly data from October 2020 to September 2021. The data were obtained from the China National Environmental Monitoring Center (http://106.37.208.233:20035/, accessed on 1 January 2021). The meteorological data included precipitation (Prec), relative humidity (RH), air temperature (AT), and wind speed (WS), which were provided by the China Meteorological Administration. For the verification of planetary boundary layer height (PBLH), reanalysis data with a horizontal resolution of 0.25° were downloaded from the European Centre for Medium-Range Weather Forecasts (ECMWF). Specifically, the dataset named "ERA5 monthly averaged data on a single level from 1979 to present", containing monthly average PBLHs (https://cds.climate.copernicus.eu/, accessed on 10 January 2021) from 2020 to 2021, was selected.

The ground-based polarization LiDAR (PL) (HKLiDAR-V, Jilin Hongke Photonics Corporation, Liaoyuan, China) was installed at the Northeast Institute of Geography and Agroecology, Chinese Academy of Sciences in Changchun (44°00′N, 125°24′E), as depicted in Figure 1. The PL was equipped with a 532 nm wavelength laser emitter, which has high atmospheric transmittance, with a maximum pumping-pulse energy of 12 MJ. Furthermore, the PL has a range resolution of 7.5 m, an integration time of 20 s, and a blind detection zone of 150 to 200 m. The LiDAR is installed in an analysis room in Changchun with a constant temperature of 20–25 °C and humidity below 60%. Quarterly calibration of the equipment was conducted using quadrant testing, geometric overlap factor correction, polarization channel gain ratio calibration, and signal-to-noise ratio calculations. The uncertainty range estimated for the LiDAR optical parameters remains stable at 25–35%. The aerosol extinction coefficient (EC) profiles were retrieved from October 2020 to September 2021 using the retrieval algorithm developed by Fernald [29].

In this study, to acquire the high-resolution images of the vertical and horizontal profiles of six major aerosol subtypes—namely clean marine, clean continental, polluted continental, dust, polluted dust, and smoke—the V3.41 products of CALIPSO Level 2 data were used [30], which are available at the NASA Langley Research Center (https://www-calipso.larc.nasa.gov/, accessed on 10 February 2021).

2.3. Potential Sources of Pollution

In this study, backwards trajectory analysis was utilized, and the HYSPIT4 model from the National Oceanic and Atmospheric Administration (NOAA) and the meteorological data of the Global Data Assimilation System (GDAS) provided by the National Centers for Environmental Prediction (NCEP) (https://www.ready.noaa.gov/archives.php, accessed on 18 January 2021) were used to calculate trajectories.

Using the GIS-based software (TrajStat 1.5.3, http://meteothink.org/docs/trajstat/trajstatrun.html, accessed on 10 March 2021), the 72-h backward trajectories at three dimensions were calculated over the Changchun observation site, starting at an arrival level of 1000 m above the ground level every 6 h (00:00, 06:00, 12:00, and 18:00 UTC). Then, the modified Ward's hierarchical clustering method was used to form the trajectory clusters [31]. In this study, the potential source contribution function (PSCF) model and the concentration weighted trajectory (CWT) model were used to analyse the source locations of pollutants (PM2.5 and PM10). The values of PSCF and CWT are defined as follows [32]:

$$PSCF_{ij} = m_{ij}/n_{ij}$$

$$C_{ij} = \frac{1}{\sum_{l=1}^{M} \tau_{ijl}} \sum_{l=1}^{M} C_l \tau_{ijl}$$

where $PSCF_{ij}$ is the PSCF value in the ijth cell; n_{ij} is the number of endpoints that fall in the ijth cell; m_{ij} is the number of endpoints for the same cell having arrival times at the sampling site corresponding to PM concentrations higher than an arbitrarily set criterion; C_{ij} is the average weighted concentration ($\mu g/m^3$) in the ijth cell; l is the index of the trajectory; M is the total number of trajectories; C_l is the concentration observed on arrival of trajectory l; and τ_{ijl} is the time spent in the ijth cell by trajectory l. A high value of C_{ij} implies that air parcels travelling over the ijth cell would be, on average, associated with high concentrations at the receptor.

Moreover, the arbitrary weighting function W_{ij} was applied to minimize the uncertainty in the PSCF and CWT values, which are referred to as the WPSCF and WCWT, respectively. The W_{ij} values are defined as follows [33]:

$$W_{ij} = \begin{cases} 1.00 & n_{ij} > 3 \cdot Avg \\ 0.70 & Avg < n_{ij} \leq 3 \cdot Avg \\ 0.42 & 0.5 \cdot Avg < n_{ij} \leq Avg \\ 0.17 & 0 < n_{ij} \leq 0.5 \cdot Avg \end{cases}$$

where Avg is the average number of trajectory segment endpoints in all cells.

2.4. Statistical Analysis

A characteristic radar chart is a relatively intuitive method for analysing the spatial and temporal types (e.g., sand dust, biomass burning, and coal combustion) of air pollutants [21]. Characteristic values, standard values, upper-limit values, and lower-limit values were used to plot the characteristic radar charts. The pollution types can be determined by the magnitude of the characteristic values. Furthermore, the extent of pollution can be determined based on standard values, upper-limit values, and lower-limit values. The standard value was set as a natural number 1, and the characteristic values were calculated from the pollutant concentration data as follows:

$$Z_{kl} = C_{kl} / \sum_{l=1}^{n} C_{kl} \quad (1)$$

$$CV_{kl} = Z_{kl} / \overline{Z_l} \quad (2)$$

where C_{kl} is the concentration of type l pollutant at time k; Z_{kl}, which can eliminate the influence of pollutant concentration fluctuations, is the component ratio of type l pollutant at time k; CV_{kl} is the characteristic value of type l pollutant at time k; and $\overline{Z_l}$ is the average concentration of type l pollutant in a study period.

Upper-limit values and lower-limit values were calculated based on the following equations:

$$Max_l = \left(\overline{Z_l} + S_l\right)/\overline{Z_l} \tag{3}$$

$$Min_l = \left(\overline{Z_l} - S_l\right)/\overline{Z_l} \tag{4}$$

where Max_l, Min_l, and S_l are the upper limit, lower limit, and standard deviation of the type l pollutant in the study period, respectively.

3. Results

3.1. Variation in Aerosol Vertical Distribution

Throughout the study period, the ECs within the near surface layer (below 1 km) consistently demonstrated the highest values in the observed range and decreased with increasing altitude (Figure 2). The average ECs within the near surface layer were greater in winter than that in other seasons, reaching their peak in January at approximately 0.45 km^{-1}. From May to June, the near-surface layer ECs averaged approximately 0.2 km^{-1}, followed by a subsequent increase from July to September, reaching approximately 0.3 km^{-1}. Additionally, during the heating period, the EC ranged from 0.3 to 0.4 km^{-1} under 1 km height, while during the non-heating period, it varied from 0.2 to 0.3 km^{-1}. In July, the extinction coefficient (EC) showed a 7.2% increase per 100 m below 1 km, while in January, it exhibited a significantly higher increase of 31.9% per 100 m below 1 km. Above 1.5 km, the average ECs remained below 0.15 km^{-1} during the study period. At an altitude of 2.5 km, a peak structure was observed in August, with the EC reaching 0.1 km^{-1}. During autumn, the mean EC within the 0–3 km range was the lowest in that year, measuring less than 0.13 km^{-1}.

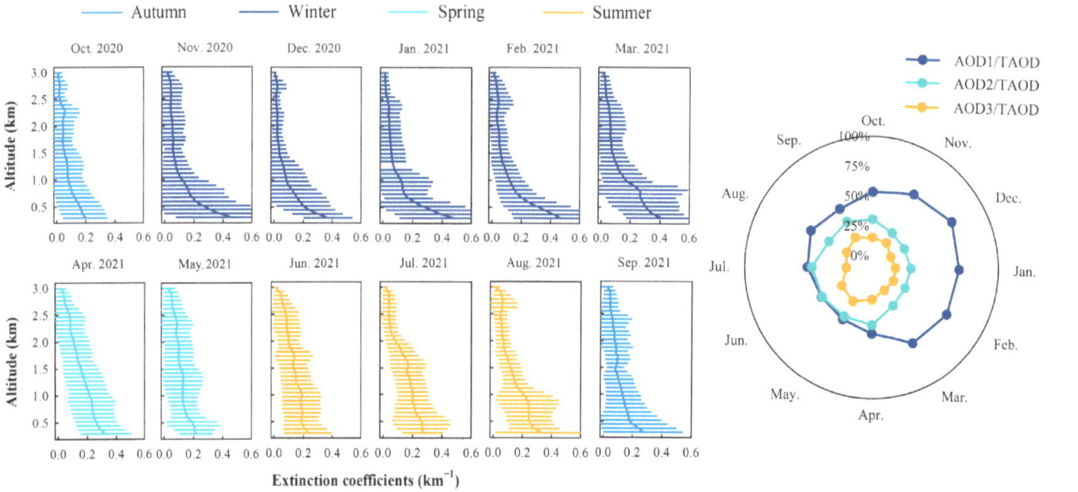

Figure 2. Monthly averaged profiles of aerosol extinction coefficients and monthly ratios of the aerosol optical depth (AOD) in different layers to the TAOD retrieved from LiDAR in the cold region of China from October 2020 to September 2021 during the day (6:00–18:00). Note: AOD1, AOD2, AOD3, and TAOD were used to indicate the AODs within the height ranges of 0–1 km, 1–2 km, 2–3 km, and 0–3 km, respectively.

The AOD1/TAOD ratio exhibited smaller values in summer (46%) and larger values in winter (65%) (Figure 2). This ratio peaked at 70% in December and reached a minimum of 40% in May. Furthermore, the AOD1/TAOD and AOD2/TAOD ratios were nearly equal from May to July. In contrast to AOD1/TAOD, AOD2/TAOD showed the opposite seasonal variation, with higher values in summer (38%) and lower values in winter (24%) (Table 1). The AOD2/TAOD ratio peaked at 42% in July and decreased to a minimum of 22% in December throughout the year. The AOD3/TAOD ratio was lowest in winter at 11% compared to other seasons, where it ranged from 16% to 19%.

Table 1. The AOD of each height range and their ratio to the TAOD in different seasons.

Season	AOD1 [1]	AOD2 [2]	AOD3 [3]	TAOD [4]	AOD1/ TAOD	AOD2/ TAOD	AOD3/ TAOD
Spring	0.155	0.136	0.069	0.359	43.13%	37.78%	19.09%
Summer	0.169	0.141	0.058	0.368	45.83%	38.42%	15.75%
Autumn	0.117	0.078	0.042	0.237	49.36%	32.86%	17.77%
Winter	0.180	0.067	0.032	0.279	64.59%	24.09%	11.32%

[1] AOD1, the integral of the extinction coefficients between the heights of 0.27 km and 1 km; [2] AOD2, the integral of the extinction coefficients between the heights of 1 km and 2 km; [3] AOD3, the integral of the extinction coefficients between the heights of 2 km and 3 km; [4] TAOD, the integral of the extinction coefficients between the heights of 0.27 km and 3 km.

Based on Pearson correlation analysis, the ECs between AOD1 and the concentrations of $PM_{2.5}$ and PM_{10} throughout the year were 0.28 ($p < 0.05$) and 0.24 ($p < 0.05$), respectively (Table 2). In this table, we present the correlation matrix between PM2.5, PM10, AOD1, AOD2, AOD3, TAOD, PBLH, AT, and RH. The correlation coefficients range from -1 to 1, where 1 represents a perfect positive correlation, -1 represents a perfect negative correlation, and 0 indicates no linear correlation. From the observations, it can be noted that $PM_{2.5}$ and PM_{10} exhibit a relatively high correlation coefficient of 0.736. AOD1 shows a moderate positive correlation with AOD3, with a coefficient of 0.8. Additionally, some variables exhibit negative correlations, such as AOD1 and AOD2 with a correlation coefficient of -0.048. The correlation coefficient between AOD2 and the $PM_{2.5}$ concentration was 0.17 ($p < 0.05$), while the correlation between AOD2 and the PM_{10} concentration was not statistically significant.

Table 2. Correlation coefficients among the PM concentrations, AOD, PBLH, air temperature, and relative humidity in Changchun.

	$PM_{2.5}$	PM_{10}	AOD1	AOD2	AOD3	TAOD	PBLH	AT	RH
$PM_{2.5}$	1	0.736	0.277	−0.169	−0.245	0.032 *	−0.245	−0.358	−0.133
PM_{10}	-	1	0.24	−0.029 *	−0.1 *	0.112	−0.159	−0.166	−0.343
AOD1	-	-	1	0.387	−0.048 *	0.8	−0.304	−0.015 *	0.264
AOD2	-	-	-	1	0.403	0.808	0.071 *	0.306	0.241
AOD3	-	-	-	-	1	0.44	0.419	0.174	0.139
TAOD	-	-	-	-	-—	1	−0.041 *	0.176	0.317
PBLH	-	-	-	-	-	-	1	0.262	0.053 *
AT	-	-	-	-	-	-	-	1	0.144
RH	-	-	-	-	-	-	-	-	1

Note: The significance was investigated using the independent-samples t-test. A value with an asterisk (*) ($p > 0.05$) is not significant at a confidence level of 95%. PBLH, planetary boundary layer height; AT, air temperature; RH, relative humidity.

3.2. *Variations in Meteorological Parameters and Air Pollutants*

There was a strong positive correlation of the PBLH between the retrieved data from LiDAR in this study and the data released by the ECMWF ($R = 0.74$, $p < 0.05$). Throughout the study period, the highest PBLH was observed in May with a mean value of 1.9 km, while the lowest was observed in December with a mean value of 1 km (Figure 3a). In spring and summer, the PBLH was relatively higher, with averages of 1.73 km and 1.64 km, respectively, compared to 1.47 km and 1.32 km in autumn and winter. Over the study period, a negative correlation coefficient was found between the PBLH and the concentrations of $PM_{2.5}$ (R = −0.27, $p < 0.05$) and PM_{10} (R = −0.16, $p < 0.05$) (Table 2).

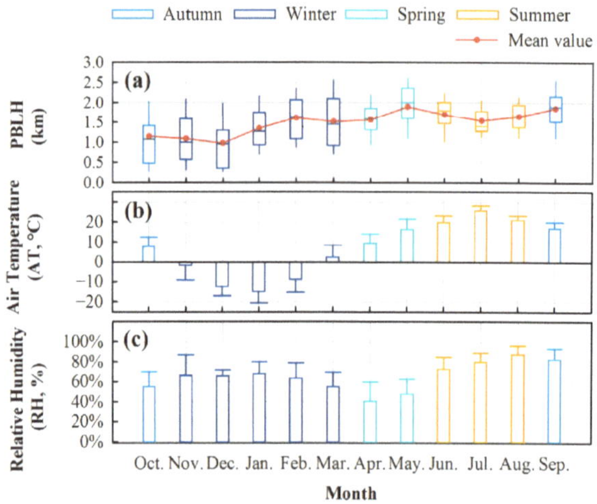

Figure 3. Monthly variations in (**a**) planetary boundary layer height (PBLH), (**b**) air temperature, and (**c**) relative humidity in the cold region of China from October 2020 to September 2021.

The study period covered a range of temperatures in the Changchun region, with January being the coldest month at −15 °C and July being the warmest month at 26 °C (Figure 3b). The average temperature in winter was −7 °C, while in summer, it was 22 °C. August had the highest relative humidity, reaching 88%, while April had the lowest at 41% (Figure 3c). In spring, the relative humidity was recorded at 44%, while in summer, it increased to 80%. These temperature and humidity variations have significant implications for air quality in the region.

Between October 2020 and September 2021, the study recorded a total of 33 days where the air-quality index (AQI) exceeded the threshold of 100, as defined by the Technical Regulation on Ambient Air-Quality Index (HJ633-2012) (Figure 4). High AQI values indicate a significant increase in air pollution levels during these periods, which could pose health risks to the population.

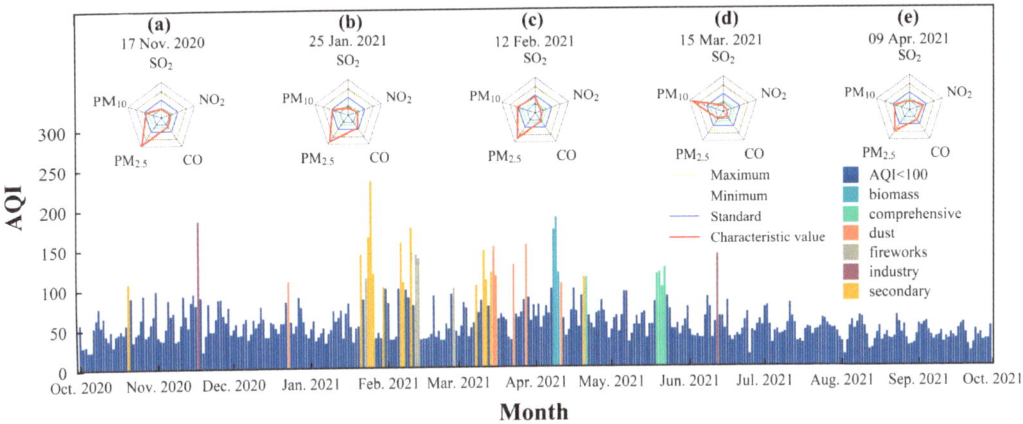

Figure 4. Daily variation in AQI with aerosol pollution type (when daily AQI value is over 100) and (**a–e**) characteristic radar chart for pollution cases in the cold region of China from October to November 2020 (Note: Columns under five aerosol pollution scenarios are marked with grey circles).

Further analysis revealed that PM dominated most pollution events, as shown in the radar charts in Figure 4a–e. This finding is consistent with previous studies, indicating that particulate matter is a major contributor to air pollution in many regions of China, including Changchun. The presence of high PM concentrations can lead to respiratory issues, cardiovascular diseases, and other health problems, highlighting the urgent need to address particulate matter pollution in the region.

In accordance with the ambient air-quality standards in China (GB 3095-2012) [34], Grade-II classification was adopted for this study (Table A1). Within the October 2020 to September 2021 period, Changchun experienced 22 days in which the Grade-II limits for $PM_{2.5}$ were exceeded (Table 3). Notably, during winter, the concentration of $PM_{2.5}$ exceeded the Grade-II limit for 17 days, indicating that winter was the primary period for haze pollution in the Changchun area. According to the WHO 2021 AQG standards [35], it was found that the pollution situation of $PM_{2.5}$ in the winter in Changchun area was severe. Additionally, there are 29 and 27 days in December and January, respectively, when NO_2 exceeds the health standards set by WHO.

Table 3. The number of days in which the concentrations of $PM_{2.5}$, PM_{10}, SO_2, CO, NO_2, and O_3 (8 h) exceeded Grade II and the WHO 2021 AQG from October 2020 to September 2021.

Period	$PM_{2.5}$ (GB/WHO)	PM_{10} (GB/WHO)	SO_2 (GB/WHO)	NO_2 (GB/WHO)	CO (GB/WHO)	O_3 (GB/WHO)
Jan. 2021	6/31	1/20	0/0	1/27	0/0	0/0
Feb. 2021	6/28	1/16	0/0	0/15	0/0	0/2
Mar. 2021	4/29	4/25	0/0	0/25	0/0	0/4
Apr. 2021	4/28	2/28	0/0	0/22	0/0	1/14
May 2021	0/28	0/25	0/0	0/20	0/0	4/15
June 2021	0/11	0/6	0/0	0/19	0/0	2/14
July 2021	0/18	0/6	0/0	0/11	0/0	0/11
Aug. 2021	0/10	0/5	0/0	0/8	0/0	0/7
Sept. 2021	0/11	0/8	0/0	0/20	0/0	0/3
Oct. 2020	1/21	0/17	0/0	0/24	0/0	0/4
Nov. 2020	1/25	0/20	0/0	0/24	0/0	0/0
Dec. 2020	0/31	1/20	0/0	0/29	0/0	0/0

3.3. *Typical Aerosol Pollution Scenarios*

Using multisource data, we analysed the vertical distributions of aerosol particles during typical aerosol pollution cases. The main information of these cases is summarized in Table 4. In northeast China during winter, pollution events exhibit a higher extinction coefficient. Moreover, based on the extinction coefficient and air-quality index (AQI) of China, it can be inferred that biomass combustion constitutes a significant pollution event in this region.

3.3.1. Firework Aerosols

In China, it is a long-standing tradition to ignite an abundance of firecrackers on Chinese New Year's Eve (11 February) and Lunar New Year's Day (12 February). At 14:00 on February 11, the average EC below 0.5 km was 0.37 km^{-1}, while at 22:00 on the same day, it increased to 0.88 km^{-1} (Figure 5a). Additionally, the ECs within a height range of 0.5 km from 22:00 on 11 February to 13:00 on 12 February remained elevated, averaging between 0.82 and 1.07 km^{-1}. In contrast, the depolarization ratio within 0.5 km is approximately 0.05 (Figure 5b), indicating a low level of coarse particles [36]. The concentration of $PM_{2.5}$ exhibited a rapidly increased after 20:00 on 11 February, reaching 118 μg/m^3 at midnight that night (Figure 5c). In recent years, as a result of the Chinese government's efforts to control fireworks, the use of fireworks in urban areas has decreased significantly. However, the tradition of setting off fireworks unavoidably persists, driven by people's pursuit of spiritual fulfilment. During 11–12 February, there was a simultaneous occurrence of high extinction and elevated concentrations of $PM_{2.5}$. Furthermore, the PBLH on that day was exceptionally lower than in the rest of that month. As a result, aerosols from firework setoffs still have a significant impact when there is an extremely low boundary layer.

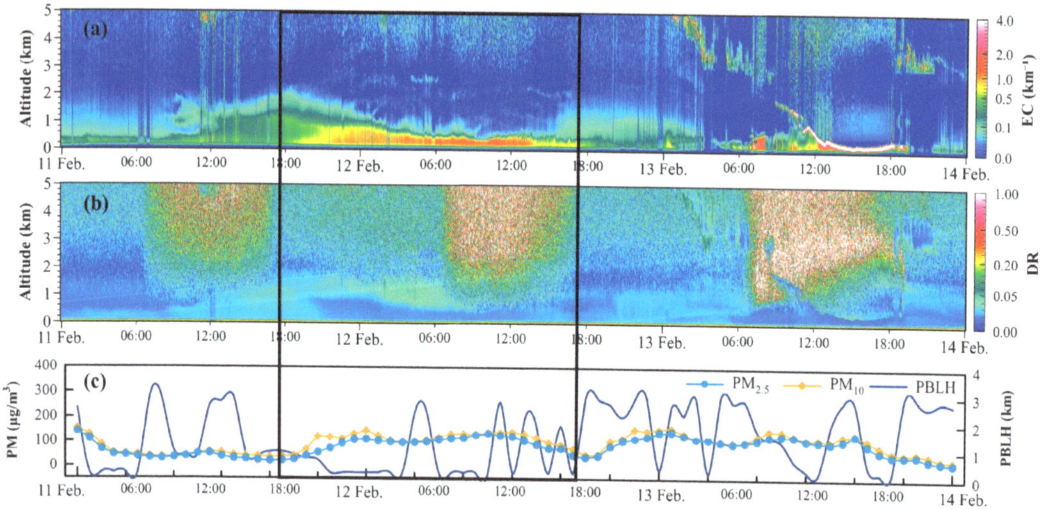

Figure 5. Variations in the (**a**) extinction coefficient (EC, 532 nm), (**b**) depolarization ratio, and (**c**) concentrations of $PM_{2.5}$, PM_{10}, and planetary boundary layer height (PBLH) in the cold region of China during a firework event (11–14 February 2021).

Table 4. Characteristics of aerosol optical parameters during haze pollution (i.e., fireworks pollution, dust, and biomass-burning) in northeast China.

Researcher	Study Area	Period	EC (0–500 m)	AQI	Uncertainty	Aerosol Types
Zhao et al., 2018 [37]	Shenyang	31 March 2016	0.1–0.5	182	None	Dust
Zhao et al., 2018 [37]	Shenyang	5 November 2016	0.5–1.1	259	None	Biomass-burning
Liu et al., 2017 [38]	Shenyang	7 January 2017	0.3–1.1	250	None	None
Zhao et al., 2020 [39]	Siping	12 January 2019	0.9–1.3	210	None	None
This study	Changchun	12 February 2021	0.1–0.7	143	25–35%	Firework
This study	Changchun	15 March 2021	0.2–0.6	152	25–35%	Dust
This study	Changchun	9 April 2021	0.3–0.7	188	25–35%	Biomass-burning

3.3.2. Dust Aerosols

At 22:00 on 14 March, an aerosol air mass was observed with its base located approximately 1 km above ground level and a thickness that could extend up to 2 km (Figure 6a). However, by 4:00 on 15 March, the air mass had mostly dissipated. Furthermore, at 12:00 on 15 March, the average depolarization ratio within the 0–1 km range increased to 0.11 (Figure 6b). At 19:00 on 15 March, the profiles of the EC and depolarization ratio both indicated that the top of the aerosol air mass reached a maximum altitude of 2.2 km. Additionally, between 10:00 on 15 March and 4:00 on 16 March, the $PM_{2.5}$ concentration ranged from 16 to 69 $\mu g/m^3$, while the PM_{10} concentration ranged from 91 to 838 $\mu g/m^3$ (Figure 6c).

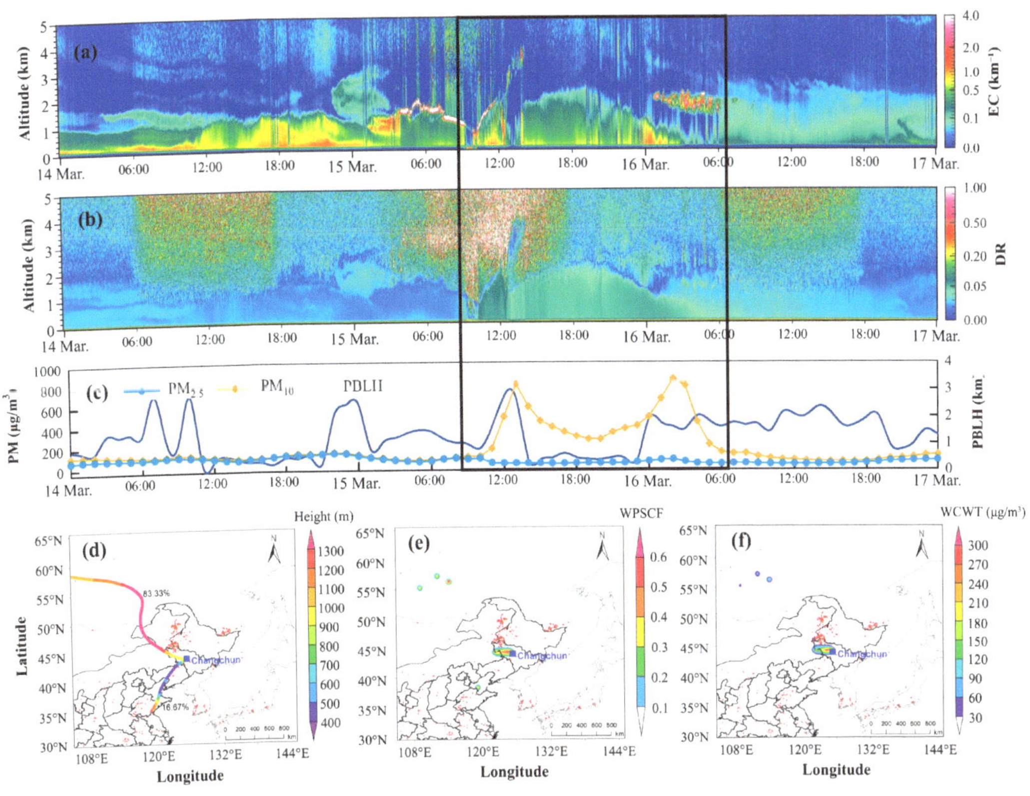

Figure 6. Variations in the (**a**) extinction coefficient (EC, 532 nm); (**b**) depolarization ratio; (**c**) concentrations of $PM_{2.5}$, PM_{10}, and planetary boundary layer height (PBLH); (**d**) mean 48-hour backwards trajectories; (**e**) spatial distributions of WPSCF values for PM_{10}; and (**f**) spatial distributions of WCWT values for PM_{10} in the cold region of China during a dust event (14–16 March 2021).

Based on the backwards trajectories, the sources of PM_{10} transport in Changchun on 15 March were analysed (Figure 6d–f). The cluster analysis of the 24 trajectories revealed two primary clusters. The first cluster originated from the desertified area in the western part of Jilin Province, accounting for 83% of the trajectories, while the second cluster originated from the Liaoning and Shandong provinces, accounting for 17%. Additionally, based on PSCF and CWT calculations, it was determined that the desert areas in the western CRC had the highest potential for the source of dust aerosol pollution. The CALIPSO data from 17 March at 02:27 revealed that dust aerosols had accumulated below 2 km and between 4–5 km, creating two distinct zones of dust accumulation (Figure 7a,b). This indicates that dust aerosols were the predominant aerosol subtype in Changchun on 17 March. Despite the lack of CALIPSO data for the Changchun region on 15 March, the observation of dust aerosols on 17 March still provides evidence of the impact of natural source transport in the CRC.

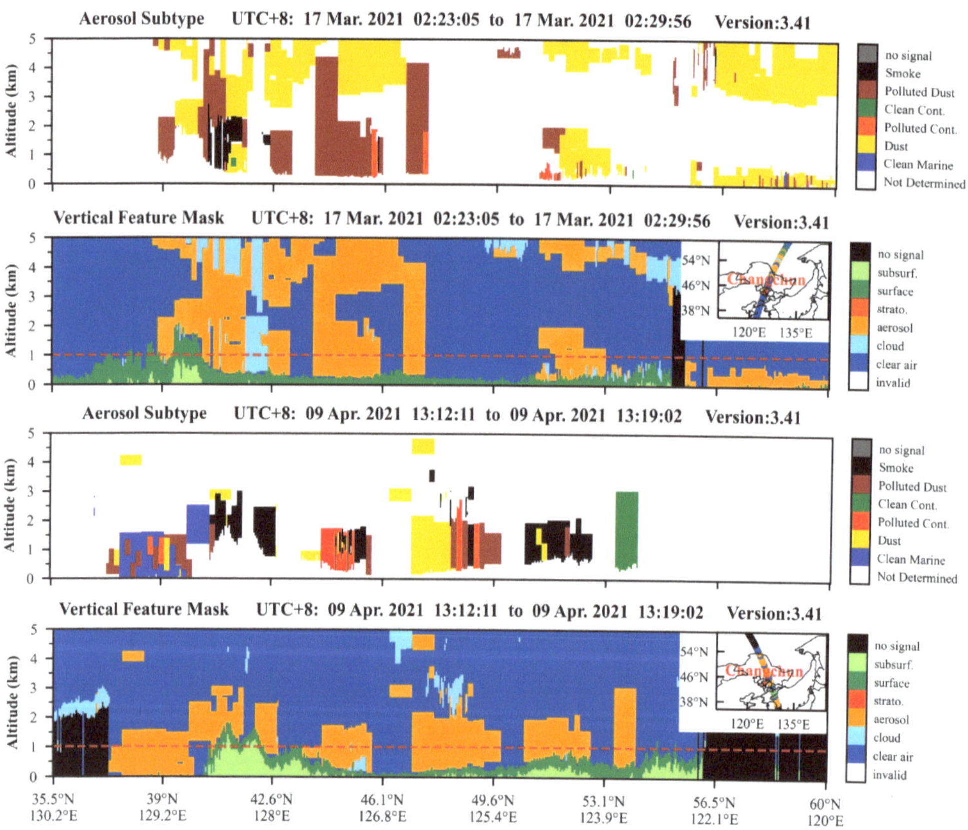

Figure 7. CALIPSO L2 products of (**a**,**b**) aerosol subtype, (**c**,**d**) vertical feature mask, and their ground track in the cold region of China on 17 March 2021 and 9 April 2021 (Note: The vertical feature mask at a height of 1 km is shown on the track).

3.3.3. Biomass-Burning Aerosols

In the Changchun region, straw burning is regulated by the government and is typically scheduled before the spring tilling period, which occurs in April each year. Consequently, in April 2021, a total of 5070 fire points was recorded by the MODIS product in Changchun (Figure 8).

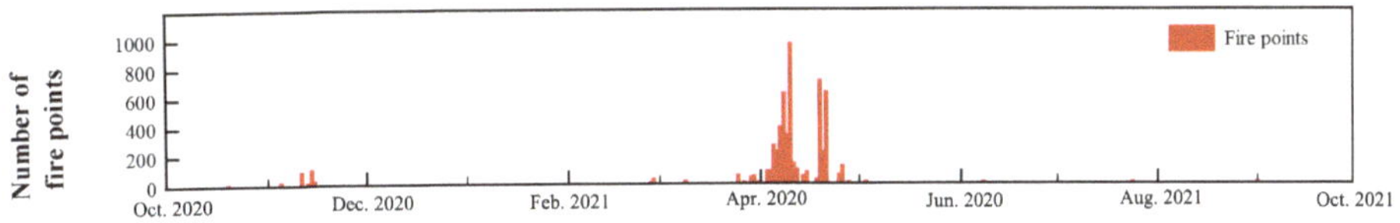

Figure 8. Daily variations of total fire points in Changchun from October 2020 to September 2021.

At 04:00 on 8 April, the EC of the air mass at a height of 0.6 km was 1.5 km^{-1}, and it decreased by 7:00 (Figure 9a). At approximately 20:00 on 8 April, the average EC within 1 km increased to 1.11 km^{-1} and then decreased to 0.31 km^{-1} by 15:00 on 9 April. Nevertheless, on 9 April, the mean depolarization ratio below 1 km was less than 0.01 (Figure 9b). At 04:00 on 8 April, the $PM_{2.5}$ concentration peaked at 598 $\mu g/m^3$ and subsequently decreased to 21 $\mu g/m^3$ by 08:00 (Figure 9c). Moreover, between 16:00 on 8 April and 08:00 on 9 April, the $PM_{2.5}$ concentration increased from 55 to 251 $\mu g/m^3$.

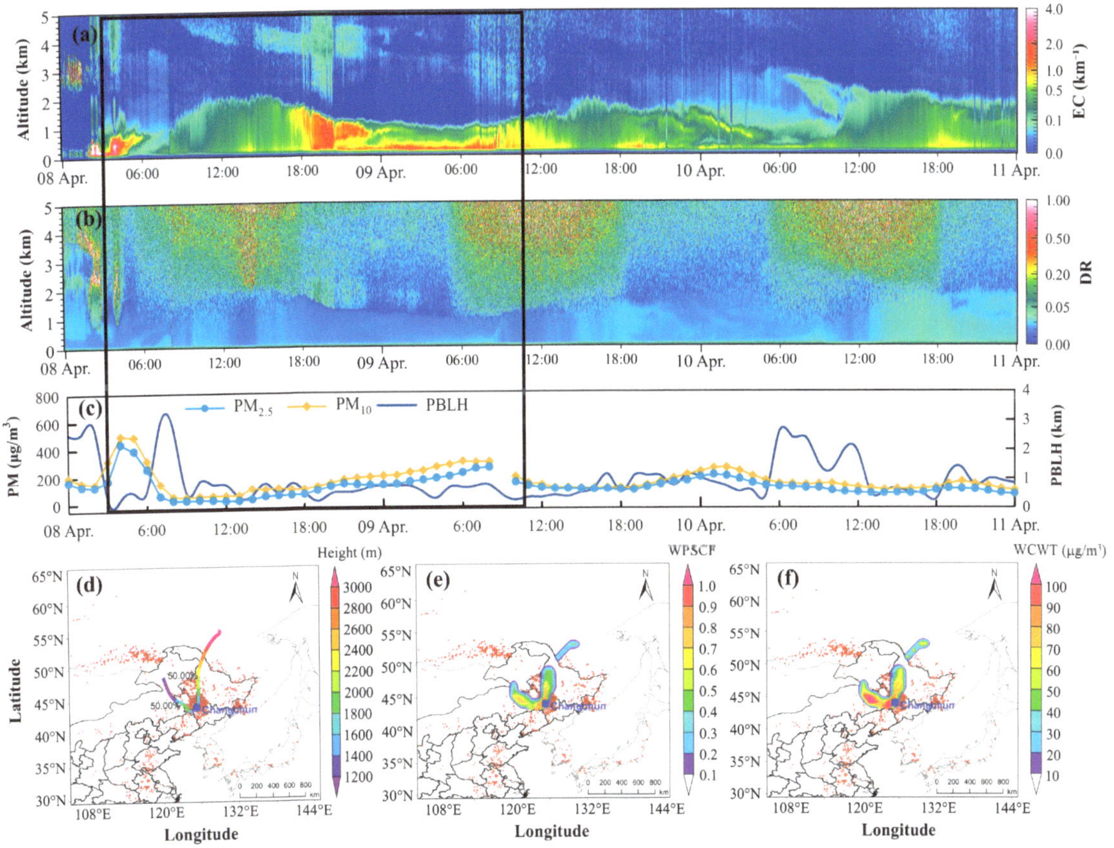

Figure 9. Variations of the (**a**) extinction coefficient (EC, 532 nm), (**b**) depolarization ratio, (**c**) concentrations of $PM_{2.5}$ and PM_{10} and planetary boundary layer height (PBLH), (**d**) mean 48-hour backward trajectories, (**e**) spatial distributions of WPSCF values for $PM_{2.5}$, and (**f**) spatial distributions of WCWT values for $PM_{2.5}$ (**f**) in the cold region of China during a biomass-burning event (8–10 April 2021).

CALIPSO dataset analysis indicated that smoke and polluted continents were the predominant aerosol subtypes observed in the Changchun area (Figure 7c,d). To determine the source of $PM_{2.5}$ transport on 9 April, a backwards trajectory analysis was conducted

(Figure 9d–f). The trajectory clustering analysis revealed that the backwards trajectories could be primarily separated into two clusters, each accounting for 50% of the trajectories. These clusters originated from the western and northern regions of the Changchun region, respectively. The PSCF analysis indicated that these regions were the primary potential sources of $PM_{2.5}$. Additionally, the CWT analysis suggested that these regions, with a significant number of fire points, contributed more than 90 $\mu g/m^3$ of $PM_{2.5}$ to Changchun.

4. Discussion

4.1. Seasonal Vertical Distribution of Aerosols

Previous investigations have demonstrated that in warm regions, such as the Yangtze River Delta region of China, the near-surface aerosol ECs were comparable between summer and winter in most years, with the maximum extinction sometimes occurring during the period from June to September [40]. However, in CRC, the near-surface layer within 1 km exhibits greater ECs during winter than during other months. Furthermore, there were significant differences in the vertical variation trend of EC with height among the different seasons. This is mainly manifested by a significant increase in the nonuniformity of the gradient in the vertical profile of aerosols as the temperature decreases. This difference may be attributed to several factors. First, during winter, a lower PBLH of 1.32 km in Changchun and a higher proportion of AOD within 1 km (65%) were observed, indicating that aerosols were primarily concentrated at lower altitude. Second, the low temperatures during winter may attenuate the atmospheric thermal turbulence in northern China [41], thereby facilitating the accumulation of aerosol particles. Finally, the concentrations of $PM_{2.5}$ and PM_{10} in the atmosphere were significantly greater during winter compared to other seasons, primarily due to material combustion for heating purposes. These factors collectively contribute to the increase in ECs near the surface in CRC during the winter. It is worth noting that, due to the inconvenience of data acquisition, the impact of motor vehicles was not fully considered. Further investigation into the impact of motor vehicles is still needed. Additionally, in spring, the extinction characteristics may be influenced by fine particles emitted from centralized biomass burning and wind erosion from bare agricultural land surfaces in CRC. The atmospheric diffusion conditions during spring are more favourable compared to other seasons [42], which could explain the relatively lower EC in spring compared to winter in this region.

The study period revealed a negative correlation coefficient between the PBLH and the concentrations of $PM_{2.5}$ and PM_{10}. The negative correlation coefficient suggests that as the PBLH decreases, the concentrations of $PM_{2.5}$ and PM_{10} increase, indicating a strong link between atmospheric mixing depth and particulate matter concentration. These findings are consistent with similar studies conducted across many regions in China [40,43]. However, the correlation between the PBLH and PM concentration in CRC was weak, which could be due to the impact of various complex meteorological and anthropogenic emissions in the area. For example, the Changchun region has a unique climate, with cold temperatures prevailing throughout the year. Additionally, the region has a high density of industries and factories, leading to significant anthropogenic emissions, which could have influenced the observed weak correlation coefficient. Furthermore, the topography of the region could also play a role in the weak correlation, as it may contribute to the formation of local circulations or wind patterns that could further influence the dispersion and accumulation of particulate matter.

Although the weak correlation between the PBLH and PM concentration in CRC may indicate the complexity of the factors affecting air pollution in the region, it also highlights the need for targeted measures to mitigate the adverse effects of particulate matter pollution. Effective strategies that address the unique circumstances of the region are necessary to promote healthy living environments for the population. Such measures could include the implementation of stricter emission controls for industrial facilities in the area, the promotion of cleaner transportation modes, and the adoption of sustainable agricultural practices, among others. In conclusion, this study provides valuable insights into the relationship between the PBLH and particulate matter concentrations in the Changchun region. Although the weak correlation coefficient observed in CRC underscores the complexity of the factors influencing air pollution in the area, it also highlights the importance of adopting targeted measures to mitigate the adverse effects of particulate matter pollution in the region.

4.2. Typical Aerosol Pollution

To gain a comprehensive understanding of aerosol pollution in urban atmospheres during firework displays, scientists conducted analyses utilizing MODIS data [44]. The analysis revealed that during festival days, India experienced a significant increase of 56–121% in aerosol surface mass loading in the middle Indo-Gangetic Plain compared to that under normal conditions. This indicates that firework displays have a substantial impact on aerosol pollution levels in the region. Furthermore, Joly's findings [45] indicated that the concentration of $PM_{2.5}$, which is a fine particulate matter known to have detrimental effects on human health, reached its peak during firework emissions. These concentrations were found to be nearly 1000 times higher than the background level within just one hour. Such high levels of $PM_{2.5}$ pose serious health risks to the population, highlighting the need for effective measures to mitigate the pollution caused by fireworks. Although our study primarily focused on low-intensity fireworks emissions that did not result in a vertical expansion of the affected area, they still had a considerable impact on the enhancement of EC within the boundary layer. EC is a key component of aerosols and is associated with various adverse health effects. Our research revealed that the emissions from low-intensity fireworks contributed to a significant 76.4% increase in EC concentrations within the boundary layer, further emphasizing the implications of pollution from fireworks.

During the spring season, East Asia frequently experiences the presence of dust aerosols, as indicated by previous research [46]. To identify these dust aerosols in East Asia, past studies have utilized a depolarization ratio threshold of 0.1 [47,48]. For instance, on 15 March 2021, an observed aerosol air mass exhibited a depolarization ratio of 0.11, surpassing the established threshold. This air mass extended vertically up to 3 km, indicating the significant spread and coverage of the dust aerosols in the atmosphere. Furthermore, the concentration of PM_{10}, which includes inhalable particles with diameters that are generally 10 μm and smaller, peaked for the entire year on this day. Additionally, the EC values exceeded the average level for the month of March. These findings suggest a substantial presence of dust aerosols and associated particulate matter pollution during this period, underscoring the environmental and public health implications of such atmospheric conditions in East Asia.

In warmer regions of China, farmers implement crop rotation practices, which results in biomass-burning aerosols being released into the atmosphere during the summer season [49]. However, in colder regions such as Changchun, only one cultivation cycle is possible each year, and as a result, anthropogenic biomass burning is mainly conducted during the spring. Prior research has indicated that biomass-burning aerosols exhibit a lower depolarization ratio and a moderate-to-significant level of EC [47,50]. On 8–9 April, an aerosol air mass with a vertical extent of up to 2 km was observed in the CRC area, showing high values of EC. This indicates that the intensity of biomass burning in this region is higher compared to warmer regions in China. Furthermore, the depolarization ratio of this air mass was found to be lower than 0.1, indicating the presence of biomass-burning aerosols. These findings suggest that anthropogenic biomass-burning activities significantly contribute to the atmospheric pollution in this region. Additionally, on 9 April, an increase in the concentration of $PM_{2.5}$ was detected, further indicating the presence of pollutants in the atmosphere. High levels of $PM_{2.5}$ pose serious health risks to the population, especially for those with respiratory issues, highlighting the urgent need for effective measures to mitigate the pollution caused by biomass-burning activities. Overall, the study suggested that the air pollution observed on 8–9 April in the CRC area was primarily associated with the transportation of biomass-burning aerosols. The results illustrate the importance of understanding the impact of different agricultural practices on the environment and the need to adopt sustainable farming methods to minimize the negative consequences of biomass burning on air quality. While the implementation of crop rotation practices may have benefits in some regions, it is essential to consider the associated environmental impacts carefully. By adopting sustainable agricultural practices, we can reduce the adverse effects of biomass burning on air quality and promote healthier living environments for all.

4.3. Implication for Local Pollution Control

During the implementation of air pollution control measures, it is essential to understand the long-term variations in local atmospheric aerosols [51]. However, traditional aerosol monitoring networks are primarily located in developed areas and are designed to capture atmospheric composition characteristics near the surface only. Fortunately, ground-based LiDAR technology allows for the acquisition of long-term and high-resolution vertical profiles of aerosols [52]. In this study, utilizing ground-based LiDAR technology, we found that the ECs in the near-surface layer of in CRC were significantly greater during the winter season than during other seasons. Therefore, it is advisable to implement specific air pollution control measures tailored to the distinct characteristics of winter and summer seasons. A case in point is the observation made on Chinese New Year's Eve in Changchun, where the daytime and night-time discharge of firecrackers resulted in an elevation of EC levels at a height of 500 m to 0.37 km^{-1} and 0.88 km^{-1}, respectively. As a potential solution, it is proposed to ease restrictions on firecracker usage during the day while intensifying regulations during night-time hours. Furthermore, through an analysis of data obtained from CALIPSO and backwards trajectory assessments, it was determined that two instances of dust–biomass-burning composite pollution occurred in March and April. The findings revealed that in March, the predominant aerosol component consisted of dust originating from western Changchun, while in April, biomass-burning aerosols stemming from the north and east of Changchun were identified as the primary aerosol component. Consequently, a reduction in the intensity of spring biomass-burning activities could serve as an effective measure to alleviate the prevalence of dust–biomass-burning composite pollution. These insights could offer valuable guidance for formulating emission policies aimed at addressing similar challenges related to the transmission of biomass burning pollutants and firework emissions in other regions.

To effectively control air pollution, it is crucial to establish a comprehensive monitoring network that enables the government to develop informed policies and manage air quality [53]. However, it is important to note that atmospheric pollutants are primarily generated, dispersed, and degraded in the upper troposphere, typically within altitudes ranging from 2 to 8 km above ground level [54]. Consequently, an air pollution episode caused by aerosol transport from the free troposphere may not be effectively addressed only by reducing local anthropogenic sources [55]. This highlights the need for a well-established three-dimensional characterization of air pollutants to ensure informed decision making in policy development. Therefore, we propose the implementation of three-dimensional monitoring in CRC to analyse and forecast the extent and types of air pollution. This approach will not only enhance local pollution control efforts but also safeguard public health.

As for the limitations, although certain vertical structures of aerosol pollution have been emphasized, decades of monitoring are required to reduce the impact of uncertainty and consider other factors, such as chemical generation and natural emissions. In the future, we will continue to conduct long-term monitoring in cold regions and develop aerosol pollution recognition models based on identification elements such as aerosol pollution recognition to provide valuable policy recommendations.

5. Conclusions

In this study, the properties and impacts of atmospheric aerosols in the megacity of Changchun in the CRC were investigated. The nonuniformity of the gradient in the vertical profile of aerosols significantly increased with decreasing temperature. In contrast, in warmer regions, the differences in the aerosol vertical distribution remained relatively consistent across the different seasons. The correlation between the temperature and ground-level PM concentration was greater than that between the planetary boundary layer height and PM concentration. The EC concentration in the near-surface layer (below 1 km) in winter in the CRC was significantly greater than that during the other seasons, which could be attributed to the low PBLH and anthropogenic emissions in winter. The ignition of fireworks in suburban regions on Chinese New Year's Eve, despite the government's prohibition in urban areas, led to a substantial increase in aerosol concentrations below the 3 km altitude. Additionally, during the scheduled biomass-burning period, the EC concentration of polluted aerosols originating from the western and northern areas of Changchun below 3 km exceeded the annual average in April, indicating the persistence of high aerosol levels despite the prevailing monsoon winds. The analysis using the CWT method indicated that these regions, characterized by a substantial number of fire points, contributed over 90 $\mu g/m^3$ of $PM_{2.5}$ to the air pollution levels in Changchun. Sand–dust activity and biomass burning in March and April could cause severe pollution in this region, even under the conditions of elevated boundary layers. These findings are beneficial for understanding the characteristics of the aerosol variations in cold climates and substantial anthropogenic emissions.

Author Contributions: Conceptualization, L.D., L.G., and Z.X.; data curation, L.D.; formal analysis, Y.Y.; funding acquisition, W.C. and L.G.; investigation, J.F.; methodology, H.Y. and G.S.; project administration, W.C.; resources, W.C.; software, L.D.; supervision, W.C.; validation, Y.Y.; visualization, G.S.; writing—original draft, L.D.; writing—review and editing, W.C. and L.G. All authors have read and agreed to the published version of the manuscript.

Funding: This study was funded by the National Natural Science Foundation of China (No. U23A6001), the Education Department of Jilin Province (No. JJKH20211079KJ), the National Natural Science Foundation of China (No. 41775116), the Strategic Priority Research Program of the Chinese Academy of Sciences (XDA28020400), the Research Foundation of Science and Technology Department of Jilin Province, China (20230508032RC), the Joint Funds of the Key Research and Development Program Project of Jilin Province (20220202028NC) and the Government Project of the Ecology and Environment Department of Jilin Province (2023-11).

Data Availability Statement: Data will be made available on request.

Conflicts of Interest: The Author Hongwu Yang was employed by the Hongke Photonics Company. The remaining authors declare that the research was conducted in the absence of any commercial or financial relationships that could be construed as a potential conflict of interest. The funders had no role in the design of the study; in the collection, analyses, or inter-pretation of data; in the writing of the manuscript; or in the decision to publish the results.

Appendix A

Table A1. Concentration limits of $PM_{2.5}$, PM_{10}, SO_2, CO, NO_2, and O_3 (8 h) within the air-quality standards of China (GB 3095-2012) [34].

Standard	$PM_{2.5}$ ($\mu g/m^3$)	PM_{10} ($\mu g/m^3$)	SO_2 ($\mu g/m^3$)	CO (mg/m^3)	NO_2 ($\mu g/m^3$)	O_3 (8 h) ($\mu g/m^3$)
Grade II	75	150	150	4	80	160

References

1. Prather, K.A.; Hatch, C.D.; Grassian, V.H. Analysis of Atmospheric Aerosols. *Annu. Rev. Anal. Chem.* **2008**, *1*, 485–514. [CrossRef] [PubMed]
2. Burnett, R.T.; Pope, C.A.; Ezzati, M.; Olives, C.; Lim, S.S.; Mehta, S.; Shin, H.H.; Singh, G.; Hubbell, B.; Brauer, M.; et al. An Integrated Risk Function for Estimating the Global Burden of Disease Attributable to Ambient Fine Particulate Matter Exposure. *Environ. Health Perspect.* **2014**, *122*, 397–403. [CrossRef] [PubMed]

3. Menon, S. Current uncertainties inassessing aerosol effects on climate. *Annu. Rev. Environ. Resour.* **2004**, *29*, 1–30. [CrossRef]
4. Mhawish, A.; Banerjee, T.; Sorek-Hamer, M.; Lyapustin, A.; Broday, D.M.; Chatfield, R. Comparison and evaluation of MODIS Multi-angle Implementation of Atmospheric Correction (MAIAC) aerosol product over South Asia. *Remote Sens. Environ.* **2019**, *224*, 12–28. [CrossRef]
5. Chen, W.; Duanmu, L.; Qin, Y.; Yang, H.W.; Fu, J.; Lu, C.W.; Feng, W.; Guo, L. Lockdown-induced urban aerosol change over Changchun, China during COVID-19 outbreak with polarization LiDAR. *Chin. Geogr. Sci.* **2022**, *32*, 824–833. [CrossRef] [PubMed]
6. She, L.; Mei, L.L.; Xue, Y.; Che, Y.H.; Guang, J. SAHARA: A Simplified Atmospheric Correction Algorithm for Chinese gaofen Data: 1. Aerosol Algorithm. *Remote Sens.* **2017**, *9*, 21. [CrossRef]
7. Sorribas, M.; Andrews, E.; Ogren, J.A.; del Aguila, A.; Fraile, R.; Sheridan, P.; Yela, M. Climatological study for understanding the aerosol radiative effects at southwest Atlantic coast of Europe. *Atmos. Environ.* **2019**, *205*, 52–66. [CrossRef]
8. Yu, H.B.; Chin, M.; Winker, D.M.; Omar, A.H.; Liu, Z.Y.; Kittaka, C.; Diehl, T. Global view of aerosol vertical distributions from CALIPSO lidar measurements and GOCART simulations: Regional and seasonal variations. *J. Geophys. Res. Atmos.* **2010**, *115*, 19. [CrossRef]
9. Amiridis, V.; Wandinger, U.; Marinou, E.; Giannakaki, E.; Tsekeri, A.; Basart, S.; Kazadzis, S.; Gkikas, A.; Taylor, M.; Baldasano, J.; et al. Optimizing CALIPSO Saharan dust retrievals. *Atmos. Chem. Phys.* **2013**, *13*, 12089–12106. [CrossRef]
10. Cheng, Y.M.; Dai, T.; Goto, D.; Schutgens, N.A.J.; Shi, G.Y.; Nakajima, T. Investigating the assimilation of CALIPSO global aerosol vertical observations using a four-dimensional ensemble Kalman filter. *Atmos. Chem. Phys.* **2019**, *19*, 13445–13467. [CrossRef]
11. Li, J.W.; Han, Z.W. Aerosol vertical distribution over east China from RIEMS-Chem simulation in comparison with CALIPSO measurements. *Atmos. Environ.* **2016**, *143*, 177–189. [CrossRef]
12. Wang, Y.; Zhuang, G.; Xu, C.; An, Z. The air pollution caused by the burning of fireworks during the lantern festival in Beijing. *Atmos. Environ.* **2007**, *41*, 417–431. [CrossRef]
13. Lai, Y.; Brimblecombe, P. Changes in Air Pollutants from Fireworks in Chinese Cities. *Atmosphere* **2022**, *13*, 1388. [CrossRef]
14. Yao, W.; Zhao, Y.; Chen, R.; Wang, M.; Song, W.; Yu, D. Emissions of Toxic Substances from Biomass Burning: A Review of Methods and Technical Influencing Factors. *Processes* **2023**, *11*, 853. [CrossRef]
15. Jiang, K.; Xing, R.; Luo, Z.; Huang, W.; Yi, F.; Men, Y.; Zhao, N.; Chang, Z.; Zhao, J.; Pan, B.; et al. Pollutant emissions from biomass burning: A review on emission characteristics, environmental impacts, and research perspectives. *Particuology* **2023**, *85*, 296–309. [CrossRef]
16. Yang, T.; Gbaguidi, A.; Yan, P.Z.; Zhang, W.D.; Zhu, L.L.; Yao, X.F. Model elucidating the sources and formation mechanisms of severe haze pollution over Northeast mega-city cluster in China. *Environ. Pollut.* **2017**, *230*, 692–700. [CrossRef] [PubMed]
17. Chen, S.; Jiang, L.; Liu, W.; Song, H. Fireworks regulation, air pollution, and public health: Evidence from China. *Reg. Sci. Urban Econ.* **2022**, *92*, 103722. [CrossRef]
18. Tomlin, A.S. Air quality and climate impacts of biomass use as an energy source: A review. *Energy Fuels* **2021**, *35*, 14213–14240. [CrossRef]
19. Chen, J.; Li, C.; Ristovski, Z.; Milic, A.; Gu, Y.; Islam, M.S.; Wang, S.; Hao, J.; Zhang, H.; He, C.; et al. A review of biomass burning: Emissions and impacts on air quality, health and climate in China. *Sci. Total Environ.* **2017**, *579*, 1000–1034. [CrossRef]
20. Yin, S.; Wang, X.F.; Zhang, X.R.; Zhang, Z.X.; Xiao, Y.; Tani, H. Exploring the effects of crop residue burning on local haze pollution in Northeast China using ground and satellite data. *Atmos. Environ.* **2019**, *199*, 189–201. [CrossRef]
21. Li, L.L.; Wang, K.; Chen, W.W.; Zhao, Q.L.; Liu, L.J.; Liu, W.; Liu, Y.; Jiang, J.Q.; Liu, J.M.; Zhang, M.D. Atmospheric pollution of agriculture-oriented cities in Northeast China: A case in Suihua. *J. Environ. Sci.* **2020**, *97*, 85–95. [CrossRef] [PubMed]
22. Li, C.; Liu, M.; Hu, Y.; Wang, H.; Xiong, Z.; Wu, W.; Liu, C.; Zhang, C.; Du, Y. Investigating the vertical distribution patterns of urban air pollution based on unmanned aerial vehicle gradient monitoring. *Sustain. Cities Soc.* **2022**, *86*, 104144. [CrossRef]
23. Dickerson, R.R.; Li, C.; Li, Z.; Marufu, L.T.; Stehr, J.W.; McClure, B.; Krotkov, N.; Chen, H.; Wang, P.; Xia, X.; et al. Aircraft observations of dust and pollutants over northeast China: Insight into the meteorological mechanisms of transport. *J. Geophys. Res.* **2007**, *112*, D24S90. [CrossRef]
24. Ma, Y.; Zhao, H.; Dong, Y.; Che, H.; Li, X.; Hong, Y.; Li, X.; Yang, H.; Liu, Y.; Wang, Y.; et al. Comparison of two air pollution episodes over Northeast China in winter 2016/17 using ground-based lidar. *J. Meteorol. Res.* **2018**, *32*, 313–323. [CrossRef]
25. Zhao, H.; Gui, K.; Ma, Y.; Wang, Y.; Wang, Y.; Wang, H.; Zheng, Y.; Li, L.; Zhang, L.; Che, H.; et al. Seasonal evolution of aerosol loading and its vertical distribution in northeastern China from long-term satellite observations and model reanalysis. *Atmos. Environ.* **2023**, *302*, 119720. [CrossRef]
26. Li, X.; Ma, Y.; Wang, Y.; Wei, W.; Zhang, Y.; Liu, N.; Hong, Y. Vertical Distribution of Particulate Matter and its Relationship with Planetary Boundary Layer Structure in Shenyang, Northeast China. *Aerosol Air Qual. Res.* **2019**, *19*, 2464–2476. [CrossRef]
27. Wang, F.; Li, Z.; Ren, X.; Jiang, Q.; He, H.; Dickerson, R.R.; Lv, F. Vertical distributions of aerosol optical properties during the spring 2016 ARIAs airborne campaign in the North China Plain. *Atmos. Chem. Phys.* **2018**, *18*, 8995–9010. [CrossRef]
28. Zhang, M.D.; Zhang, S.C.; Bao, Q.Y.; Yang, C.J.; Qin, Y.; Fu, J.; Chen, W.W. Temporal Variation and Source Analysis of Carbonaceous Aerosol in Industrial Cities of Northeast China during the Spring Festival: The Case of Changchun. *Atmosphere* **2020**, *11*, 18. [CrossRef]
29. Fernald, F.G. Analysis of atmospheric lidar observations: Some comments. *Appl. Optics* **1984**, *23*, 652. [CrossRef]

30. Han, Y.; Wu, Y.H.; Wang, T.J.; Zhuang, B.L.; Li, S.; Zhao, K. Impacts of elevated-aerosol-layer and aerosol type on the correlation of AOD and particulate matter with ground-based and satellite measurements in Nanjing, southeast China. *Sci. Total Environ.* **2015**, *532*, 195–207. [CrossRef]
31. Sirois, A.; Bottenheim, J.W. Use of backward trajectories to interpret the 5-year record of PAN and O3 ambient air concentrations at Kejimkujik National Park, Nova Scotia. *J. Geophys. Res. Atmos.* **1995**, *100*, 2867–2881. [CrossRef]
32. Han, Y.J.; Holsen, T.M.; Hopke, P.K. Estimation of source locations of total gaseous mercury measured in New York State using trajectory-based models. *Atmos. Environ.* **2007**, *41*, 6033–6047. [CrossRef]
33. Xin, Y.J.; Wang, G.C.; Chen, L. Identification of Long-Range Transport Pathways and Potential Sources of PM10 in Tibetan Plateau Uplift Area: Case Study of Xining, China in 2014. *Aerosol. Air Qual. Res.* **2016**, *16*, 1044–1054. [CrossRef]
34. GB 3095-2012*Ambient Air Quality Standards of the People's Republic of China*; The Ministry of Environmental Protection of the People's Republic of China: Beijing, China, 2016. Available online: http://www.mee.gov.cn/ywgz/fgbz/bz/bzwb/dqhjbh/dqhjzlbz/201203/W020120410330232398521.pdf (accessed on 21 February 2021).
35. World Health Organization. WHO Global Air Quality Guidelines: Particulate Matter (PM2.5 and PM10), Ozone, Nitrogen Dioxide, Sulfur Dioxide and Carbon Monoxide. 2021. Available online: https://iris.who.int/handle/10665/345329. (accessed on 21 February 2021).
36. Freudenthaler, V.; Esselborn, M.; Wiegner, M.; Heese, B.; Tesche, M.; Ansmann, A. Depolarization ratio profiling at several wavelengths in pure Saharan dust during SAMUM 2006. *Tellus B Chem. Phys. Meteorol.* **2009**, *61*, 165–179. [CrossRef]
37. Zhao, H.; Che, H.; Wang, Y.; Dong, Y.; Ma, Y.; Li, X.; Hong, Y.; Yang, H.; Liu, Y.; Wang, Y.; et al. Aerosol vertical distribution and typical air pollution episodes over northeastern China during 2016 analyzed by ground-based lidar. *Aerosol. Air Qual. Res.* **2018**, *18*, 918–937. [CrossRef]
38. Liu, Y.; Zhao, H.; Ma, Y.; Yang, H.; Wang, Y.; Wang, H.; Zhang, Y.; Zou, X.; Wang, H.; Wen, R.; et al. Characteristics of particulate matter and meteorological conditions of a typical air-pollution episode in Shenyang, northeastern China, in winter 2017. *Atmos. Pollut. Res.* **2021**, *12*, 316–327. [CrossRef]
39. Zhao, H.; Che, H.; Zhang, L.; Gui, K.; Ma, Y.; Wang, Y.; Wang, H.; Zheng, Y.; Zhang, X. How aerosol transport from the North China plain contributes to air quality in northeast China. *Sci. Total Environ.* **2020**, *738*, 139555. [PubMed]
40. Fan, W.Z.; Qin, K.; Xu, J.; Yuan, L.M.; Li, D.; Jin, Z.; Zhang, K.F. Aerosol vertical distribution and sources estimation at a site of the Yangtze River Delta region of China. *Atmos. Res.* **2019**, *217*, 128–136. [CrossRef]
41. Tao, M.H.; Chen, L.F.; Li, R.; Wang, L.L.; Wang, J.; Wang, Z.F.; Tang, G.Q.; Tao, J.H. Spatial oscillation of the particle pollution in eastern China during winter: Implications for regional air quality and climate. *Atmos. Environ.* **2016**, *144*, 100–110. [CrossRef]
42. Fu, J.; Song, S.T.; Guo, L.; Chen, W.W.; Wang, P.; Duanmu, L.J.; Shang, Y.J.; Shi, B.W.; He, L.Y. Interprovincial Joint Prevention and Control of Open Straw Burning in Northeast China: Implications for Atmospheric Environment Management. *Remote Sens.* **2022**, *14*, 17. [CrossRef]
43. Miao, Y.C.; Liu, S.H. Linkages between aerosol pollution and planetary boundary layer structure in China. *Sci. Total Environ.* **2019**, *650*, 288–296. [CrossRef] [PubMed]
44. Kumar, M.; Singh, R.K.; Murari, V.; Singh, A.K.; Singh, R.S.; Banerjee, T. Fireworks induced particle pollution: A spatio-temporal analysis. *Atmos. Res.* **2016**, *180*, 78–91. [CrossRef]
45. Joly, A.; Smargiassi, A.; Kosatsky, T.; Fournier, M.; Dabek-Zlotorzynska, E.; Celo, V.; Mathieu, D.; Servranckx, R.; D'Amours, R.; Malo, A.; et al. Characterisation of particulate exposure during fireworks displays. *Atmos. Environ.* **2010**, *44*, 4325–4329. [CrossRef]
46. Kim, S.W.; Yoon, S.C.; Kim, J.; Kang, J.Y.; Sugimoto, N. Asian dust event observed in Seoul, Korea, during 29–31 May 2008: Analysis of transport and vertical distribution of dust particles from lidar and surface measurements. *Sci. Total Environ.* **2010**, *408*, 1707–1718. [CrossRef] [PubMed]
47. Fan, S.D.; Liu, C.; Xie, Z.Q.; Dong, Y.S.; Hu, Q.H.; Fan, G.Q.; Chen, Z.Y.; Zhang, T.S.; Duan, J.B.; Zhang, P.F.; et al. Scanning vertical distributions of typical aerosols along the Yangtze River using elastic lidar. *Sci. Total Environ.* **2018**, *628–629*, 631–641. [CrossRef] [PubMed]
48. Liu, J.J.; Zheng, Y.F.; Li, Z.Q.; Flynn, C.; Cribb, M. Seasonal variations of aerosol optical properties, vertical distribution and associated radiative effects in the Yangtze Delta region of China. *J. Geophys. Res. Atmos.* **2012**, *117*, 16. [CrossRef]
49. Sun, T.Z.; Che, H.Z.; Qi, B.; Wang, Y.Q.; Dong, Y.S.; Xia, X.G.; Wang, H.; Gui, K.; Zheng, Y.; Zhao, H.J.; et al. Characterization of vertical distribution and radiative forcing of ambient aerosol over the Yangtze River Delta during 2013–2015. *Sci. Total Environ.* **2019**, *650*, 1846–1857. [CrossRef] [PubMed]
50. Burton, S.P.; Ferrare, R.A.; Hostetler, C.A.; Hair, J.W.; Rogers, R.R.; Obland, M.D.; Butler, C.F.; Cook, A.L.; Harper, D.B.; Froyd, K.D. Aerosol classification using airborne High Spectral Resolution Lidar measurements—Methodology and examples. *Atmos. Meas. Tech.* **2012**, *5*, 73–98. [CrossRef]
51. Lu, X.; Hong, J.Y.; Zhang, L.; Cooper, O.R.; Schultz, M.G.; Xu, X.B.; Wang, T.; Gao, M.; Zhao, Y.H.; Zhang, Y.H. Severe Surface Ozone Pollution in China: A Global Perspective. *Environ. Sci. Technol. Lett.* **2018**, *5*, 487–494. [CrossRef]
52. Badarinath, K.V.S.; Kharol, S.K.; Sharma, A.R. Long-range transport of aerosols from agriculture crop residue burning in Indo-Gangetic Plains-A study using LIDAR, ground measurements and satellite data. *J. Atmos. Sol. Terr. Phys.* **2009**, *71*, 112–120. [CrossRef]

53. Jin, Y.N.; Andersson, H.; Zhang, S.Q. Air Pollution Control Policies in China: A Retrospective and Prospects. *Int. J. Environ. Res. Public Health* **2016**, *13*, 22. [CrossRef]
54. Liu, C.; Gao, M.; Hu, Q.H.; Brasseur, G.P.; Carmichael, G.R. Stereoscopic Monitoring A Promising Strategy to Advance Diagnostic and Prediction of Air Pollution. *Bull. Am. Meteorol. Soc.* **2021**, *102*, E730–E737. [CrossRef]
55. Colette, A.; Menut, L.; Haeffelin, M.; Morille, Y. Impact of the transport of aerosols from the free troposphere towards the boundary layer on the air quality in the Paris area. *Atmos. Environ.* **2008**, *42*, 390–402. [CrossRef]

remote sensing

MDPI

Article

Impact of Atmospheric Correction Methods Parametrization on Soil Organic Carbon Estimation Based on Hyperion Hyperspectral Data

Prajwal Mruthyunjaya [1,*], Amba Shetty [1], Pruthviraj Umesh [1] and Cécile Gomez [2,3]

1 Department of Water Resources and Ocean Engineering, National Institute of Technology Karnataka, Surathkal 575025, India
2 IRD, UMR LISAH (INRA-IRD-SupAgro), 34060 Montpellier, France
3 Indo-French Cell for Water Sciences, IRD, Indian Institute of Science, Bangalore 560012, India
* Correspondence: prajwal0021@gmail.com or prajwal.187am501@nitk.edu.in

Abstract: Visible Near infrared and Shortwave Infrared (VNIR/SWIR, 400–2500 nm) remote sensing data is becoming a tool for topsoil properties mapping, bringing spatial information for environmental modeling and land use management. These topsoil properties estimates are based on regression models, linking a key topsoil property to VNIR/SWIR reflectance data. Therefore, the regression model's performances depend on the quality of both topsoil property analysis (measured on laboratory over-ground soil samples) and Bottom-of-Atmosphere (BOA) VNIR/SWIR reflectance which are retrieved from Top-Of-Atmosphere radiance using atmospheric correction (AC) methods. This paper examines the sensitivity of soil organic carbon (SOC) estimation to BOA images depending on two parameters used in AC methods: aerosol optical depth (AOD) in the FLAASH (Fast Line-of-Sight Atmospheric Analysis of Spectral Hypercubes) method and water vapor (WV) in the ATCOR (ATmospheric CORrection) method. This work was based on Earth Observing-1 Hyperion Hyperspectral data acquired over a cultivated area in Australia in 2006. Hyperion radiance data were converted to BOA reflectance using seven values of AOD (from 0.2 to 1.4) and six values of WV (from 0.4 to 5 cm), in FLAASH and ATCOR, respectively. Then a Partial Least Squares regression (PLSR) model was built from each Hyperion BOA data to estimate SOC over bare soil pixels. This study demonstrated that the PLSR models were insensitive to the AOD variation used in the FLAASH method, with R^2_{cv} and $RMSE_{cv}$ of 0.79 and 0.4%, respectively. The PLSR models were slightly sensitive to the WV variation used in the ATCOR method, with R^2_{cv} ranging from 0.72 to 0.79 and $RMSE_{cv}$ ranging from 0.41 to 0.47. Regardless of the AOD values, the PLSR model based on the best parametrization of the ATCOR model provided similar SOC prediction accuracy to PLSR models using the FLAASH method. Variation in AOD using the FLAASH method did not impact the identification of bare soil pixels coverage which corresponded to 82.35% of the study area, while a variation in WV using the ATCOR method provided a variation of bare soil pixels coverage from 75.04 to 84.04%. Therefore, this work recommends (1) the use of the FLAASH AC method to provide BOA reflectance values from Earth Observing-1 Hyperion Hyperspectral data before SOC mapping or (2) a careful selection of the WV parameter when using ATCOR.

Keywords: Hyperion; hyperspectral imagery; atmospheric corrections; soil organic carbon; ATCOR; FLAASH; mapping

Citation: Mruthyunjaya, P.; Shetty, A.; Umesh, P.; Gomez, C. Impact of Atmospheric Correction Methods Parametrization on Soil Organic Carbon Estimation Based on Hyperion Hyperspectral Data. *Remote Sens.* **2022**, *14*, 5117. https://doi.org/10.3390/rs14205117

Academic Editors: Jing Wei, Xingfa Gu and Shuaiyi Shi

Received: 12 September 2022
Accepted: 10 October 2022
Published: 13 October 2022

1. Introduction

Soil is a heterogeneous composite material with a wide range of physical and chemical properties. It supports life and vegetation on earth and plays a central role in many of today's environmental challenges. A solid understanding of soil is required for improved soil management. Composition and properties such as clay, sand, silt, calcium carbonate, free iron, organic carbon, and pH in the spatial and temporal domains. Soil organic carbon

(SOC) is one of the major soil constituents because of its ability to influence plant growth as a source of energy and a trigger for nutrient availability via mineralization [1]. SOC is also important for the carbon cycle in terrestrial environments [2]. The determination of SOC content is also an obvious requirement to determine the rate and/or extent of Carbon soil sequestration. In agronomy, soil science, and climate change research, information on geographical and temporal fluctuations in SOC content is needed [3,4]. Traditional laboratory procedures for determining SOC content are time-consuming, expensive, and labor-intensive. The VNIR/SWIR spectroscopy (350–2500 µm) can be used to estimate the SOC in lab, field and from remote sensing conditions [5]. The soil reflectance spectra in the 1100–2500 nm range generally includes three distinct absorption peaks around 1400, 1900, and 2200 nm with a few small absorption peaks between 2200 and 2500 nm [6]. The effect of SOC containing biochemical constituents like chlorophyll, oil, cellulose, pectin, starch, lignin, and humic acids [7] is dominant in the visible and near infra-red portion of the spectrum (400–1100 nm), and some parts of the SWIR region (1100–2500 nm) [8], and soil reflectance generally decreases with organic matter content [9]. Simple Linear Regression [10], Principal Component Analysis [11], Partial Least Squares regression (PLSR) [12], and Artificial Neural Networks [13] are some of the statistical techniques used for SOC estimation. Out of these methods, PLSR is the most common method. VNIR/SWIR hyperspectral remote sensing data were becoming a potential tool to extend SOC mapping [14,15]. However, this technique is affected by the presence of green and dry vegetation [16,17], topsoil moisture [18] and atmospheric composition [19]. Concerning the latter, since the mid-1980s some Atmospheric Correction (AC) methods have been developed which were applied on VNIR/SWIR data obtained via airborne and spaceborne multispectral and hyperspectral sensors. AC methods are used to convert the Top-Of-Atmosphere radiance to the surface reflectance (also called Bottom-of-Atmosphere reflectance). By reducing external atmospheric components, reflectance can be improved. Empirical models, radiative transfer models, and hybrid models are the three categories of AC models. They have evolved from an empirical approach to rigorous radiative transfer modeling [20].

Empirical models mainly rely on image statistics to estimate surface reflectance. A few of the empirical methods are Internal Average Reflectance (IAR), Flat Field (FF), and Empirical line method [21]. Darkest Pixel principle (DP) is also an empirical method, where studies [22] indicated the best results for accurate surface reflectance estimation are achieved using a method that adds the effect of Rayleigh scattering to conventional dark object subtraction. For atmospheric correction and aerosol retrieval, semi-empirical approaches such as QUick atmospheric correction (QUAC) [23] are used. Radiative transfer methods simulate the absorption and scattering effects of atmospheric components. To mimic the atmosphere, several algorithms have been created, notably ATREM (ATmospheric REMoval algorithm), ATCOR (ATmospheric CORrection [24], FLAASH (Fast Line-of-sight Atmospheric Analysis of Spectral Hypercubes [25], ISDAS (Imaging Spectrometer Data Analysis System [26], HATCH (High-accuracy ATmosphere Correction for Hyperspectral data [27], and ACORN (Atmospheric CORrection Now) [28]. For the derivation of surface reflectance from hyperspectral imaging data, hybrid models commonly combine radiative modeling with empirical methodologies. With physical models, Gordon [29] has summarized the evolution of Ocean color Atmospheric corrections right from its early years to the current scenarios. Other physical models are also developed and used for various applications [30,31]

Some studies compared performances of AC models for identifying tree species [32], mineral mapping [33], littoral environments studies [34], and SOC content mapping [35]. Beyond the type of AC methods and the choice of the AC method, the atmospheric parameters (e.g., Aerosol Optical Depth, Water vapor, Ozone, Visibility) used in AC methods might be uncertain and the choice of these atmospheric parameters may impact the reflectance quality. Only a few studies focused on an evaluation of the impacts of these atmospheric parameters on reflectance quality such as ozone, water vapor, Rayleigh scattering, aerosol scattering on the spectral reflectance of NOAA-AVHRR and derived NDVI data [36,37].

Mannschatz et al. [38] studied sensitivity of LAI estimation from satellite imaging to atmospheric correction using ATCOR, and highlighted that the ATCOR parameter 'visibility' has the strongest influence on LAI estimation. Griffin et al. [39] studied the sensitivity of Visibility, Aerosol model type, Atmospheric model and solar zenith on surface reflectance using AVIRIS data in FLAASH and ATREM and concluded the sensitivity to input visibility generally varied between +0.03 for both AC models. Davaadorj. A. [40] studied the sensitivity of different aerosol models using Worldview 3 data with FLAASH and ATCOR AC and concluded that maritime aerosol has the least sensitivity compared to rural and urban models. The variation of surface reflectance caused by different surface types such as sand, artificial turf, grass, bright and dark targets with respect to AOD was studied by Nazer et al. and Bassani et al. [41,42].

In this context, the aim of the study is to evaluate the uncertainty of SOC prediction due to two atmospheric parameters used in AC methods: the water vapor (WV) using ATCOR and the Aerosol Optical Depth (AOD) using FLAASH. A Hyperspectral Hyperion image was corrected by ATCOR and FLAASH AC methods to convert the at-sensor radiance spectrum to the ground reflectance spectrum, with a feasible range of WV and AOD values, respectively. And the PLSR method was selected as the classical chemometric method for SOC content estimation based on each corrected Hyperion reflectance data. The data and methods used are presented in Section 2 and the results in Section 3. Finally, the results are discussed in Section 4.

2. Materials and Methods

2.1. Study Area

The Narrabri site is in Australia's northwestern state of New South Wales (NSW) (Figure 1a). The study area is divided into two subsites: the cotton fields of Namoi (referred to as Narrabri site#1 henceforth) (149°37′E, 30°10′S), which cover 70 km^2, and the pastures in the town of Narrabri (referred to as Narrabri site#2), which covers 69 km^2 (Figure 1b). The Australian fields are extensive and continuous (approximately 500 m × 900 m). Vertisols predominate in this area. The terrain is flat, with plane regions rising to an average elevation of 210 m above sea level.

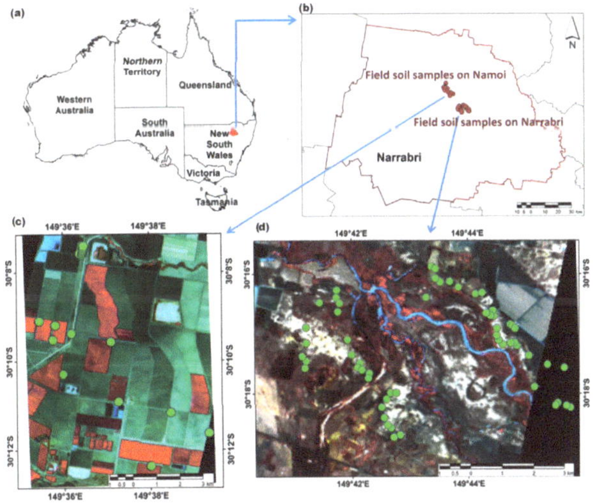

Figure 1. Location of the study area (**a**) in Australia, (**b**) on the Narrabri region; hyperion image acquired over (**c**) the site#1 and (**d**) site#2 with sampling points plotted in red dots.

2.2. Raw Hyperion Data

The EO-1 satellite's Hyperion sensor monitors radiation from 400 to 2500 nm in 242 spectral bands with a spectral resolution of about 10 nm and a spatial resolution of 30 m. The Hyperion images have a sweep width of 7.6 km. Folkman et al. [43] provided a full description of Hyperion's characteristics, operations, and uses.

Two Hyperion cloud-free pictures were obtained over the Narrabri study area in this study: the first on 13 December 2006 at 23:51 UT over Narrabri pasture soils (site#2), and the second on 17 January 2007 at 23:47 UT over cropping soils (site#1) and traveling stock routes of Namoi. The Hyperion images were georeferenced Hyperion images in L1T format (Figure 1c,d).

2.3. Soil Sampling

Over the two Narrabri locations, a total of 98 soil samples were collected (Figure 1c,d) from researchers of The University of Sydney in Australia [12]. It includes 46 samples from Narrabri site#1's cropping soils and stock routes collected in October 2006 (during the same semester as Hyperion data) (Figure 1c) and 52 samples from Narrabri site#2's pastures collected in December 2006 (during the same month as Hyperion data) (Figure 1d). During the field campaign, soil samples were gathered from bare soil areas. Because there was no organic matter added or consumed in the field due to the absence of vegetation during this period at the sampling locations, SOC content may be deemed steady between sampling and picture collection. Climate change, particularly changes in temperature and rainfall, have an impact on SOC by hastening its decomposition. However, it was reliably concluded that SOC was very consistent between measurements in Narrabri site#1 during a short period of three months from October 2006 to January 2007. Soil samples were taken at a depth of 0 to 10 cm. This depth corresponds to the ploughed soil horizon, and the Hyperion measurement may be deemed indicative of this tested depth because the soil samples were gathered on cultivated areas. Soil samples are the average of sub-samples obtained at the center and four corners of a 20 × 20 m area's perimeter.

2.4. SOC Laboratory Measurements

Laboratory mid-infrared (MIR) spectroscopy was used to determine the SOC content of soil samples. Samples were crushed to 200 µm for MIR analysis and analyzed as neat powders. Each soil sample's MIR spectral reflectance was measured using a Bruker Optics Tensor 37 Fourier Transform Infrared spectrometer with an 8 cm-1 resolution and 64 scans per second in the range 2500 to 25,000 nm. PLSR from MIR calibrations was used to predict the SOC content of each soil sample [12]. The calibration used 13 factors, and the model's root mean squared error in the test set validation was 0.15 dag/kg, with an R^2 of 0.91 [12]. The SOC ranges from 0.002% to 3.6%, with a mean of 1.56%, and a standard deviation of 0.97%.

2.5. Atmospheric Parameters

2.5.1. Water Vapor

Water vapor (WV) absorption has a strong effect within specific absorption bands, but spectral bands for land surface observations are typically designed to avoid strong absorption lines. Hence water vapor absorption must be accurately corrected in the selected bands. From a joint project between the National Center for Environmental Prediction (NCEP, formerly "NMC") and the National Center for Atmospheric Research (NCAR) [44], the total column WV over the study area ranged between 0.9 cm to 3.7 cm in the last 15 years (from 2000 to 2015, Figure 2, Table 1). These WV values were estimated by reanalysis models built from historic meteorological data collected over 15 years.

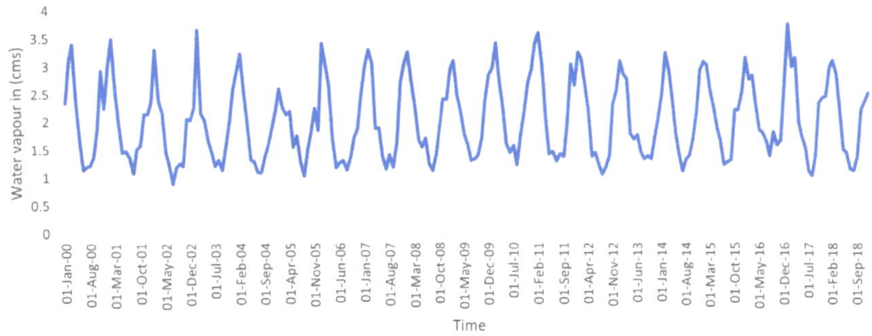

Figure 2. The WV distribution over Narrabri over 15 years estimated by the joint project between the National Centers for Environmental Prediction (NCEP, formerly "NMC") and the National Center for Atmospheric Research (NCAR).

Table 1. WV and AOD estimations, estimated by the joint project between the National Centers for Environmental Prediction (NCEP, formerly "NMC"), the National Center for Atmospheric Research (NCAR), and the atmosphere global product based on Modis MOD08_M3 V6, respectively.

	Water Vapor (WV) (in cm)	**Aerosol Optical Depth (AOD)**
Minimum	0.9	0.02
Maximum	3.7	1.50
Mean	2	0.48
Median	1.9	0.42
Standard Deviation	0.7	0.25

At the date of the Hyperion images acquisition, the total column WV value would be estimated between 2.8 cm and 3.1 cm. Finally, considering the "Mid Latitude Summer" climatic condition over the study area, the default WV value usually selected in the ATCOR model would be 2–3 cm (Table 3, [12]). The Hyperion data were corrected from atmospheric effects using six WV values, preselected by the ATCOR software: 0.4, 1, 2, 2.9, 4, and 5 cm.

2.5.2. Aerosol Optical Depth

Molecular scattering and absorption can be adequately explained, but not the optical properties of aerosols for the portion of the spectrum dominated by scattering processes rather than thermal emission. Kaufman [45] demonstrated how aerosols reduce apparent spatial resolution, alter the apparent spectral characteristics, resulting in a loss of classification accuracy. From MODIS MOD08_M3 V6, which is an atmosphere global product that contains monthly 1 × 1-degree grid average values of atmospheric parameters, the AOD values over the study area ranged from 0.02 to 1.5 over the last 15 years (from 2000 to 2015, Figure 3, Table 1) [46].

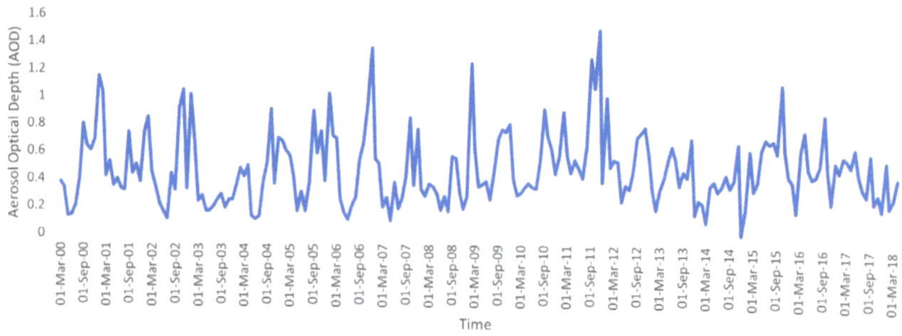

Figure 3. The aerosol optical depth distribution over Narrabri over 15 years estimated by the atmosphere global product based on Modis MOD08_M3 V6.

At the date of Hyperion images acquisition, over the study area, the AOD value would be estimated between 0.25 and 0.26. Finally, considering the "Mid Latitude Summer" climatic condition over the study area, the default AOD value in FLAASH would be 1. The Hyperion data were corrected from atmospheric effects using seven AOD values in FLAASH: 0.2, 0.4, 0.6, 0.8, 1, 1.2, and 1.4.

2.6. Methodology

The FLAASH and ATCOR atmospheric correction was performed in ENVI and ATCOR software, respectively. Bands selection and bare soil pixels selection were performed in QGIS, an open-source GIS software. The regression model, including the calibration data preparation, the model building, and the evaluation of the performances were performed in Matlab(R). Historical AOD and WV data were extracted using the Google Earth Engine platform.

2.7. Atmospheric Correction Models

2.7.1. ATCOR Atmospheric Correction Model

ATCOR uses the MODTRAN5 radiative transfer model [24] with the newest high-resolution transmission molecular absorption (HITRAN) database for atmospheric correction. MODTRAN5 is an updated version with greater spectral resolution and supplementary atmospheric gas treatment [47,48]. In the 400–2500 nm range, ATCOR uses a variable wavenumber grid to achieve a constant spectral sample distance of 0.4 nm [24]. In this study, the band model of MODTRAN was used.

For each image, ATCOR assumes a Lambertian surface and requires a sensor type, flight date, solar zenith angle, sensor view geometry, average ground height, and adjacency range. The initial visibility estimate was 30 km. The spectral specifications for the sensor Hyperion are stored in a calibration file called "hyperion 167.cal" provided by ATCOR. There is an atmospheric database, which consists of look-up tables for radiative transfer calculations that span a wide range of weather situations and sun angles. The user has the option of using the standard MODTRAN model for atmosphere and aerosol types to depict the scene, and each image receives its own MODTRAN solution.

In this study, both Hyperion images were corrected from atmospheric effects with the ATCOR model (Figure 4) using the six WV values, preselected by the ATCOR software: 0.4, 1, 2, 2.9, 4, and 5 cm. The default values of other parameters were selected for running the ATCOR model as described in Table 2. Six atmospherically corrected Hyperion images were created for both sites (site#1 and site#2).

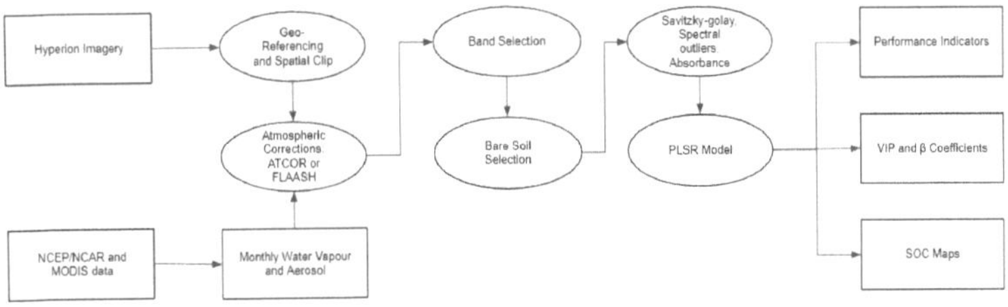

Figure 4. Process flow.

Table 2. ATCOR parameters.

Parameter	Selected Value
Atmospheric Model	Mid-Latitude Summer
Adjacency correction	No
Aerosol Model	Rural
Visibility	30 km
Region for water vapor retrieval	820 nm
Spectral polishing	No
CO_2	390 ppm
Water vapor	Ranges from 0.9 cm to 3.7 cm

2.7.2. Fast Line-of-Sight Atmospheric Analysis of Spectral Hypercubus (FLAASH) Atmospheric Correction Model

FLAASH is an atmospheric adjustment approach based on physics [49]. It calculates the parameters needed for RT equations using the MODTRAN4 code, which results in surface reflectance. A scaled DISORT (DIScrete Ordinate Radiative Transfer) algorithm is utilized to calculate atmospheric multiple scattering, and the correlated k technique is employed to represent absorption for places with significant absorption effects. FLAASH, with a resolution of 5 cm^{-1}, results in SSD ranging from 0.08 to 3.12 nm depending on wavelength in the 400–2500 nm range.

Hyperspectral sensors, such as HyMAP, AVIRIS, HYDICE, Hyperion, Probe-1, CASI, and AISA, and multispectral sensors, such as HyMAP, AVIRIS, HYDICE, Hyperion, Probe-1, CASI, and AISA), are supported by FLAASH (such as Landsat, SPOT, IRS, and ASTER). Retrieval of water vapor and aerosol are only possible when the image comprises bands with incorrect wavelength locations. A method for extracting an estimated aerosol/haze amount from chosen dark land pixels in the picture is included in the FLAASH model.

In this study, both Hyperion images were corrected from atmospheric effects with the FLAASH model (Figure 4) using the seven AOD values selected based on the Modis MOD08_M3 V6 product (Section 2.5.2): 0.2, 0.4, 0.6, 0.8, 0.1, 0.12, and 0.14. The default values of other parameters were selected for running the FLAASH model as described in Table 3. Seven atmospherically corrected Hyperion images were created for both sites (site#1 and site#2).

Table 3. FLAASH parameters.

Parameter	Selected Value
Atmospheric Model	Mid-Latitude Summer
Adjacency correction	No
Aerosol Model	Rural
Visibility	30 kms
Region for water vapor retrieval	820 nm
Spectral polishing	No
CO_2	390 ppm
Aerosol optical depth	Ranges from 0.02 to 0.15

2.8. Bands Selection

The HYPERION spectral bands with a very low SNR and those located in the atmospheric absorption bands are removed, as previously done by Gomez et al. [12], Lu et al. [50], and Minu et al. [35]. A total of 122 spectral bands, belonging to spectral ranges from 467 to 854 nm, 1063 to 1336 nm, 1548 to 1790 nm, and 1981 to 2355 nm, were finally kept to estimate SOC content from PLSR models. Different degrees of vertical stripping were indicated by visual observation of individual reflectance bands. It was discovered that excluding certain bands from the study was beneficial [50,51]. This is done after FLAASH or ATCOR has applied atmospheric adjustment (Figure 4). Bands surrounding large atmospheric absorption areas (e.g., at the 950, 1400, 1900, and 2500 nm water absorption bands) that appear as over/under-corrected reflectance values were also removed from the reflectance spectrum. After using the FLAASH and ATCOR algorithms, a few bands with negative or zero reflectance values near the 450 nm spectral area that were not actual reflectance values were also removed.

Finally, a total of 122 spectral bands were used in both ATCOR and FLAASH and then used to estimate SOC content from PLSR models. This is in accordance with the bands used by Minu et al. [35]. These 122 spectral bands belong to the spectral ranges from 467 to 854, 1063 to 1336, 1548 to 1790, and 1981 to 2355 nm.

2.9. Bare Soil Pixels Selection

The Normalized Differential Vegetative Index (NDVI) was used to identify and mask green vegetation pixels [52], based on the following equation:

$$\mathrm{NDVI} = (\mathrm{NIR} - \mathrm{Red})/(\mathrm{NIR} + \mathrm{Red}) \quad (1)$$

where NIR and Red are the reflectance measured at 834 nm and 661 nm, respectively. Pixels with NDVI values over 0.2 were masked.

The Modified Normalized Difference Water Index (MNDWI) was used to identify and mask water pixels. The MNDWI is calculated for the enhancement of open water features [53], as follows:

$$\mathrm{MNDWI} = (\mathrm{Green} - \mathrm{SWIR}_1))/(\mathrm{Green} + \mathrm{SWIR}_1) \quad (2)$$

where Green and SWIR_1 are the reflectance measured at 559 nm and 1609 nm, respectively. It also diminishes built-up area features that are often correlated with open water in other indices. Pixels with MNDWI values greater than 0 were masked.

The Normalized Burned Ratio 2 (NBR2) index was used to mask the dry vegetation [54] as follows:

$$\mathrm{NBR2} = (\mathrm{SWIR}_1 - \mathrm{SWIR}_2)/(\mathrm{SWIR}_1 + \mathrm{SWIR}_2) \quad (3)$$

where SWIR_1 and SWIR_2 are the reflectance measured at 1609 nm and 2193 nm, respectively. Pixels with an NBR2 value greater than 0.2 were masked [54].

2.10. Regression Model

2.10.1. Data Preparation

Before performing quantitative statistical analysis, a noise reduction was achieved through standard pre-treatments: a Savitzky–Golay filter with third-order polynomial smoothing and window widths of 61 nm [55], mean centering and variance scaling. A generalised moving average filter, the Savitzky Golay filter [55], is also known as a digital smoothing polynomial filter or least-squares smoothing filter. The filter coefficients are calculated by fitting a polynomial of a particular degree with an unweighted linear least-squares fit. It requires equidistant bandwidth and fits a local polynomial regression on the signal. It is feasible to get a high amount of smoothing without attenuating data features using a higher degree polynomial. Mathematically, it's nothing more than a weighted sum of nearby values. The Hyperion spectra were filtered with a third order Savitzky Golay filter.

$$\overline{x}_i = \frac{\sum_{j=1}^{n} Xij}{n} \tag{4}$$

Standard normal variate (SNV) calculates the standard deviation of all the pooled variables for the given sample [56]. The entire sample is then normalized by this value, thus giving the sample a unit standard deviation (s = 1).

$$w_i = \sqrt{\frac{\sum_{j=1}^{n} (Xij - \overline{x}_i)2}{n-1}} + \delta^{-1} \tag{5}$$

where n is the number of variables, x_{ij} is the value of the jth variable for the ith sample, and δ is a user-definable offset. The user-definable offset can be used to avoid over-normalizing samples which have near-zero standard deviation. The default value for this offset is zero, indicating that samples will be normalized by their unweighted standard deviation. The selection of δ is dependent on the scale of the variables. A setting near the expected noise level (in the variables' units) is a good approximation.

Outliers are observations that differ significantly from normal values and are not compatible with the rest of the data [57,58]. Spectra can be defined as a spectral outlier when the sample is spectrally different from the rest of the samples. Out of the 98 samples for which SOC values are present, spectral outliers were identified using Mahalanobis distance [59]. This class of methods only uses distance space to flag outlier observations. The Mahalanobis distance (MD) for the ith observation is given by:

$$\begin{array}{c} MD_i = \sqrt{(x_i - \overline{x})tC^{-1} \, (x_i - \overline{x})} \\ C = \frac{1}{n-1} X'X \end{array} \tag{6}$$

where X is the data matrix of size n × p, where p is the number of variables and n is the number of observations. x_i is an observation (a row of X), $\overline{x}$ is the mean vector, and C is the sample covariance matrix which gives information about the covariance structure of the data. In the analysis, Mahalanobis distance cutoff was set to 3.

2.10.2. PLSR with Leave-One-Out Cross-Validation (LOOCV)

In this study, Partial Least Squares Regression (PLSR) is carried out to find predictive models for soil organic carbon data and discusses the effectiveness of these methods in predicting soil carbon using spaceborne Hyperspectral data. Here, the PLSR method is used to regress a relationship between response variables to many predictor variables [60]. Care should be taken whilst choosing the number of latent variables. Using many components will lead to a good fitting model but is a strategy that leads to overfitting. In this case, cross-validation, which is a widely used method, was used in choosing the optimal number of variables. The maximum number of latent variables was defined as 10, then the optimal number of latent variables was determined using the prediction residual error sum of squares (PRESS) analysis. This allows avoiding under- and over-fitting.

Cross validations also avoid overfitting by not using the same data to both fit a model and to estimate prediction error. In this study, 10-fold cross-validation was used. Here the original data was divided into subsets or folds, which in this case is 10. One of the folds is not used in the model building and is used to estimate the error. This process is replicated for all the folds for multiple iterations. This leads to a minimized prediction error for different numbers of latent variables. Hence there is no fixed validation dataset.

2.10.3. Model Evaluation

The coefficient of determination (R^2_{cv}) is the percent of the total variation in the response variable that is explained by the regression line [61]. The R^2_{cv} was obtained for the cross-validation dataset as follows:

$$R^2 = 1 - \frac{SSE}{SST} \quad (7)$$

where SSE is the sum squared differences between the predicted and observed value, and SST is the sum of squared differences between the observed and overall mean value.

The root means squared error ($RMSE_{cv}$) is the average prediction error (square root of mean squared error) [62].

$$RMSE = \sqrt{\frac{\sum_{i=1}^{n}(y_i - \hat{y}_i)2}{n}} \quad (8)$$

RPD is the ratio of the standard deviation (SD) of analyzed data to $RMSE_{cv}$. It is given by the following equation:

$$RPD = \frac{SD}{RMSE} \quad (9)$$

Chang and Laird [63] defined three classes of RPD: category A ($RPD > 2$) describes models that can accurately predict the soil property, category B ($2 > RPD > 1.4$) describes models with limited predictive power, and category C ($RPD < 1.4$) describes models that have no prediction ability.

The ratio of performance to interquartile ($RPIQ$) value is more objective than RPD [64] as it is based on quartiles, which better represents the spread of the population. The quartiles are milestones in the population range: $Q1$ is the value below which 25% of the samples are found; $Q3$ is the value below which 75% of the samples are found; and $Q2$, commonly called the median, is the value under which 50% of samples are found. $RPIQ$ is the ratio of IQ to $RMSE$, where IQ is the difference between the third quartile $Q3$ and the first quartile $Q1$. A larger $RPIQ$ value indicates improved model performance. The formula is shown as follows:

$$\begin{aligned} IQ &= Q3 - Q1 \\ RPIQ &= \tfrac{IQ}{RMSE} \end{aligned} \quad (10)$$

The Variable Importance in Projections (VIP) is a weighted sum of squares of PLS weights, with weights generated from each PLS component's Y-Variance [65,66]. Multiplying the VIP and the absolute value of PLS regression coefficients from the PLSR model yields the wavelengths important for modeling. Each PLSR model was examined for a higher value for the product of absolute regression coefficient and variable importance for projection (VIP) to identify significant wavelengths in determining SOC content from spectral reflectance data rather than the full spectrum.

2.10.4. SOC Mapping

The calibrated PLSR models were applied to bare soil pixels. This would produce an estimated SOC map for each model. In case of a negative estimated SOC value, this value was replaced by 0 g/kg. So, a total of 13 SOC maps were obtained: six from the Hyperion images corrected by FLAASH and different AOD values and seven from the Hyperion images corrected by ATCOR and different WV values. A map of the standard deviation of the estimated SOC along the six SOC maps obtained from the Hyperion images corrected by FLAASH was produced. A map of the standard deviation of the estimated SOC along the seven SOC maps obtained from the Hyperion images corrected by ATCOR was produced.

3. Results

3.1. Bare Soil Coverage Analysis

Similar histograms of NDVI, MNDWI and NBR2 values were obtained whatever be the AOD value used in the FLAASH model to correct the Hyperion images (Figure 5(A1,B1,C1)). Moreover, a similar bare soil coverage of 82.35% was obtained, whatever be the used AOD value (Table 4). So, a variation of the AOD value used in the FLAASH model did not impact the reflectance at the spectral bands used to calculate the NDVI, MNDWI, and NBR2 spectral indexes (Equations (1)–(3)).

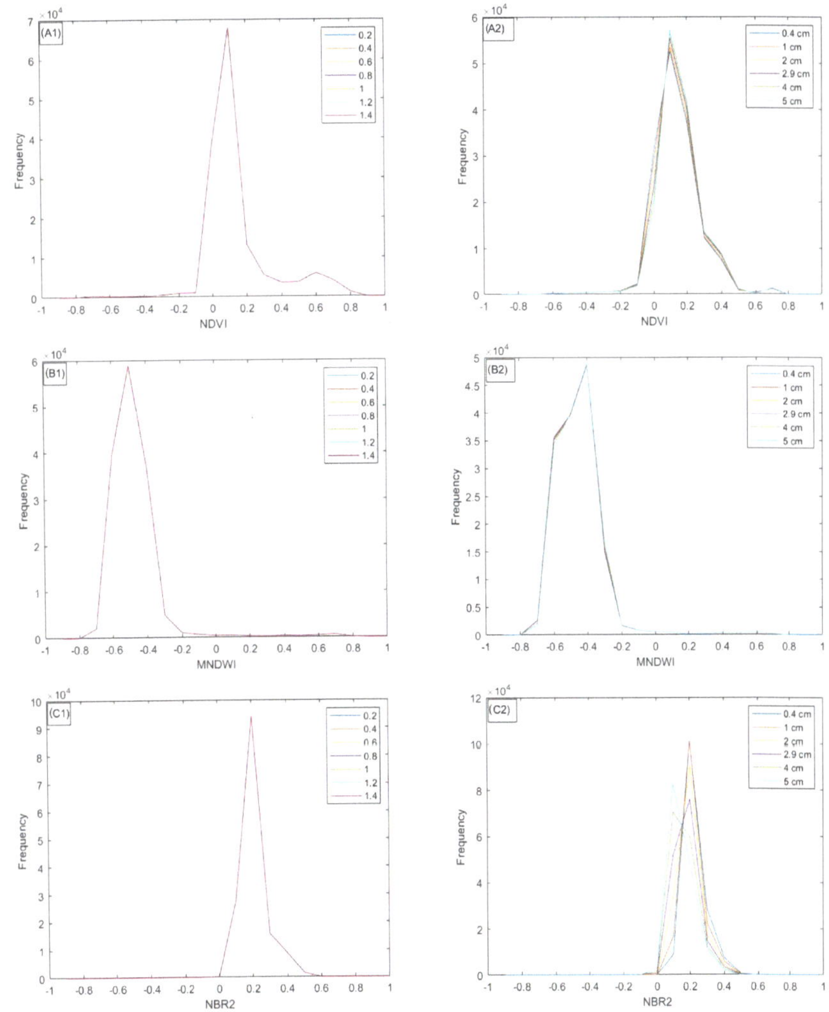

Figure 5. Histograms of NDVI (overall the area) obtained from the Hyperion images corrected with several (**A1**) AOD and (**A2**) WV; Histograms of MNDWI (overall the area) obtained from the Hyperion images corrected with several (**B1**) AOD and (**B2**) WV; Histograms of NBR2 (overall the area) obtained from the Hyperion images corrected with several (**C1**) AOD and (**C2**) WV.

Table 4. Figures of merit obtained from PLSR models, based on Hyperion data corrected by the FLAASH method, and depending on 7 values of AOD.

AOD	0.2	0.4	0.6	0.8	1	1.2	1.4
Bare soil coverage (%)	82.35	82.35	82.35	82.35	82.35	82.35	82.35
R^2_{cv}	0.79	0.79	0.79	0.79	0.79	0.79	0.79
$RMSE_{cv}$ (%)	0.40	0.41	0.40	0.40	0.40	0.40	0.40
RPD	2.23	2.21	2.22	2.23	2.23	2.23	2.23
RPIQ	3.26	3.24	3.25	3.26	3.27	3.26	3.26
bias (%)	0.32	0.33	0.33	0.33	0.33	0.33	0.33
Number of Latent Variables	2	2	2	2	4	2	2

By contrast, the selection of the WV value used in the ATCOR model to correct the Hyperion images appeared to impact marginally the spectral bands used to calculate the NDVI, MNDWI, and NBR2 spectral indexes (Equations (1)–(3)). Histograms of NDVI and MNDWI values displayed the same shapes irrespective of the WV value used in the ATCOR model, with the same modes at 0.1 and −0.4 for NDVI and MNDWI, respectively (Figure 5(A2,B2)). Histograms of NBR2 values displayed the same shapes with a slight shift of modes (Figure 5(C2)). Consequently, the bare soil coverage varied depending on the used WV value from 75 % (obtained with a WV of 0.4) to 84 % (obtained with a WV of 2) (Table 5).

Table 5. Figures of merit obtained from PLSR models, based on Hyperion data corrected by the ATCOR method, and depending on 6 values of WV. The best performances are highlighted in bold.

Water Vapor (in cm)	**0.4**	**1**	**2**	**2.9**	**4**	**5**
Bare soil coverage (%)	75.04	80.54	84.04	83.50	83.17	82.85
R^2_{cv}	0.75	0.78	0.79	0.75	0.72	0.72
$RMSE_{cv}$ (%)	0.44	0.41	0.41	0.44	0.47	0.46
RPD	2.04	2.17	2.19	2.03	1.91	1.93
RPIQ	2.98	3.17	3.21	2.97	2.80	2.83
bias (%)	0.35	0.33	0.31	0.34	0.38	0.35
Number of Latent Variables	5	2	3	4	2	6

For the same pixel, it is observed that there is no variation of the spectrum along with the AOD range, but variations can be observed of the spectrum along with the WV range (Figure 6). No absorption bands of clay or carbonate are observed in the spectrum with FLAASH AC. Meanwhile, for the spectrum with ATCOR AC, the higher the WV the higher the albedo of the spectra.

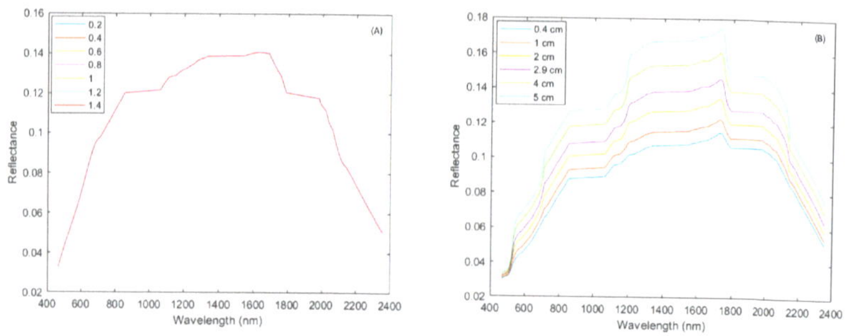

Figure 6. Spectral signature of the same bare soil pixel corrected by (**A**) FLAASH with varying AOD levels and (**B**) ATCOR with varying WV levels.

3.2. SOC Prediction Models Performances Using Hyperion Data Corrected by FLAASH

A PLSR model was built from the Hyperion spectra corrected by the FLAASH model and from SOC associated with the available soil samples. Four spectral outliers were removed from the calibration database, regardless of the study area, and the number of latent variables were determined following the rule of the first local minimum of the $RMSE_{cv}$ and varied between 2 and 4 (Table 4). The performance of the PLSR models was correct, with $R^2{}_{cv}$ of 0.79, $RMSE_{cv}$ of 0.40 and RPD of 2.2 whatever be the AOD values (Table 4). So, the selection of the AOD for correcting the Hyperion images using the FLAASH model did not affect the model's performance for SOC prediction.

3.3. SOC Prediction Models Performances Using Hyperion Data Corrected by ATCOR

A PLSR model was built from the Hyperion spectra corrected by the ATCOR model and from SOC associated with the available soil samples. Four spectral outliers were removed from the calibration database, regardless of the study area, and the number of latent variables were determined following the rule of the first local minimum of the RMSE_{cv}, which varied between 2 and 6 (Table 5). The performances of the PLSR models were correct, with $R^2{}_{cv}$ from 0.72 to 0.79, $RMSE_{cv}$ from 0.41 to 0.47 and RPD from 1.9 to 2.2 (Table 5).

The best performance was obtained from the model based on spectra corrected by the ATCOR model and WV of 2 cm (Table 5). The lowest performance was obtained from the model based on spectra corrected by the ATCOR model and WV of 4 cm (Table 5).

3.4. Significant Wavelengths for SOC Estimation

Between 41 to 43 wavelengths were identified as significant in the PLSR models built from the 6 Hyperion images corrected by FLAASH and different AOD values. 41 of these significant wavelengths are common between the six PLSR models. 48% of the significant wavelengths belonged to the VNIR (400–1100 nm) and 42% of the significant wavelengths belonged to the spectral domain from 1100–2000 nm (Figure 7A). In the region, 2000–2400 nm fewer significant wavelengths (14%) were found corresponding to some minerals (Carbonates and Clay) (Figure 7A).

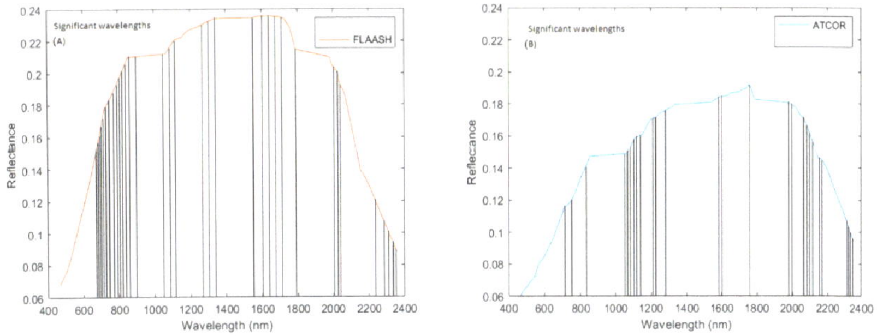

Figure 7. Spectral signature of the same bare soil pixel corrected by (**A**) FLAASH for an AOD level of 80 and the significant wavelengths for the associated PLSR model (highlighted by gray vertical lines); (**B**) ATCOR for an WV of 2 cm and the significant wavelengths for the associated PLSR model (highlighted by gray vertical lines).

Between 28 to 42 wavelengths were identified as significant in the PLSR models built from the two Hyperion images corrected by ATCOR and different WV values, and only 1 of these significant wavelengths are common between the thirteen PLSR models. Most of the significant wavelengths (36%) belonged to the spectral domain from 1100 to 2000 nm (Figure 7B). 33% of significant wavelengths belonged to the spectral domain from 2000–2500 nm (Figure 7B).

3.5. SOC Maps Using Hyperion Data Corrected by FLAASH

PLSR models built from Hyperion images corrected with different AOD values using the FLAASH model were applied to the bare soil pixels to produce SOC maps over both sites. The percentage of negative predictive SOC values over both sites varied from 2.64 to 6.31% depending on the AOD value used to correct the Hyperion images with the FLAASH model (Table 6). The percentage of negative predictive SOC increases from an AOD of 0.4 to AOD 1.4 (Table 6) while the highest percentage was observed with an AOD of 0.2 (Table 6).

Table 6. Percentage of negative SOC estimated using Hyperion data corrected by FLAASH, from both the sites.

AOD	Negative SOC Pixels %
0.2	6.31
0.4	1.93
0.6	2.64
0.8	2.65
1.0	3.27
1.2	4.98
1.4	5.50

The distributions of SOC contents estimated by PLSR over both sites varied slightly when using an AOD value from 0.4 to 1.4 with a mode around 1.8%, while the distribution of SOC contents estimated by PLSR when using an AOD value of 0.2 has a mode around 1.6% (Figure 8). The SOC map using AOD 0.2 has a greater number of pixels which have SOC of greater than 3%. This is not observed in the other SOC maps. The SOC prediction map with AOD 80 is visualized in the Figure 9.

Variation in the SOC content of the pixels adjacent and in between vegetation pixels can also be observed. This can be clearly seen in the Standard Deviation map (Figure 10). A higher standard deviation along the riverbed in site#2 is observed. This may be due to no bare soil and a mixture of water, green, and dry vegetation. High standard deviation is observed in areas that have negative values with AOD = 0.2 and AOD > 1.0.

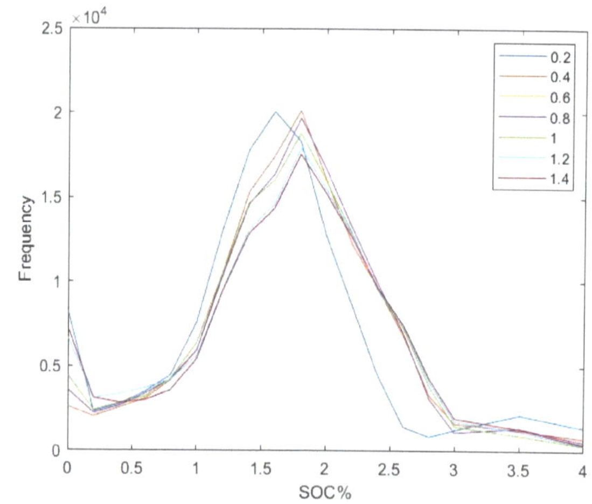

Figure 8. Histogram of the estimated SOC content from both the sites with varying AOD.

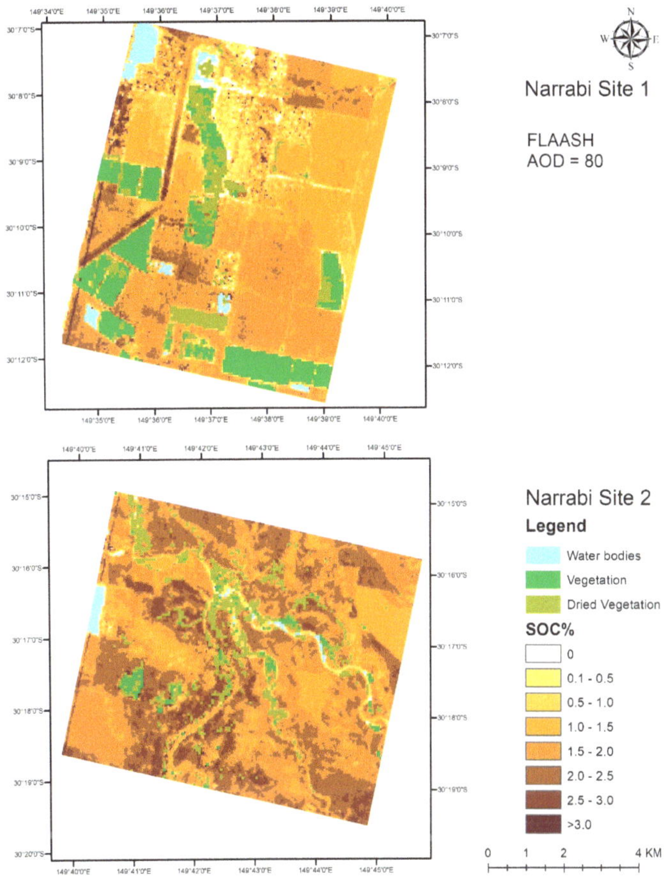

Figure 9. SOC prediction maps using an AOD of 80 with FLAASH method.

It is observed that there is a slight deviation in the spectral signatures with varying AOD in the VNIR region of 400 to 1100 nm and no deviation in the rest of the spectrum up to 2400 nm (Figure 6). The percentage of bare soil pixel is very consistent 82.35% and good consistency is observed for the significant wavelengths for the prediction model with FLAASH atmospheric correction (Table 4). Hence due to these observations, consistency can be observed in the model performance across varying AOD levels.

The range of AOD in the Narrabri region varies from 0.02 to 1.5 as per Table 1. Comparing the SOC maps obtained by varying the AOD level with the exception of AOD level 0.2, good consistency prediction from AOD 0.4 to AOD 1.4 is observed. This is also confirmed by the histogram as per (Figure 8). The standard deviation map of SOC variation is fairly consistent with high variation at a few places. These places match with farm boundaries indicating inconsistencies at a farm level.

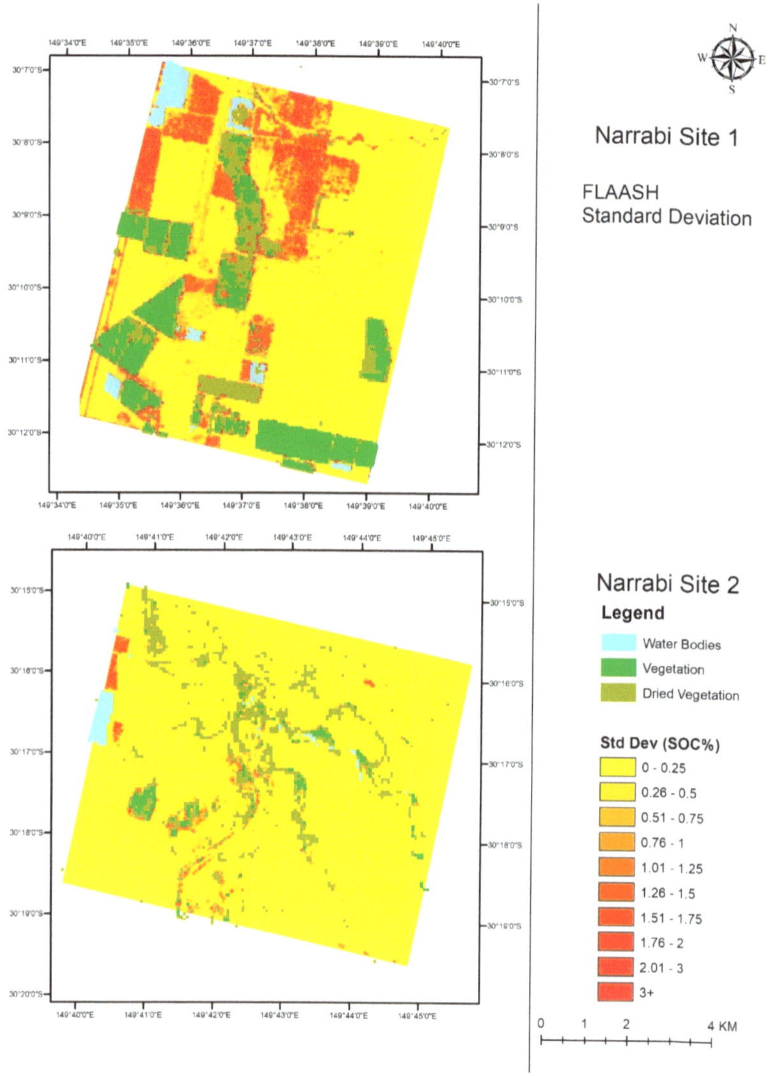

Figure 10. Standard deviation map of estimated SOC obtained along the seven AOD values used with FLAASH method.

3.6. SOC Maps Using Hyperion Data Corrected by ATCOR

PLSR models built from Hyperion images corrected with different WV values using the ATCOR model were applied to the bare soil pixels to produce SOC maps over both sites.

The percentage of negative predictive SOC values over both areas varied from 1.78% (obtained with a WV of 5 cm) to 5.51% (obtained with a WV of 0.4 cm) depending on the WV value used to correct the Hyperion images with the ATCOR model (Table 7).

Table 7. Percentage of negative SOC estimated using Hyperion data corrected by ATCOR, from both the sites.

Water Vapor (in cms)	Negative SOC Pixels %
0.4	5.51
1.0	4.60
2.0	4.63
2.9	3.22
4.0	4.53
5.0	1.78

The distributions of SOC contents estimated by PLSR over both sites varied significantly when using a WV value from 0.4 to 5 cm (Figure 11), with different modes from 1.2% (obtained with a WV of 2 cm) to 1.8% of SOC (obtained with a WV of 0.4 cm). The PLSR model built from Hyperion spectra corrected by ATCOR and a WV value of 2 cm provided the best performances (Table 5) while the estimated SOC contents over the sites displayed the lower SOC values (Figure 11). The SOC prediction map with WV 2 cm is visualized in the Figure 12.

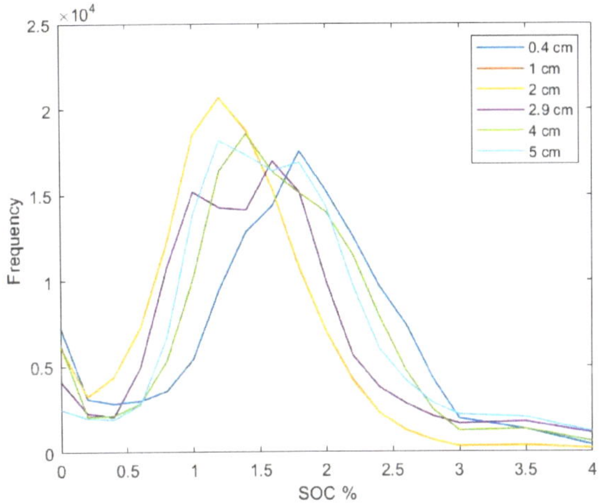

Figure 11. Histogram of the estimated SOC content from both the sites with varying WV.

There is variation in the SOC map using WV of 0.4, 4, and 5 cm, respectively (Figure 11), compared to the rest of the maps, which are fairly consistent. Variation in the pixels around and between vegetation can also be observed. This can be clearly seen in the standard deviation map (Figure 13). High standard deviation is observed along the riverbed due to no bare soil pixel and the mixture of water and vegetation pixels in site#2. A couple of prominent stripes in site#1 and site#2 are observed. In site#1, high standard deviations are observed along the farm boundaries.

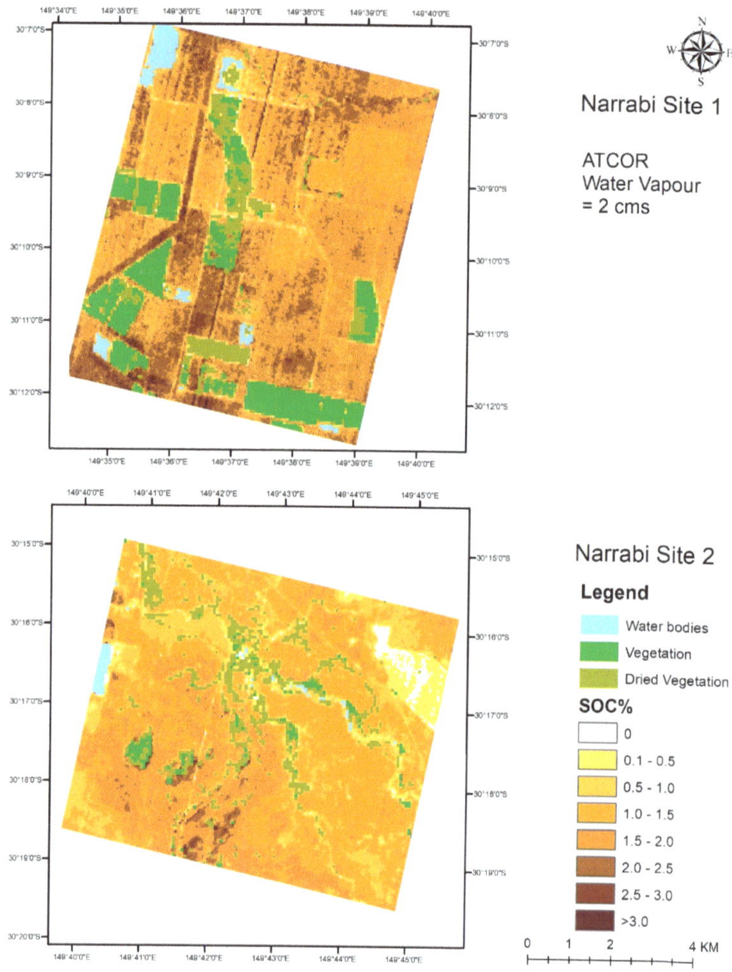

Figure 12. SOC prediction maps using WV of 2cm with ATCOR method.

The distribution of SOC estimates varied with different WV conditions though similarity in the modes is observed between 1 cm, 2 cm, and 2.9 cm (Figure 11). The best results statistically are obtained for the models in this range. A higher percentage of SOC is estimated in the 0.4 cm, 4 cm, and 5 cm (Figure 11). It is to be noted these ranges are not in the range of WV over the study area (Figure 2, extracted from NCEP). Small deviations in SOC can be observed throughout the study area and stronger deviations in SOC can be observed where there is uncertainty associated with surface types like riverbeds and boundaries of farm plot (Figure 13).

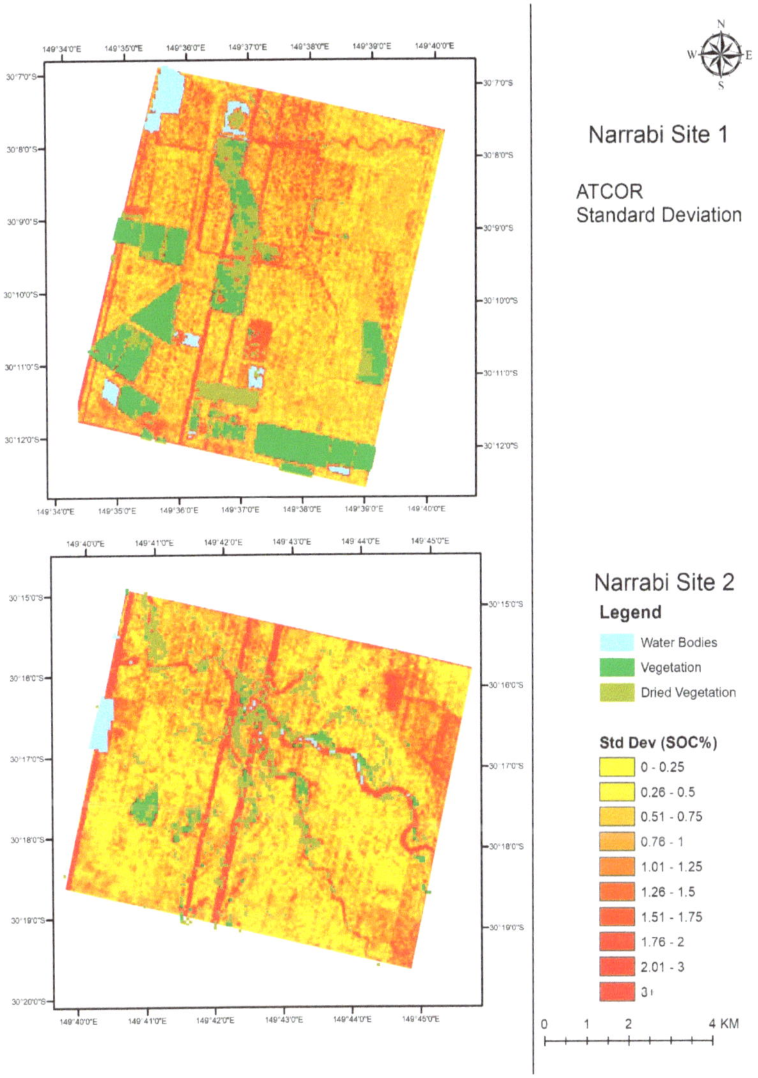

Figure 13. Standard deviation map of estimated SOC obtained along the six WV values used with ATCOR method.

4. Discussions

4.1. Bare Soil Identification Variations Based on Atmospheric Parameters

The choice of the AOD parameter in the FLAASH AC method seems to have no impact on the three tested spectral indexes (NDVI, MNDWI, and NBR2; Figure 5(A1,B1,C1)). Therefore, the percentage of bare soil pixels, based on NDVI indexes, remained constant along the AOD parameters in FLAASH (Table 4). On the other hand, the WV parameter in the ATCOR AC method affects the three tested spectral indexes (NDVI, MNDWI, and NBR2; Figure 5(A2,B2,C2)). Therefore, the percentage of bare soil pixels identification varied depending on the WV parameter (Table 5).

As soil properties estimations are affected by the presence of green and dry vegetation on spectra [16,17], bare soil pixel identification is a major step in soil properties mapping by remote sensing. As this step of bare soil pixels identification depends on spectral indices calculations (e.g., NDVI and NBR2) [67,68], the choice of the AOD parameter may impact the soil properties mapping. Additionally, the topsoil moisture conditions may also affect reflectance spectra and therefore spectral indices calculation, such as the NBR2 index [67], and then soil properties mapping [18].

4.2. Performance Analysis of PLSR Models after FLAASH AC Method

The SOC predictions obtained by PLSR models using Hyperion data corrected by the FLAASH AC method were accurate within the range of the AOD parameter (Tables 4 and 5). These PLSR performances are in accordance with Gomez et al. [12] using the same data and the Atmospheric Removal Program' (ATREM) model, and with Minu et.al [35] using the same data and the FLAASH method, who obtained R^2_{cv} of 0.71 and $RMSE_{cv}$ of 0.46% for SOC prediction. These PLSR performances for mapping SOC are slightly better than those obtained by Lu et al. [50], who also used the FLAASH AC method to derive reflectance from Hyperion radiance.

The performance results in the SOC prediction model using Hyperion data corrected by the FLAASH method show a constant R^2_{cv} and $RMSE_{cv}$ of 0.79 and 0.4% whatever the AOD parameter. This is in line with the study of Griffin et al. [39] where for the dry and relatively clear rural aerosol case, the effect of varying the aerosol optical depth produces errors < 0.02 with sensitivity. Davaadorj A. [40] also stated that FLAASH produced less variability in reflectance with varying AOD compared to other AC models (ATCOR). Most of the significant wavelengths (48%, Figure 7A) used in the PLSR models belonged to the VNIR spectral range (400–1100 nm) and more precisely around 600 to 750 nm which has been also referred by Ben Dor et al. [69] confirming that PLSR models are based on spectral features. The remaining 42% of significant wavelengths belonged to the spectral domain from 1100–2000 nm (Figure 7), including 1358 nm, probably due to OH ions in water [70].

The distribution of SOC estimation over bare soils were similar with Hyperion images corrected with FLAASH and AOD of 0.4 to 1.4 (Figure 8). While constant PLSR performances were observed whatever the AOD parameters (Table 4), differences in the SOC mapping can be observed in the maps (Figure 10). The differences in SOC estimations over some fields (Figure 10) due to the choice of AOD in FLAASH may be due to dry vegetation residue [71]. Studies by Nazeer et al. [41] and Bassani et al. [42] showed that influences of AOD vary with respect to the surface types, such as grass, water, artificial turf, sand etc. The current study deals with only one surface type i.e., bare soil, whereas compared to Nazeer et al. [41] and Bassani et al. [42] who accounted for multiple surface types.

4.3. Performance Analysis of PLSR Models after ATCOR AC Method

The SOC predictions obtained by PLSR models after the ATCOR AC method were accurate within the range of the WV (Table 5). These PLSR performances are in accordance with Gomez et al. [12] using ATREM AC method [20,72] and with Minu et.al [35] using the same data and the ATCOR AC method and obtained R^2_{cv} of 0.76, respectively.

In the ATCOR method, the R^2_{cv} lies in between 0.72 and 0.79 and variations in *RPD* and *RPIQ* values can be observed. A RPD higher than 2 is observed for WV of 0.4 to 2.9 cm while *RPD* values in between 1.91 to 2 were observed for WV 4 to 5 cm. An increase in the reflectance values of the spectral signatures for higher values of WV i.e., 4 cm, and 5 cm in the VNIR region of 400–1100 nm, as well as SWIR region of 1200 to 2400 nm (Figure 6B), is observed. Consistency in the significant wavelengths of the models using 1, 2, and 2.9 cm of WV is seen. 33% of significant bands in predicting SOC are in the region 2000–2400 nm (Figure 7). Henderson et al. [8] showed that soils generated from various parent materials had the best correlation with SOC concentration at wavelengths of 1955–1965, 2215, 2265, 2285–2295, and 2315–2495 nm. A combination of metal–OH bonding and O–H stretching causes an absorption band in the soil signature (Figure 6B) around 2200 nm [73]. In an ATCOR adjusted signal, this absorption band is retrieved. At wavelengths between 640 and 690 nm, Bartholomeus et al. [16] highlighted the strongest association between the inverse of reflectance and SOC (Figure 6B). They also discovered a negative relationship with respect to SOC between the area of the absorption feature between 2050 and 2200 nm.

Best SOC estimation performances were obtained with Hyperion images corrected with ATCOR and WV of 2 cm (Table 5), which is the default parameter in for "Mid-latitude summer" context (Table 3, [24]), while at the date of Hyperion images acquisition, the WV value would be estimated between 2.8 cm and 3.1 cm (Figure 2, extracted from NCEP). Inversely, worst SOC estimation performances were obtained with Hyperion images corrected with ATCOR and WV of 4 and 5 cm (Table 5) which are not in the range of WV over the study area from NCEP data (Figure 2).

4.4. *ATCOR versus FLAASH for SOC Predictions*

The best SOC prediction model after the ATCOR method was obtained with a WV of 2 cm (Table 5) and gave similar performance than the one after FLAASH method and whatever the AOD. Though the overall percentage of the bare soil pixel selection using spectral indices between ATCOR method and FLAASH method was similar (Tables 4 and 5), FLAASH method showed more consistency than the ATCOR method. The study conducted by Yusuf et al. 2018 [74] showed that for rural aerosol model FLAASH and ATCOR showed similar Standard Error of Estimate for surface types such as vegetation, sand, and water bodies. In the study conducted by Marcello et al. [75] on soil using worldview imagery using for a rural aerosol model, FLAASH gave an *RMSE* of 0.0398 and ATCOR gave an *RMSE* of 0.0406, which is line with the superior *RMSE* of FLAASH over ATCOR (Tables 4 and 5).

The reflectance obtained by FLAASH spectra has a slightly higher albedo compared to ATCOR spectra (Figure 6). Similar results have been also observed by Manakos et al. [76] where FLAASH and ATCOR spectra of rural aerosol model were studied with respect to Asphalt and Gravel surface types. The comparison of Spectral angle mapper results for FLAASH and ATCOR were similar for the Asphalt surface type, but FLAASH performed slightly better with the Gravel surface type [76]. Majority of the significant wavelengths were found after 2000 nm in the ATCOR method (Figure 7B), whereas the majority of the significant wavelengths were found in the visible region in the FLAASH method (Figure 7A). Similar results were arrived at by Minu et al. [35] while comparing FLAASH and ATCOR for the soil reflectance. FLAASH method showed better consistency overall in the SOC distribution than the ATCOR method (Figures 8 and 11). With respect to mapping of SOC, FLAASH method showed high standard deviations at a few farms whereas salt and pepper standard deviation distribution was found throughout the study area (Figures 10 and 13).

5. Conclusions

This study highlights the influence of AOD and WV input parameters in AC methods. The AOD parameter was found to have no impact on performance of the FLAASH method and the spectral reflectance. Some amount of variability of estimated SOC was found at low values of AOD i.e., 0.2 and it was found to be fairly consistent at other values of AOD. Variation of the WV had an effect on the performance of the ATCOR method. The best performance of the ATCOR method was observed at 2 cm. With increase in WV, there was increase in the reflectance values between 1000 nm and 2000 nm. The estimation of SOC was consistent around the default value of 2 cm (i.e., from 1 cm to 2.9 cm) but inconsistency is observed at WV < 1 cm and >2.9 cm. Hence in terms of atmospheric parameters, it is found that AOD is a more robust parameter to WV in terms of uncertainty in selection. Caution must be exercised when selecting values of WV in ATCOR method which are not close to the default parameter. As Hyperion is a moderate resolution satellite of 30 m spatial resolution, the bare soil pixel selection is an important step. Wrong bare soil classification especially at the locations when different surface types meet has a high impact and uncertainty on the estimated SOC.

Finally, the next hyperspectral sensors (EnMAP, PRISMA, and HyspIRI) will generate an increasing amount of VNIR/SWIR data around the planet, providing a new chance for mapping topsoil features. If the soil spectroscopy community tries to produce frequently topsoil characteristics maps, the selection of a robust and appropriate AC method, input AC parameters, regression method, and bare soil identification process will become crucial phases.

Author Contributions: Conceptualization, P.M., A.S., P.U. and C.G.; methodology, P.M., A.S., C.G.; software, P.U.; validation, P.M., & C.G.; writing—original draft preparation, P.M. & C.G.; writing—review and editing, A.S. & C.G.; funding acquisition, C.G. All authors have read and agreed to the published version of the manuscript.

Funding: This research was funded by Programme National de Télédétection Spatiale (PNTS, http://www.insu.cnrs.fr/pnts (accessed on 11 September 2022)), grant no. PNTS-2019-8.

Conflicts of Interest: The authors declare no conflict of interest.

References

1. Lal, R. Soil health and carbon management. *Food Energy Secur.* **2016**, *5*, 212–222. [CrossRef]
2. Xiao, J.; Chevallier, F.; Gomez, C.; Guanter, L.; Hicke, J.A.; Huete, A.R.; Ichii, K.; Ni, W.; Pang, Y.; Rahman, A.F.; et al. Remote sensing of the terrestrial carbon cycle: A review of advances over 50 years. *Remote Sens. Environ.* **2019**, *233*, 111383. [CrossRef]
3. Sanchez, P.A.; Ahamed, S.; Carré, F.; Hartemink, A.E.; Hempel, J.; Huising, J.; Lagacherie, P.; McBratney, A.B.; McKenzie, N.J.; de Lourdes Mendonça-Santos, M.; et al. Digital soil map of the world. *Science* **2009**, *325*, 680–681. [CrossRef] [PubMed]
4. Zaehle, S. Terrestrial nitrogen–carbon cycle interactions at the global scale. *Philos. Trans. R. Soc. B Biol. Sci.* **2013**, *368*, 20130125. [CrossRef]
5. Gholizadeh, A.; Borůvka, L.; Saberioon, M.; Vašát, R. Visible, near infrared, and mid-infrared spectroscopy applications for soil assessment with emphasis on soil organic matter content and quality: State-of-the-art and key issues. *Appl. Spectrosc.* **2013**, *67*, 1349–1362. [CrossRef]
6. Chang, C.W.; Laird, D.A. Near-infrared reflectance spectroscopic analysis of soil C and N. *Soil Sci.* **2002**, *167*, 110–116. [CrossRef]
7. Beyer, L.; Kahle, P.; Kretschmer, H.; Wu, Q. Soil organic matter composition of man-impacted urban sites in North Germany. *J. Plant Nutr. Soil Sci.* **2001**, *164*, 359–364. [CrossRef]
8. Henderson, T.L.; Baumgardner, M.F.; Franzmeier, D.P.; Stott, D.E.; Coster, D.C. High dimensional reflectance analysis of soil organic matter. *Soil. Sci. Soc. Am. J.* **1992**, *56*, 865–872. [CrossRef]
9. Stoner, E.R.; Baumgardner, M.F. Characteristic variations in reflectance of surface soils. *Soil Sci. Soc. Am. J.* **1981**, *45*, 1161–1165. [CrossRef]
10. Shelukindo, H.B.; Semu, E.; Singh, B.R.; Munishi, P.K.T. Predictor variables for soil organic carbon contents in the Miombo woodlands ecosystem of Kitonga forest reserve, Tanzania. *Int. J. Agric. Sci.* **2014**, *4*, 222–231.
11. Abdelhakim, B.; Tahar, G. Land use effect on soil and particulate organic carbon, and aggregate stability in some soils in Tunisia. *Afr. J. Agric. Res.* **2010**, *5*, 764–774.
12. Gomez, C.; Viscarra Rossel, R.A.; McBratney, A.B. Soil organic carbon prediction by hyperspectral remote sensing and field vis-NIR spectroscopy: An Australian case study. *Geoderma* **2008**, *146*, 403–411. [CrossRef]
13. Ayoubi, S.; Shahri, A.P.; Karchegani, P.M.; Sahrawat, K.L. Application of artificial neural network (ANN) to predict soil organic matter using remote sensing data in two ecosystems. *Biomass Remote Sens. Biomass* **2011**, 181–196. [CrossRef]
14. Stevens, A.; van Wesemael, B.; Vandenschrick, G.; Touré, S.; Tychon, B. Detection of Carbon Stock Change in Agricultural Soils Using Spectroscopic Techniques. *Soil Sci. Soc. Am. J.* **2006**, *70*, 844–850. [CrossRef]
15. Hbirkou, C.; Pätzold, S.; Mahlein, A.-K.; Welp, G. Airborne hyperspectral imaging of spatial soil organic carbon heterogeneity at the field-scale. *Geoderma* **2012**, *175*, 21–28. [CrossRef]
16. Bartholomeus, H.; Schaepman-Strub, G.; Blok, D.; Sofronov, R.; Udaltsov, S. Spectral estimation of soil properties in siberian tundra soils and relations with plant species composition. *Appl. Environ. Soil Sci.* **2012**, *2012*, 241535. [CrossRef]
17. Ouerghemmi, W.; Gomez, C.; Naceur, S.; Lagacherie, P. Applying blind source separation on hyperspectral data for clay content estimation over partially vegetated surfaces. *Geoderma* **2011**, *163*, 227–237. [CrossRef]
18. Chabrillat, S.; Ben-Dor, E.; Cierniewski, J.; Gomez, C.; Schmid, T.; van Wesemael, B. Imaging spectroscopy for soil mapping and monitoring. *Surv. Geophys.* **2019**, *40*, 361–399. [CrossRef]
19. Thompson, D.R.; Guanter, L.; Berk, A.; Gao, B.C.; Richter, R.; Schläpfer, D.; Thome, K.J. Retrieval of atmospheric parameters and surface reflectance from visible and shortwave infrared imaging spectroscopy data. *Surv. Geophys.* **2019**, *40*, 333–360. [CrossRef]
20. Gao, B.-C.; Montes, M.J.; Davis, C.O.; Goetz, A.F.H. Atmospheric correction algorithms for hyperspectral remote sensing data of land and ocean. *Remote Sens. Environ.* **2009**, *113*, S17–S24. [CrossRef]
21. Souri, A.H.; Sharifi, M.A. Evaluation of scene-based empirical approaches for atmospheric correction of hyperspectral imagery. In Proceedings of the 33rd Asian Conference of Remote Sensing, Ambassador City Jomtien Hotel, Pattaya, Thailand, 26 November 2012.
22. Song, C.; Woodcock, C.E.; Seto, K.C.; Lenney, M.P.; Macomber, S.A. Classification and change detection using Landsat TM data: When and how to correct atmospheric effects? *Remote Sens. Environ.* **2001**, *75*, 230–244. [CrossRef]

23. Bernstein, L.S. Quick atmospheric correction code: Algorithm description and recent upgrades. *Opt. Eng.* **2012**, *51*, 111719. [CrossRef]
24. Richter, R.; Schläpfer, D. *Atmospheric/Topographic Correction for Satellite Imagery (ATCOR-2/3 User Guide, Version 8.3. 1, February 2014)*; ReSe Applications Schläpfer, Langeggweg: Wil, Switzerland, 2013.
25. Adler-Golden, S.; Berk, A.; Bernstein, L.S.; Richtsmeier, S.; Acharya, P.K.; Matthew, M.W.; Chetwynd, J.H. FLAASH, a MODTRAN4 atmospheric correction package for hyperspectral data retrievals and simulations. In Proceedings of the 7th Annual JPL Airborne Earth Science Workshop, Pasadena, CA, USA, 1 February 1988; Volume 97, pp. 9–14.
26. Mazer, A.S.; Martin, M.; Lee, M.; Solomon, J.E. Image processing software for imaging spectrometry data analysis. *Remote Sens. Environ.* **1998**, *24*, 201–210. [CrossRef]
27. Zheng, Q.; Kindel, B.C.; Goetz, A.F. The high accuracy atmospheric correction for hyperspectral data (HATCH) model. *IEEE Trans. Geosci. Remote Sens.* **2003**, *41*, 1223–1231. [CrossRef]
28. Katkovsky, L.V.; Martinov, A.O.; Siliuk, V.A.; Ivanov, D.A.; Kokhanovsky, A.A. Fast atmospheric correction method for hyperspectral data. *Remote Sens.* **2018**, *10*, 1698. [CrossRef]
29. Gordon, H.R. Evolution of Ocean Color Atmospheric Correction: 1970–2005. *Remote Sens.* **2021**, *13*, 5051. [CrossRef]
30. Rahman, H. Influence of atmospheric correction on the estimation of biophysical parameters of crop canopy using satellite remote sensing. *Int. J. Remote Sens.* **2001**, *22*, 1245–1268. [CrossRef]
31. El Alem, A.; Lhissou, R.; Chokmani, K.; Oubennaceur, K. Remote Retrieval of Suspended Particulate Matter in Inland Waters: Image-Based or Physical Atmospheric Correction Models? *Water* **2021**, *13*, 2149. [CrossRef]
32. Cetin, M.; Musaoglu, N.; Kocal, O.H. A comparison of Atmospheric correction methods on Hyperion imagery in forest areas. *Uludag Universit. J. Fac. Eng.* **2017**, *22*, 103–114. [CrossRef]
33. Rani, N.; Mandla, V.R.; Singh, T. Evaluation of atmospheric corrections on hyperspectral data with special reference to mineral mapping. *Geosci. Front.* **2017**, *8*, 797–808. [CrossRef]
34. Montes, M.J.J.; Fusina, R.A.; Donato, T.F.; Bachmann, C.M.; Gao, B.-C. The effects of atmospheric correction schemes on the hyperspectral imaging of littoral environments, IGARSS 2004. In Proceedings of the 2004 IEEE International Geoscience and Remote Sensing Symposium, Online. 26 September–2 October 2004; Volume 6, pp. 4187–4190. [CrossRef]
35. Minu, S.; Shetty, A.; Minasny, B.; Gomez, C. The role of atmospheric correction algorithms in the prediction of soil organic carbon from Hyperion data. *Int. J. Remote Sens.* **2017**, *38*, 6435–6456. [CrossRef]
36. Holzer-Popp, T. Retrieving aerosol optical depth and type in the boundary layer over land and ocean from simultaneous GOME spectrometer and ATSR-2 radiometer measurements 2. Case study application and validation. *J. Geophys. Res.* **2002**, *107*, AAC 16-1–AAC 16-17. [CrossRef]
37. Bhatia, N.; Tolpekin, V.A.; Reusen, I.; Sterckx, S.; Biesemans, J.; Stein, A. Sensitivity of Reflectance to Water Vapor and Aerosol Optical Thickness. *IEEE J. Sel. Top. Appl. Earth Obs. Remote Sens.* **2015**, *8*, 3199–3208. [CrossRef]
38. Mannschatz, T.; Pflug, B.; Borg, E.; Feger, K.H.; Dietrich, P. Uncertainties of LAI estimation from satellite imaging due to atmospheric correction. *Remote Sens. Environ.* **2014**, *153*, 24–39. [CrossRef]
39. Griffin, M.K.; Burke, H.H.K. Compensation of hyperspectral data for atmospheric effects. *Linc. Lab. J.* **2013**, *14*, 29–54.
40. Davaadorj, A. Evaluating Atmospheric Correction Methods Using Worldview-3 Image. Master's Thesis, University of Twente, Enschede, The Netherlands, 2019.
41. Nazeer, M.; Nichol, J.E.; Yung, Y.K. Evaluation of atmospheric correction models and Landsat surface reflectance product in an urban coastal environment. *Int. J. Remote Sens.* **2014**, *35*, 6271–6291. [CrossRef]
42. Bassani, C.; Cavalli, R.M.; Antonelli, P. Influence of aerosol and surface reflectance variability on hyperspectral observed radiance. *Atmos. Meas. Tech.* **2012**, *5*, 1193–1203.
43. Folkman, M.A.; Pearlman, J.; Liao, L.B.; Jarecke, P.J. EO-1/Hyperion hyperspectral imager design, development, characterization, and calibration. *Hyperspectral Remote Sens. Land Atmos.* **2001**, *4151*, 40–51. [CrossRef]
44. Kalnay, E.; Kanamitsu, M.; Kistler, R.; Collins, W.; Deaven, D.; Gandin, L.; Iredell, M.; Saha, S.; White, G.; Woollen, J.; et al. The NCEP/NCAR 40-year reanalysis project. *Bull. Am. Meteorol. Soc.* **1996**, *77*, 437–472. [CrossRef]
45. Kaufman, Y.J. Atmospheric effect on spatial resolution of surface imagery. *Appl. Opt.* **1984**, *23*, 3400–3408. [CrossRef]
46. Platnick, S.T.; King, M.D.; Meyer, K.G.; Wind, G.A.; Amarasinghe, N.A.; Marchant, B.E.; Arnold, G.T.; Zhang, Z.H.; Hubanks, P.A.; Ridgway, B.I.; et al. MODIS cloud optical properties: User guide for the Collection 6 Level-2 MOD06/MYD06 product and associated Level-3 Datasets. Version. 2015, 1. Available online: https://modis-images.gsfc.nasa.gov/_docs/C6MOD06OPUserGuide.pdf (accessed on 11 September 2022).
47. Berk, A.; Cooley, T.W.; Anderson, G.P.; Acharya, P.K.; Bernstein, L.S.; Muratov, L.; Lewis, P.E. MODTRAN5: A reformulated atmospheric band model with auxiliary species and practical multiple scattering options. In Proceedings of the Remote Sensing of Clouds and the Atmosphere IX, Maspalomas, Gran Canaria, Spain, 13–15 September 2004; Volume 5571, pp. 78–85.
48. Berk, A.; Anderson, G.P. Impact of MODTRAN® 5.1 on atmospheric compensation. In Proceedings of the IGARSS 2008—2008 IEEE International Geoscience and Remote Sensing Symposium, Boston, MA, USA, 6–11 July 2008; Volume 3, pp. III-127.
49. Cooley, T.; Anderson, G.P.; Felde, G.W.; Hoke, M.L.; Ratkowski, A.J.; Chetwynd, J.H.; Lewis, P. FLAASH, a MODTRAN4-based atmospheric correction algorithm, its application and validation. *IEEE Int. Geosci. Remote Sens. Symp.* **2002**, *3*, 1414–1418.
50. Lu, P.; Wang, L.; Niu, Z.; Li, L.; Zhang, W. Prediction of Soil Properties Using Laboratory VIS–NIR Spectroscopy and Hyperion Imagery. *J. Geochem. Explor.* **2013**, *132*, 26–33. [CrossRef]

51. Jaber, S.M.; Lant, C.L.; Al-Qinna, M.I. Estimating spatial variations in soil organic carbon using satellite hyperspectral data and map algebra. *Int. J. Remote Sens.* **2011**, *32*, 5077–5103. [CrossRef]
52. Rouse, J.W.; Haas, R.H.; Schell, J.A.; Deering, D.W. Monitoring vegetation systems in the Great Plains with ERTS. *NASA Spec. Publ.* **1974**, *351*, 309.
53. Singh, K.V.; Setia, R.; Sahoo, S.; Prasad, A.; Pateriya, B. Evaluation of NDWI and MNDWI for assessment of waterlogging by integrating digital elevation model and groundwater level. *Geocarto Int.* **2015**, *30*, 650–661. [CrossRef]
54. Castaldi, F.; Chabrillat, S.; Don, A.; van Wesemael, B. Soil organic carbon mapping using LUCAS topsoil database and Sentinel-2 data: An approach to reduce soil moisture and crop residue effects. *Remote Sens.* **2019**, *11*, 2121. [CrossRef]
55. Savitzky, A.; Golay, M.J. Smoothing and differentiation of data by simplified least squares procedures. *Anal. Chem.* **1964**, *36*, 1627–1639. [CrossRef]
56. Barnes, R.J.; Dhanoa, M.S.; Lister, S.J. Standard normal variate transformation and de-trending of near-infrared diffuse reflectance spectra. *Appl. Spectrosc.* **1989**, *43*, 772–777. [CrossRef]
57. Chiang, J.H.; Hao, P.Y. A new kernel-based fuzzy clustering approach: Support vector clustering with cell growing. *IEEE Trans. Fuzzy Syst.* **2003**, *11*, 518–527. [CrossRef]
58. Pearson, R.K. Outliers in process modeling and identification. *IEEE Trans. Control. Syst. Technol.* **2002**, *10*, 55–63. [CrossRef]
59. Mark, H.L.; Tunnell, D. Qualitative near-infrared reflectance analysis using Mahalanobis distances. *Anal. Chem.* **1985**, *57*, 1449–1456. [CrossRef]
60. Tenenhaus, M.; Amato, S.; Esposito Vinzi, V. A global goodness-of-fit index for PLS structural equation modelling. In Proceedings of the XLII SIS Scientific Meeting, Padova, Italy, 9 June 2004; Volume 1, Number 2, pp. 739–742.
61. Nakagawa, S.; Schielzeth, H. A general and simple method for obtaining R^2 from generalized linear mixed-effects models. *Methods Ecol. Evol.* **2013**, *4*, 133–142. [CrossRef]
62. Chai, T.; Draxler, R.R. Root mean square error (RMSE) or mean absolute error (MAE)–Arguments against avoiding RMSE in the literature. *Geosci. Model Dev.* **2014**, *7*, 1247–1250.
63. Malley, D.F.; Martin, P.D.; Ben-Dor, E. Application in analysis of soils. *Near-Infrared Spectrosc. Agric.* **2004**, *44*, 729–784.
64. Bellon-Maurel, V.; Fernandez-Ahumada, E.; Palagos, B.; Roger, J.M.; McBratney, A. Critical review of chemometric indicators commonly used for assessing the quality of the prediction of soil attributes by NIR spectroscopy. *TrAC Trends Anal. Chem.* **2010**, *29*, 1073–1081. [CrossRef]
65. Wold, S.; Sjöström, M.; Eriksson, L. PLS-regression: A basic tool of chemometrics. *Chemom. Intell. Lab. Syst.* **2001**, *58*, 109–130. [CrossRef]
66. Mehmood, T.; Liland, K.H.; Snipen, L.; Sæbø, S. A review of variable selection methods in partial least squares regression. *Chemom. Intell. Lab. Syst.* **2012**, *118*, 62–69. [CrossRef]
67. Vaudour, E.; Gomez, C.; Fouad, Y.; Lagacherie, P. Sentinel-2 image capacities to predict common topsoil properties of temperate and Mediterranean agroecosystems. *Remote Sens. Environ.* **2019**, *223*, 21–33. [CrossRef]
68. Dvorakova, K.; Heiden, U.; van Wesemael, B. Sentinel-2 Exposed Soil Composite for Soil Organic Carbon Prediction. *Remote Sens.* **2021**, *13*, 1791. [CrossRef]
69. Ben-Dor, E. Quality assessment of several methods to recover surface reflectance using synthetic imaging spectroscopy data. *Remote Sens. Environ.* **2004**, *90*, 389–404. [CrossRef]
70. Ben-Dor, E.; Inbar, Y.; Chen, Y. The reflectance spectra of organic matter in the visible near-infrared and short wave infrared region (400–2500 nm) during a controlled decomposition process. *Remote Sens. Environ.* **1997**, *61*, 1–15. [CrossRef]
71. Wang, H.; Zhang, X.; Wu, W.; Liu, H. Prediction of Soil Organic Carbon under Different Land Use Types Using Sentinel-1/-2 Data in a Small Watershed. *Remote Sens.* **2021**, *13*, 1229. [CrossRef]
72. Gao, B.C.; Goetz, A.F.H. Column Atmospheric Water Vapor and Vegetation Liquid Water Retrievals from Airborne Imaging Spectrometer Data. *J. Geo. Res.* **1990**, *95*, 3549–3564. [CrossRef]
73. Das, S.; Pandey, P.; Mohanty, S.; Nayak, S.K. Influence of NCO/OH and transesterified castor oil on the structure and properties of polyurethane: Synthesis and characterization. *Mater. Express* **2015**, *5*, 377–389. [CrossRef]
74. Yusuf, F.R.; Santoso, K.B.; Ningam, M.U.L.; Kamal, M.; Wicaksono, P. Evaluation of atmospheric correction models and Landsat surface reflectance product in Daerah Istimewa Yogyakarta, Indonesia. *IOP Conf. Ser. Earth Environ. Sci.* **2018**, *169*, 012004. [CrossRef]
75. Marcello, J.; Eugenio, F.; Perdomo, U.; Medina, A. Assessment of Atmospheric Algorithms to Retrieve Vegetation in Natural Protected Areas Using Multispectral High Resolution Imagery. *Sensors* **2016**, *16*, 1624. [CrossRef] [PubMed]
76. Manakos, I.; Manevski, K.; Kalaitzidis, C.; Edler, D. Comparison between atmospheric correction modules on the basis of worldview-2 imagery and in situ spectroradiometric measurements. In Proceedings of the 7th EARSeL SIG Imaging Spectroscopy Workshop, Edinburgh, UK, 11–13 April 2011; pp. 11–13.

remote sensing

MDPI

Article

Construction of Aerosol Model and Atmospheric Correction in the Coastal Area of Shandong Peninsula

Kunyang Shan [1,2], Chaofei Ma [1,*], Jingning Lv [3], Dan Zhao [3] and Qingjun Song [1]

1 National Satellite Ocean Application Service, Beijing 100081, China; shankunyang21@mails.ucas.ac.cn (K.S.); kingdream@mail.nsoas.org.cn (Q.S.)
2 National Marine Environmental Forecasting Center, Beijing 100081, China
3 School of Environment, Southern University of Science and Technology, Shenzhen 518055, China; lvjn2022@mail.sustech.edu.cn (J.L.); zhaod3@sustech.edu.cn (D.Z.)
* Correspondence: mcf@mail.nsoas.org.cn

Abstract: Applying standard aerosol models for atmospheric correction in nearshore coastal waters introduces significant uncertainties due to their inability to accurately represent aerosol characteristics in these regions. To improve the accuracy of remote sensing reflectance (R_{rs}) products in the nearshore waters of the Shandong Peninsula, this study develops an aerosol model based on aerosol data collected from the Mu Ping site in the coastal area of the Shandong Peninsula, enabling tailored atmospheric correction for this specific region. Given the pronounced seasonal variations in aerosol optical properties, monthly aerosol models were developed. The monthly aerosol model is derived using the average values of aerosol microphysical properties. Compared to the standard aerosol model, this model is more effective in characterizing the absorption and scattering characteristics of aerosols in the study area. Corresponding lookup tables for the aerosol model were created and integrated into the NIR-SWIR atmospheric correction algorithm. According to the accuracy evaluation indexes of RMSD, MAE, and UPD, it can be found that the atmospheric correction results of the aerosol model established in this paper are better than those of the standard aerosol model, especially in the 547 nm band. It demonstrates that the new aerosol model outperforms the standard model in atmospheric correction performance. With the increasing availability of aerosol observational data, the aerosol model is expected to become more accurate and applicable to other satellite missions.

Keywords: atmospheric correction; aerosol model; GRASP; size distribution

Citation: Shan, K.; Ma, C.; Lv, J.; Zhao, D.; Song, Q. Construction of Aerosol Model and Atmospheric Correction in the Coastal Area of Shandong Peninsula. *Remote Sens.* **2024**, *16*, 1309. https://doi.org/10.3390/rs16071309

Academic Editors: Carmine Serio and Tamas Várnai

Received: 22 February 2024
Revised: 1 April 2024
Accepted: 3 April 2024
Published: 8 April 2024

1. Introduction

Ocean color remote sensing provides abundant observational data for studying water color constituents, including chlorophyll-a concentration, suspended sediment concentration, and water transparency [1,2]. These water color constituents can be inferred using inversion algorithms constructed with remote sensing reflectance (R_{rs}) at different spectral bands as variables. Remote sensing reflectance is typically defined as the ratio of radiance leaving the water (L_w) to the downwelling irradiance just above the water surface (E_d). The signal outside the water constitutes only a small fraction of the signal received by the satellite, with the majority originating from interactions with aerosols and Rayleigh molecules. This atmospheric signal contributes approximately 80% to 90% of the satellite signal in the visible light spectrum over the ocean [3,4]. Additionally, it is subject to subtle influences such as glint and whitecaps. Therefore, obtaining accurate water-leaving signals requires the crucial step of mitigating the impact of aerosols and Rayleigh molecules. Due to the relatively stable composition of molecules in the atmosphere, the portion of Rayleigh molecular scattering can be accurately assessed by considering the influences of polarization, surface pressure, and roughness [5,6]. It is evident from this that mitigating the impact of aerosol components is particularly crucial for obtaining accurate water-leaving signals. As is well known, atmospheric aerosol properties exhibit significant variability across both temporal and spatial scales. Differences in geographical location and time can lead to pronounced disparities in aerosol optical and microphysical characteristics [7]. Consequently, developing an appropriate aerosol model for the relevant satellite images is crucial for eliminating aerosol signals and achieving effective atmospheric correction [8].

The microphysical and optical properties of aerosols, such as particle size distribution and complex refractive index, effectively characterize the features of regional aerosols. They contribute to the successful elimination of aerosol signals during atmospheric correction processes, thus obtaining accurate R_{rs} products [9]. However, characterizing these aerosol properties proves challenging due to their significant temporal and spatial variability [10]. To improve this situation, many studies began analyzing various aerosol observational data to characterize aerosol properties. By constructing different aerosol models, they aim to characterize the absorption and scattering characteristics of aerosols under different temporal and spatial conditions [11–13].

Aerosol particle size distribution is an important indicator for studying the origin and distribution of aerosols, and it is commonly incorporated as one of the aerosol model parameters [14,15]. The aerosol size distribution is commonly characterized using the classical log-normal distribution theory, which assigns specific parameters to each component, including modal radius and standard deviation [16]. Several other studies utilize power functions, gamma functions, and analytical functions to depict aerosol particle sizes and their distribution [17–20]. Based on size distribution, aerosols are classified into five categories (rural, urban, maritime, tropospheric, and fog) by Shettle and Fenn, commonly known as SF79 [12]. In establishing relevant aerosol models, they further consider the impact of relative humidity on each group of aerosol particles. Gordon and Wang [21] were the first to apply three SF79 aerosol models (maritime, coastal, and tropospheric) in atmospheric correction for ocean color remote sensing satellite imagery. Chomko and Gordon introduced maritime, coastal, tropospheric, and urban aerosol models [17].

Ahmad et al. [22] categorized aerosol types into 10 categories and combined them with eight relative humidity levels to form 80 aerosol models (referred to as AF10). They constructed corresponding lookup tables for each model. Among these, the fine-mode particle component of the 10 aerosol types is comprised of 99.5% dust particles and 0.5% soot particles, showing excellent consistency with the average results of aerosol optical properties measured by AERONET. These 10 aerosol models have varying proportions of coarse and fine modal particles but share the same effective radius and mean radius [5,22]. These aerosols can be used to eliminate aerosol signals during the atmospheric correction process. However, due to the limitations of the models themselves, they cannot effectively account for the presence of strongly absorbing aerosols. These aerosols often occur in coastal waters where aerosols from land are transported to the sea due to wind advection effects, even extending into the open ocean. To effectively characterize the characteristics of strongly absorbing aerosols, many researchers focused their studies on this aspect [10]. However, during the atmospheric correction process, there is no reliable method to detect the presence of absorbing aerosols in the Near-Infrared (NIR) band. As a consequence, all pixels are processed using non-absorbing or weakly absorbing aerosol algorithms [5].

Current atmospheric correction algorithms often overestimate aerosol radiance in coastal marine areas, leading to the underestimation of water-leaving signals, particularly near the blue spectral bands, and even negative remote sensing reflectance values may occur [6]. This situation may arise because AERONET sites are primarily located on land, with fewer stations situated in marine areas, and their global distribution is not uniform. As a result, the observed data from AERONET sites used for constructing aerosol models are limited, with a mere 11 sites situated in open oceans and adjacent coastal waters. In coastal waters, influenced by both land and sea, aerosol sources are more complex. Merely relying on observed marine data to characterize aerosol conditions with complex sources in coastal areas poses challenges [1]. Given this limitation, other studies started to collect more aerosol observational data and integrate it into aerosol models [22,23]. Some researchers utilize continuously updated AERONET datasets to build new aerosol models for studying specific regions [24]. One study divided 27 aerosol models based on global AERONET aerosol observational data to improve the inadequacy of AF10 in effectively characterizing strongly absorbing aerosols [25]. In this research, we attempt to utilize aerosol observation data from the Yellow Sea Offshore Verification Platform established by the National Satellite Ocean Application Center. We aim to construct aerosol models and Look-Up Tables (LUTs) specifically for the nearshore region of the Shandong Peninsula in China. This endeavor is undertaken to achieve more accurate remote sensing reflectance (Rrs) inversion products for the coastal waters of China.

2. Materials and Methods

2.1. Materials

2.1.1. Sun/Sky Photometer Data

The Mu Ping site (37.681N, 121.700E) is located in the coastal area approximately 20 km north of the coastline in the Mu Ping district of Yantai City, Shandong Province, as shown in Figure 1. Situated near the junction of the Bohai Sea and Yellow Sea, this site experiences a dual influence from both the ocean and the mainland due to its proximity to the coastline. The station employs the CE318 Sun Photometer, a water color version manufactured by the French company CIMEL Electronics. It comprises 11 observation channels at wavelengths of 400 nm, 412 nm, 442 nm, 490 nm, 510 nm, 560 nm, 620 nm, 665 nm, 779 nm, 865 nm, and 1020 nm. In atmospheric aerosol research, the wavelengths of 442 nm, 665 nm, 865 nm, and 1020 nm are commonly used [26]. The instrument primarily measures solar radiance, allowing for the estimation of atmospheric aerosol characteristics and other components.

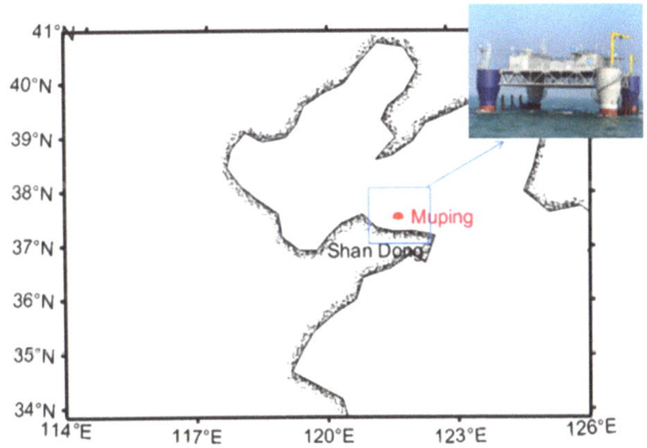

Figure 1. The geographical location of the Mu Ping station.

The CE318 multi-band sun photometer, using the PhotoGetData (v2.18.5) software, can transmit the measured data through the RS232 serial port to a computer and save it, generating binary storage files (K8 files). Each K8 file, after conversion into ASCII format, typically includes multiple file types with different suffixes. Among them, the NSU file represents data for direct solar radiation measurements, primarily used for calculating aerosol optical thickness and other parameters. The ALL and ALR files, respectively, denote data for zenith angle scans in the sky and left–right scans along the azimuth circle, mainly used for inverting sky radiance parameters. The sun photometer data at this site are sourced from the National Satellite Ocean Application Center.

2.1.2. Remote Sensing Images

MODIS is equipped with 36 spectral bands, among which 16 are primarily suitable for water color research, including 10 visible bands, 3 Near-Infrared bands, and 3 Short-Wave

Infrared bands. The high temporal resolution of MODIS sensor provides significant advantages for studying rapidly changing water color in time dynamics, making it one of the widely used image data globally. For turbid coastal waters, MODIS's Short-Wave Infrared bands can effectively serve atmospheric correction work. MODIS sensor provides various level data products, including L0, L1A, L1B, L2, and L3. L0 data are the raw data received by the sensor without processing, which is only used in a few specific applications. L1A data are reconstructed from L0 data and supplemented with auxiliary information (including radiometric and geometric calibration coefficients, as well as geolocation parameters). L1B data are radiometrically calibrated L1A data. L2 data are the geophysical products developed by combining L1B data with auxiliary data through processes such as atmospheric correction and parameter inversion. L3 data are raster data obtained by uniform mosaicking and projecting L2 products over a certain geographic grid within a certain time period. In this article, MODIS Aqua L1A data are used as the source of remote sensing image data to validate the atmospheric correction results obtained from aerosol models. The data were downloaded from NASA (https://oceancolor.gsfc.nasa.gov/, accessed on 1 October 2023).

2.1.3. R_{rs} Validation Data

This study utilizes on-site measurements of R_{rs} spectra from the Mu Ping station, established by the National Satellite Ocean Application Center on the northern coast of the Shandong Peninsula. The dataset includes measurements collected from the year 2020 to the present. When conducting measurements, the method used was that recommended by NASA for obtaining above-water remote sensing reflectance (R_{rs}) data [27], which were estimated as

$$R_{rs} = \frac{\rho_p \left(L_u - \rho_f \times L_{sky} \right)}{\pi L_{pla}} \quad (1)$$

where L_u is the upward radiance, L_{sky} denotes skylight radiance, L_{pla} represents the radiance emitted from a standard reference plaque, all of which were directly measured using a spectrometer. ρ_p represents the reflectance (approximately 10%) of a standard reference plaque provided by the producer, and ρ_f signifies the Fresnel reflection off the water surface, assumed to be 0.022 for a flat water surface.

2.2. Methods

2.2.1. Construction of Aerosol Models

The raw data used for the inversion of aerosol microphysical properties comes from the Mu Ping station's CE318 Water Color Sun/Sky Photometer. The microphysical characteristics of aerosols, including particle size distribution and complex refractive index, can be inferred from spectral sky radiance measurements conducted at 442, 665, 865, and 1020 nm, obtained at the almucantar plane, along with solar direct irradiance measurements at corresponding wavelengths [26]. The selected four bands exhibit no significant gas absorption and demonstrate high sensitivity to the typical sizes of major aerosols [28]. Before the inversion of microphysical properties, it is necessary to select data that meet certain criteria. First, the observed sky radiance data from both left and right azimuth scans (clockwise and counterclockwise) should differ by no more than 20%. The average of the left and right azimuth scan data is then taken as the sky radiance data for that specific time point. Second, under the condition of symmetric azimuth angles for left and right sky scattering radiance, the scattering radiance data used for inversion should have a symmetrical azimuthal angle count greater than 21, within the range of 3.5° to 160° [29]. Third, the selected data should have a solar zenith angle greater than 50°. The data generated under this criterion is similar to the AERONET website's version 2.0 data [29,30]. The sky radiance data were convolved using a square filter with a width of 10 nm centered at the effective wavelengths of the photometer. The standard extraterrestrial spectrum irradiance (https://oceancolor.gsfc.nasa.gov/docs/rsr/f0.txt, accessed on 1 October 2023) was used to normalize the resulting spectra [31].

The Generalized Retrieval of Aerosol and Surface Properties (GRASP) code, developed by Dubovik et al., is an open-source algorithm designed for retrieving properties of aerosols [30,32,33]. GRASP comprises two main modules: numerical inversion and forward modeling. The numerical inversion module employs statistical optimization techniques to fit observational data, while the forward model accurately simulates various atmospheric remote sensing observations [34]. Therefore, this versatility allows GRASP to be applied across passive and active satellites, as well as ground-based atmospheric observations, and it is particularly suited for synergistic retrievals when inverting different observations simultaneously [35]. The algorithm accepts diverse input data, such as satellite images, polar nephelometer measurements, sun/sky photometer data, sky camera images, and lidar data. The precise inversion of various combinations of these input data is accomplished through the use of a multi-term least squares method (LSM) [36]. This fitting algorithm can undergo statistical optimization, allowing for flexible inversions based on different observations [37]. In this study, ground-based observations from a single sun photometer were utilized for the inversion of aerosol properties. The retrieved column-integrated aerosol volume size distributions (VSDs) are adjusted using 22 logarithmically equidistant triangle bins ranging from 0.05 to 15 μm in radius [38]. While some studies suggest approximating VSDs with bimodal log-normal distributions to reduce the information required by the binned VSDs, a problem arises when the retrieved VSDs deviate from perfect log-normality, leading to asymmetrical or trimodal shapes. In such cases, the strategy based on simplified bimodal VSDs may not yield accurate retrievals, hence the preference for the former approach despite its more complex calculations. To further validate the algorithm's reliability and ensure the accuracy of the results, we selected raw data from the AERONET Beijing_PKU and Socheongcho sites as input for the GRASP algorithm [39]. We inverted the aerosol particle size distribution for the respective regions of these two sites. The Beijing_PKU site is located on land, and the data selected are from 1 January 2019, at 01:21:13 UTC. The Socheongcho site is situated in the ocean, and the data retrieved are from 3 January 2020, at 03:48:35 UTC. These two sites represent different underlying surface types. The obtained results are shown in Figure 2.

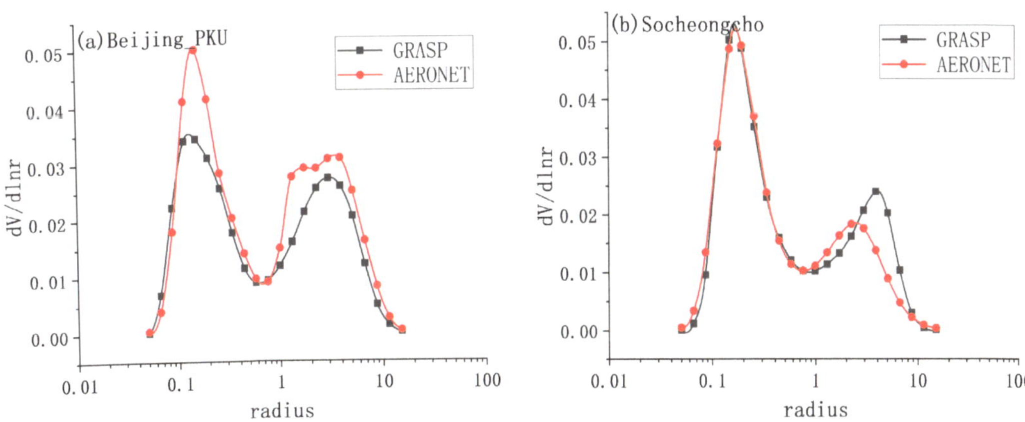

Figure 2. Comparison of aerosol size distributions obtained by the GRASP algorithm and AERONET; (**a**) Beijing_PKU, (**b**) Socheongcho.

For the Beijing_PKU site, it can be observed that the trend of aerosol size distribution obtained by the GRASP algorithm is generally consistent with the AERONET results. However, a notable difference is that the AERONET results exhibit higher peaks at the coarse and fine modal particles compared to the GRASP results. For the Socheongcho site, the consistency between GRASP and AERONET results is nearly perfect, with only minor differences observed at the peak of coarse-mode particles. This suggests that the GRASP algorithm is capable of accurately retrieving aerosol characteristics in the respective regions.

In open ocean regions, the distribution of aerosol particles in the air above the sea surface is relatively stable, with less influence from external factors, and the composition and concentration of aerosols change minimally with the seasons [22]. However, in coastal areas, due to proximity to land, the composition of aerosols in the atmosphere above undergoes rapid changes in both time and space, influenced significantly by human activities and natural factors on land, and is subject to larger variations due to terrestrial influences [40].

The AF10 model, currently adopted as the globally standardized aerosol model by NASA, incorporates relative humidity as a key parameter to adapt various regional aerosol models. However, aerosol particles originating from terrestrial sources, including continental, biomass burning, and dust sources, demonstrated limited sensitivity to variations in relative humidity, particularly under conditions where the relative humidity is below 70% [41,42]. Kinne's study of aerosol characteristics on the monthly average time scale found that large monthly differences were rare [43]. Sang Woo studied the optical properties of columnar aerosols in East Asia and found that their changes had seasonal and monthly characteristics [44]. Through studying the global aerosol particle size distribution and complex refractive index parameters, Zhao Dan found that these parameters changed slightly in the same month in different years and were relatively stable [1]. Therefore, for more accurate construction of regional aerosol models, this study builds aerosol models for the region based on monthly aerosol data. Selecting the sun photometer data from the Mu Ping station between 2020 and 2022, which underwent quality control, the GRASP algorithm is employed to invert aerosol microphysical and optical properties. Results with inversion fitting errors of less than 5% are considered valid. Valid results are then statistically analyzed, and monthly averages are calculated to construct monthly aerosol models. The microphysical properties of aerosols, including particle size distribution and complex refractive index, along with optical properties such as extinction coefficient, Ångström exponent (α), single scattering albedo (SSA), and aerosol phase function, can be simulated using a radiative transfer model predicated on these pivotal microphysical attributes [45].

The aerosol size distribution function describes how particles of different diameters are distributed within a given volume. Typically, this function is modeled as the combination of two log-normal distributions: one for fine particles and the other for coarse particles. This combined distribution is commonly known as a bimodal log-normal distribution. Mathematically, it can be represented as follows:

$$\frac{dV(r)}{dlnr} = \frac{V_f}{\sqrt{2\pi}\sigma_f}\exp\left[-\left(\frac{lnr - lnr_f}{\sqrt{2}\sigma_f}\right)^2\right] + \frac{V_c}{\sqrt{2\pi}\sigma_c}\exp\left[-\left(\frac{lnr - lnr_c}{\sqrt{2}\sigma_c}\right)^2\right] \tag{2}$$

where V denotes the volume concentration of the particles, r represents the mean volume radius of the particles, f and c represent coarse mode particles and fine mode particles, respectively. r_f and r_c represent the mean volume radius of the fine and coarse mode particles, respectively. σ_f and σ_c signify the standard deviation of the fine mode and coarse mode particle size within a model. Figure 3 shows the monthly average particle size distribution at the Mu Ping station for months 1 to 12 [43]. The complex refractive index comprises the real part (n_r) and the imaginary part (n_i), which vary depending on the chemical composition of the aerosol; n_r and n_i govern the scattering and absorption of aerosols by incident light, respectively [46,47]. Figure 4 illustrates the distribution of the real and imaginary components of the complex refractive index at 442 nm across months 1 to 12. In July, there is only one valid data point, and thus, it lacks monthly representativeness. By inputting the aforementioned particle size distribution and complex refractive index into the 6S radiative transfer model, the optical properties of aerosols can be simulated [48].

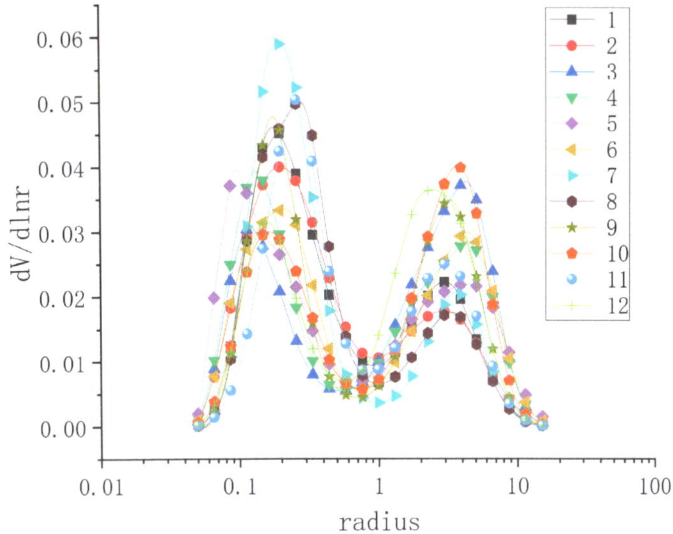

Figure 3. Monthly average particle size distribution.

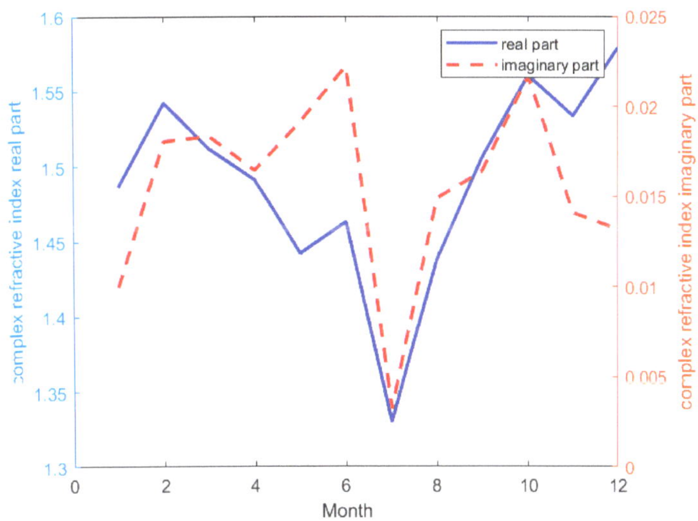

Figure 4. Real and imaginary parts of the monthly average complex refractive index at 442 nm.

2.2.2. Construction of the Lookup Table for the New Aerosol Model

These tables contain parameters identical to those used in the AF10 model. Additionally, two sets of coefficients were computed. The first set represents the correlation between single-scattering albedo and multiple-scattering albedo. The second set of coefficients relates to the relationship between diffuse transmittance from the sun to Earth's surface, diffuse transmittance from the surface to the satellite, and aerosol optical thickness.

In the case where geometric conditions and aerosol optical thickness (AOD) are known, the relationship between aerosol single-scattering reflectance (ρ_{as}) and aerosol multiple-scattering reflectance (ρ_a) can be expressed as

$$ln[\rho_a(\lambda)] = a(\lambda) + b(\lambda)ln[\rho_{as}(\lambda)] + c(\lambda)ln^2[\rho_{as}(\lambda)] \quad (3)$$

$$\rho_{as}(\lambda) = \omega_a(\lambda)\tau_a(\lambda)p_a(\theta_s, \theta_v, \phi, \lambda)/4cos\theta_v cos\theta_s \quad (4)$$

$$p_a(\theta_s, \theta_v, \phi, \lambda) = P_a(\theta_-, \lambda) + (r(\theta_s) + r(\theta_v))P_a(\theta_+, \lambda) \quad (5)$$

$$cos\theta_\pm = \pm cos\theta_s cos\theta_v - sin\theta_s sin\theta_v cos\phi \quad (6)$$

where ω_a is aerosol single scattering albedo, τ_a is aerosol optical depth, θ_s and θ_v are solar zenith angle and view zenith angle, respectively, ϕ is relative azimuth angle, $P_a(\theta_\pm, \lambda)$ is the aerosol scattering phase function for a scattering angle, $\theta_\pm$, $r(\theta)$ is the Fresnel reflectance of the interface for an incident angle θ.

Whereas the multiple-scattering reflectance is obtained through simulations using the 6S radiative transfer model [48], the single-scattering reflectance is calculated using Equations (4) and (5) [49].

By employing the method of least squares fitting regression to solve for the coefficients a, b, and c, the relationship between $ln\rho_a$ and $ln\rho_{as}$ under specific geometric conditions ($\theta_s = 36°$, $\theta_v = 66°$, $\phi = 70°$) for the aerosol model at the Mu Ping station in June is depicted in Figure 5. The regression coefficients in the regression equation correspond to the a, b, and c in Equation (3).

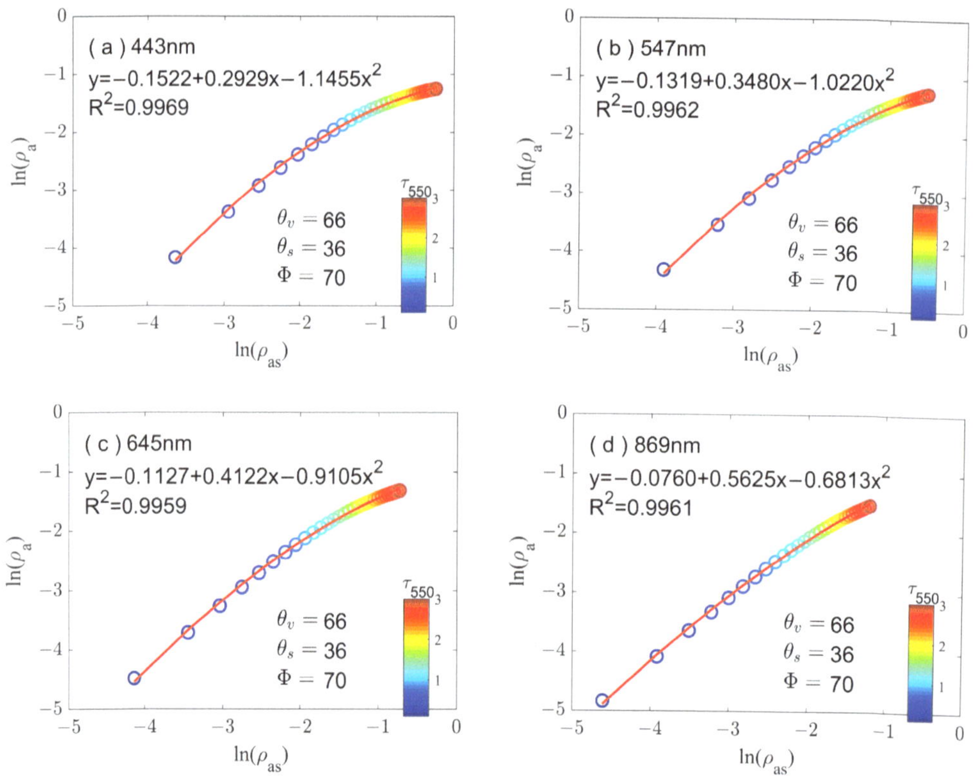

Figure 5. An example illustrating the least squares fitting relationship between variables $\ln(\rho_a)$ and $\ln(\rho_{as})$ for four MODIS bands, using the aerosol model for June under specific geometric conditions ($\theta_s = 36°$, $\theta_v = 66°$, and $\phi = 70°$). The color of each point reflects the aerosol optical depth at 550 nm (τ_{550}).

The relationship between diffuse transmittance t (t_s or t_v) and aerosol optical depth (τ_a) can be expressed as

$$t = Aexp(-B\tau_a) \tag{7}$$

Whereas the diffuse transmittance t is obtained through simulations using the 6S radiative transfer model under specific geometric parameters and AOD conditions, the coefficients (A and B) can be obtained through the least squares fitting method. Figure 6 depicts the relationship between diffuse transmittance and aerosol optical depth when the solar zenith angle is 36 degrees, and the coefficients of the fitting equation correspond to the A and B in Equation (7).

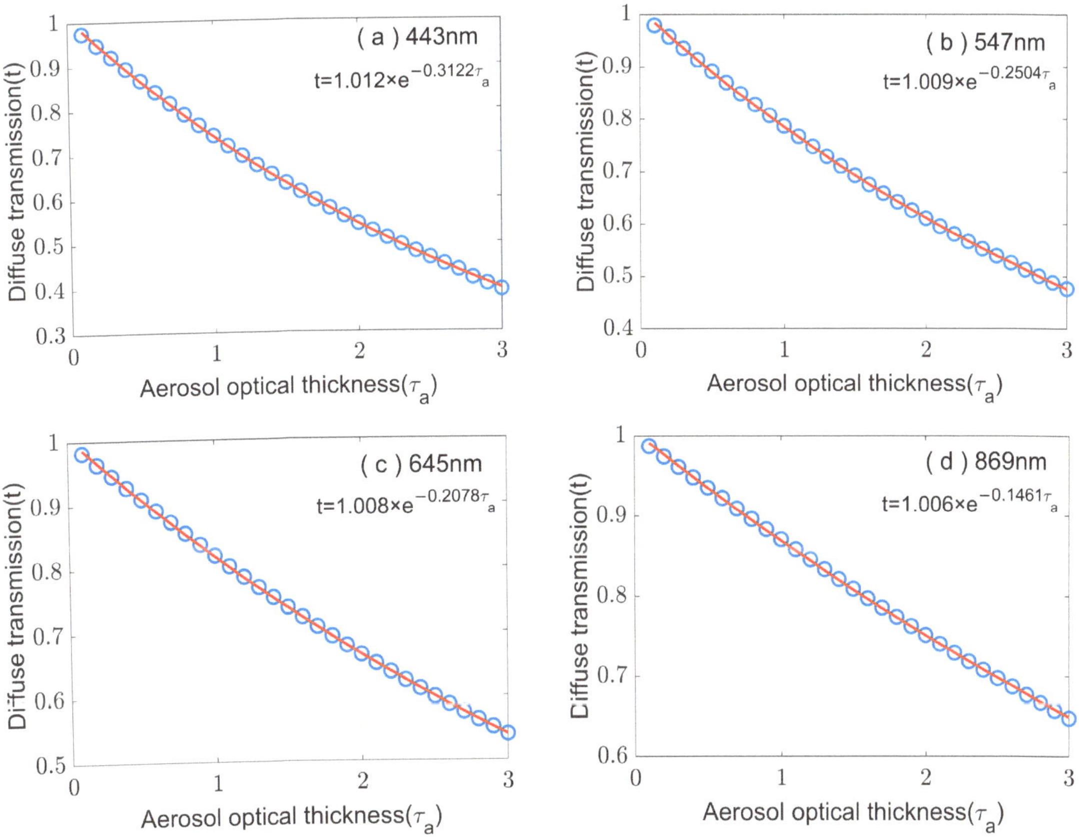

Figure 6. An example illustrating the least squares fitting relationship between variables t and τ_a for four MODIS bands, using the aerosol model for June under specific geometric conditions ($\theta_s = 36°$).

For each lookup table (LUT) associated with the aerosol model for each month, we integrated diverse combinations of aerosol multiple-scattering and diffuse transmittance coefficients correlated with aerosol optical depth (AOD) values and satellite/solar geometric parameters. In particular, the AOD varied from 0 to 3 (with an increment of 0.1), while the solar/sensor zenith angles (θ_s and θ_v) ranged from 0 to 78 degrees (with an increment of 6 degrees), and their relative azimuthal angles (ϕ) ranged from 0 to 180 degrees (with an increment of 10 degrees). Conversely, NASA's LUT employed only 8 AOD values within a significantly narrower range [4]. The format and content of the aerosol lookup table are shown in Table 1.

Table 1. Aerosol lookup table LUTs parameter list.

Parameter	Description	Dimension
wave	wavelength	1
scatt	scattering angle	1
albedo	single scattering albedo	1
extc	extinction coefficient	1
angstrom	Ångström index	1
phase	Scattering Phase Function	2
solz	Solar zenith angle	1
senz	View zenith angle	1
phi	Relative azimuth	1
accost bcost ccost	Aerosol single-multiple scattering coefficient	4
dtran_wave	Diffuse transmission wavelength	1
dtran_theta	Diffuse transmission zenith angle	1
dtran_a (a_0) dtran_b (b_0)	Diffuse transmittance coefficient	2

2.2.3. Atmospheric Correction

The radiative signals observed by satellite sensors include signals from both water bodies and the atmosphere [21,50]. To achieve more accurate retrieval of water color elements, it is essential to separate the water color signal from the atmospheric signal [51–53]. The process of removing the atmospheric signal is known as atmospheric correction [54,55].

The goal of atmospheric correction is to establish specific models that remove atmospheric path and surface influences from the Top of Atmosphere (TOA) signals. The primary difficulty involves accurately assessing the impact of aerosols, a major source of uncertainty in atmospheric correction, on atmospheric path radiance. This contribution varies considerably and must be estimated based on observations [56]. Water exhibits strong absorption in the Near-Infrared to Short-Wave Infrared (NIR-SWIR) range, providing a basis for separating atmospheric and ocean signals [57,58]. Initially, for open ocean regions, SeaWiFS and MODIS atmospheric correction was conducted using two Near-Infrared (NIR) bands under the assumption that the marine influence in these NIR bands could be considered negligible [50]. However, coastal regions, influenced by factors such as land–sea dynamics, often have turbid waters. In these areas, the radiance observed in the Near-Infrared (NIR) bands can be significant. This can pose challenges for iterative methods that rely solely on NIR, leading to underestimation or even negative values for water-leaving radiance. To tackle this issue, Wang et al. [59] proposed an atmospheric correction algorithm for MODIS that incorporates both Near-Infrared and Short-Wave Infrared bands. The key aspect of this algorithm involves introducing a turbidity index based on MODIS measurements in the Near-Infrared (748 nm) and Short-Wave Infrared bands (1240 nm, 2130 nm), as described in Equation (7). When the turbidity index meets certain conditions $T_{ind}(748,\ 1240) < 1.3$, the iterative method using Near-Infrared bands (748 nm, 869 nm) is applied; otherwise, the iterative method using Short-Wave Infrared bands (1240 nm, 2130 nm) is utilized [59,60].

$$T_{ind}(748,\ 1240) = \frac{\Delta\rho^{(RC)}(748)}{\Delta\rho^{(RC)}(1240)} exp\left\{-\frac{492}{890} ln\left(\frac{\Delta\rho^{(RC)}(1240)}{\Delta\rho^{(RC)}(2130)}\right)\right\} \tag{8}$$

$$\Delta\rho^{(RC)}(\lambda_l) = \rho_t(\lambda_l) - \rho_r(\lambda_l) \tag{9}$$

where $\rho_t(\lambda_l)$ represents the reflectance of the top atmosphere observed by satellites after gas absorption correction, and $\rho_r(\lambda_l)$ denotes the reflectance contributed by the molecules, also known as Rayleigh scattering.

3. Results

To further validate the efficacy of our new aerosol model in improving atmospheric correction, we applied the previously described NIR-SWIR atmospheric correction algorithm. This algorithm, recommended jointly by NASA for turbid water bodies, was validated utilizing datasets on a global and regional scale. The monthly Look-Up Tables (LUTs) developed for the study sites in this research were substituted for the standard LUTs in the SeaDAS (v8.3) software. During the atmospheric correction process for a specific pixel, the aerosol LUT is selected based on the location of the target pixel and the month of data collection. Subsequently, this LUT is utilized to estimate the aerosol scattering component of the signal [1].

This research utilized the same version of SeaDAS (v8.3) for the removal of gas absorption and Rayleigh scattering to assess the improvements of the new aerosol model on remote sensing reflectance (R_{rs}) products. We established specific criteria for selecting corresponding satellite and in situ observational data. Initially, the time interval between satellite observations and ground-based measurements was limited to 30 min. While NASA's standard allows a 3 h time difference in open ocean regions, our stricter criterion was necessary due to the higher variability and influence of continental factors in coastal areas. Pixels identified as being affected by cloud or ice contamination, having high sensor view zenith, high solar zenith, and stray light contamination were deemed invalid. Instead of using NASA's default Rayleigh-corrected reflectance for cloud-masked reflectance, we opted for an alternative approach [1]. This decision was made due to concerns that NASA's default threshold might incorrectly identify turbid water or thick aerosols, which are common in coastal waters. Similar methodologies were adopted in previous research studies [61,62].

Figure 7 shows the remote sensing reflectance (R_{rs}) obtained from the aerosol model developed in this study and the AF10 aerosol model processed through the NIR-SWIR algorithm. It can be observed that the R_{rs} distribution from the two models is generally consistent, with subtle differences in specific values. In certain regions, the aerosol model established in this article yields slightly higher R_{rs} than the NASA standard model. This means that the aerosol signals assessed by standard aerosol models in these regions may be higher than those in the aerosol models presented in this paper. This is due to the fact that standard aerosol models are constructed using only a small number of observations from ocean sites, especially in coastal areas. This results in the overestimation of aerosol signals in turbid water, while the measured aerosol data in this paper can better characterize the aerosol characteristics of the study area, and the aerosol signals obtained are more accurate. Figure 8 presents a further comparison between satellite-retrieved R_{rs} and ground-based measured R_{rs} spectra. For this collection of spectra, in strict adherence to the previously defined criteria, the time gap between satellite imagery and ground measurements does not exceed 30 min. It can be seen that when we use the new aerosol model, the results calculated by the NIR-SWIR algorithm are closer to the measured values.

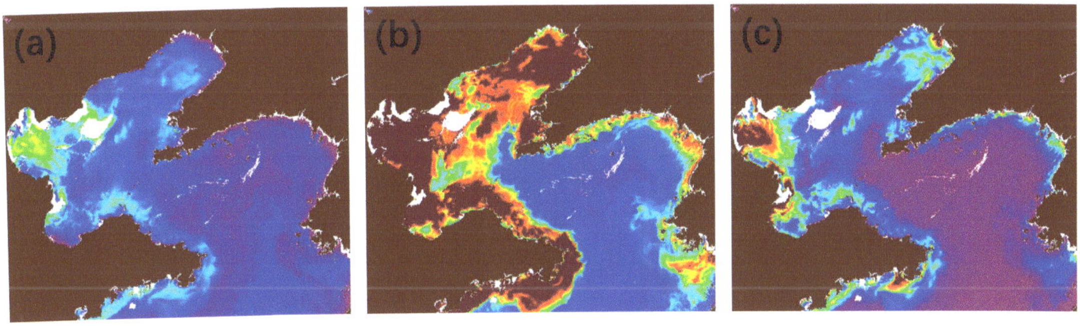

Figure 7. *Cont.*

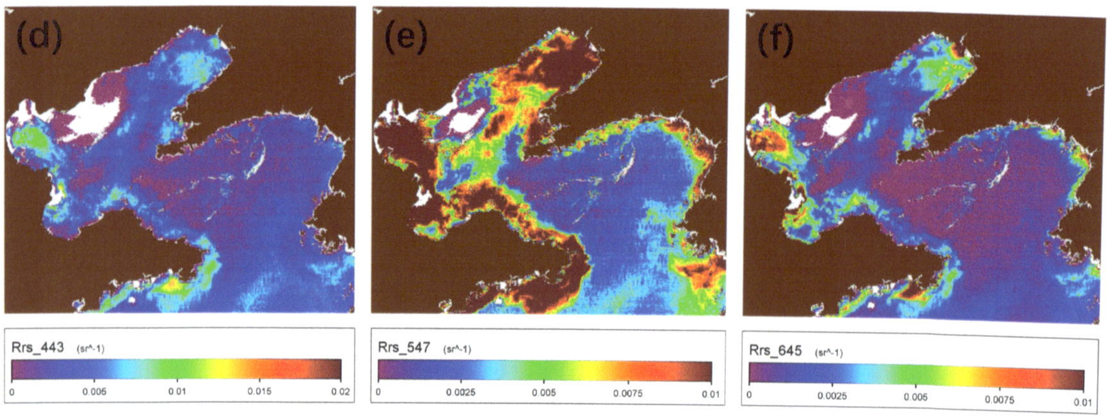

Figure 7. Atmospheric corrected R_{rs} products (443, 547, 645 nm) of MODIS-Aqua over the nearshore of Shandong Peninsula region on 26 September 2020. (new aerosol model: (**a**–**c**), NASA aerosol model: (**d**–**f**)).

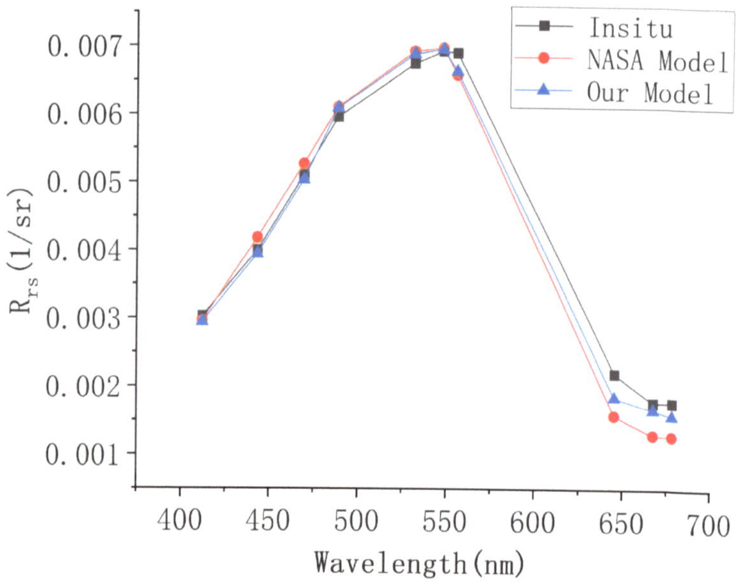

Figure 8. Comparison between satellite-retrieved spectra and in situ measurements at the Mu Ping site.

4. Discussion

We use several accuracy assessment metrics to evaluate the results of the aerosol model, encompassing the regression slope, coefficient of determination (R^2), root mean squared difference (RMSD), unbiased percentage difference (UPD), and mean absolute error (MAE).

$$RMSD = \sqrt{\frac{1}{N}\sum_{i=1}^{N}(Y_i - X_i)^2} \tag{10}$$

$$MAE = \frac{1}{N}\sum_{i=1}^{N}|Y_i - X_i| \tag{11}$$

$$UPD = \sum_{i=1}^{N} \frac{|Y_i - X_i|}{0.5 \times |Y_i + X_i|} \quad (12)$$

Figure 9 displays the satellite data alongside corresponding measurements for three visible bands (i.e., 443, 547, and 645 nm), and the satellite R_{rs} product is also calculated using the NIR-SWIR algorithm. The red points represent the results of NASA's standard aerosol model, while the blue points represent the results of the new aerosol model proposed in this study.

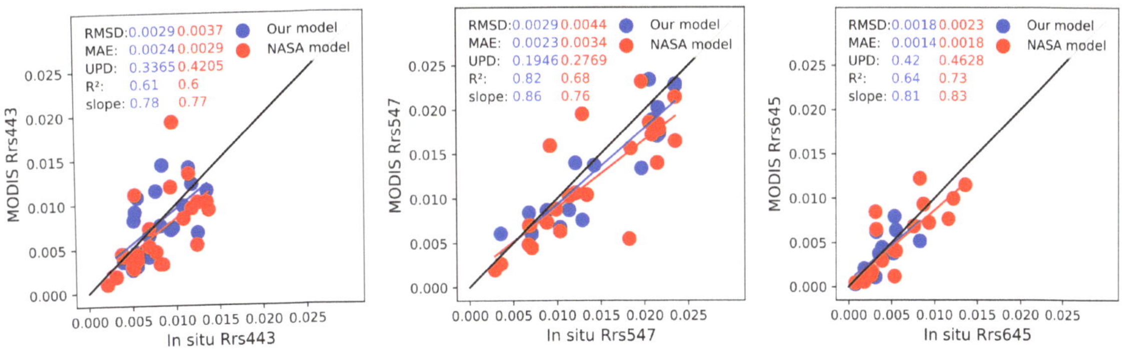

Figure 9. Scatter plots comparing the satellite-derived remote sensing reflectance (R_{rs}) using the NIR-SWIR atmospheric correction method with the in situ R_{rs} at three MODIS bands (443 nm (**left**), 547 nm (**middle**), and 645 nm (**right**)). The evaluation metrics for the atmospheric correction results of the nine MODIS bands are specifically presented in Table 1.

From the results, it is clear that for the RMSD, MAE, and UPD accuracy evaluation indicators, the results of the new model are significantly better than NASA's standard model. For R^2 and slope, except for slightly lower values at 645 nm, the new model's results at 443 nm and 547 nm are also better than the standard model. Table 2 lists the accuracy assessment statistics for MODIS's nine bands. Overall, the performance of the new aerosol model is better than the standard aerosol model, especially at 547 nm. It is important to highlight that the underestimation observed in the blue light band with the new model, relative to the standard model, was mitigated, which is likely attributed to a more accurate representation of absorbing aerosols [1]. It is notable that the slopes observed in these bands are all less than 1, possibly indicating the overcorrection of remote sensing reflectance (R_{rs}) in turbid water. Please note that the study initially matched 24 satellite situ points. However, minor discrepancies in the number of data points across different models or bands may arise from saturation or negative R_{rs} values in MODIS bands or from the atmospheric correction process.

Table 2. The evaluation metrics for the atmospheric correction results of the nine MODIS bands, evaluated using in situ R_{rs} dataset from the Mu Ping site.

	Band	R^2	**Slope**	**RMSD**	**MAE**	**UPD (%)**
	412	0.56/0.53	0.71/0.6	0.003/0.0037	0.0022/0.0029	41.68/49.13
	443	0.61/0.6	0.78/0.77	0.0029/0.0036	0.0023/0.0028	32.12/40.29
	469	0.65/0.64	0.76/0.77	0.0032/0.0039	0.0026/0.0031	28.48/38.04
	488	0.57/0.66	0.6/0.77	0.0038/0.0039	0.0031/0.0031	27.01/32.86
Our/NASA model	531	0.76/0.69	0.84/0.79	0.0033/0.0042	0.0025/0.0033	21.93/28.69
	547	0.82/0.68	0.86/0.76	0.0029/0.0044	0.0023/0.0034	19.46/27.69
	555	0.62/0.65	0.7/0.72	0.0045/0.0048	0.0037/0.0037	49.41/31.14
	645	0.64/0.73	0.81/0.83	0.0018/0.0023	0.0014/0.0018	42.00/46.28
	678	0.62/0.72	0.74/0.82	0.0014/0.0019	0.0012/0.0015	40.49/43.32

5. Conclusions

Coastal regions, characterized by complex water compositions, introduce significant uncertainties in remote sensing reflectance (R_{rs}) obtained through atmospheric correction. NASA's aerosol models are primarily developed for open ocean conditions, leading to substantial variability in coastal areas, mainly due to the inaccurate characterization of absorbing aerosols in the aerosol correction process. In this study, aerosol observation data from the Mu Ping station in the Yellow and Bohai Seas were utilized to characterize aerosol properties in the region. Aerosol properties, such as single scattering albedo (SSA) and Ångström exponent, exhibited noticeable spatial gradients and seasonal dynamics. The monthly aerosol model, based on these aerosol characteristics by taking their monthly averages, can more effectively characterize the scattering and absorption properties of aerosols in the region, facilitating the precise removal of aerosol signals from the sensor-received signals. By incorporating this aerosol model and the standard aerosol model into the NIR-SWIR algorithm for atmospheric correction, the atmospheric correction results for both models in the region are obtained. These results are then matched and validated against measured data. From the precision evaluation metrics, including RMSD, MAE, and UPD, it can be observed that the new aerosol model outperforms the standard aerosol model across the nine visible light bands of MODIS, leading to an improvement in the accuracy of atmospheric correction.

While we only applied this aerosol model to MODIS satellite imagery in this study, it can seamlessly be adapted for use with other satellite imagery. This adaptation requires constructing aerosol model lookup tables that align with the specific wavelength bands of the other satellites. This opens up the possibility of generating additional remote sensing reflectance (R_{rs}) products for coastal regions, facilitating the improved retrieval of water color-related parameters.

Certainly, there is substantial room for improvement in the current aerosol model. The model's construction was based on observational data from the Mu Ping station spanning 2020 to 2022, which is relatively limited. As time progresses, the Mu Ping station will provide more observational data, enabling the aerosol model to better capture the seasonal dynamic changes in aerosol characteristics. This will facilitate more effective atmospheric correction efforts. Depending on requirements, aerosol models can be constructed with smaller time intervals. Presently, the Huangdonghai offshore observation platform has deployed two observation stations, Mu Ping and Dongtou. In the future, more aerosol observation stations will be established in China's offshore areas, achieving comprehensive coverage of aerosol characteristic observations in the Chinese near seas. This holds extraordinary significance for obtaining precise R_{rs} products in China's nearshore regions.

Author Contributions: Conceptualization, K.S.; methodology, K.S. and J.L.; validation, K.S.; investigation, K.S.; resources, K.S.; data curation, K.S.; writing—original draft, K.S.; writing—review and editing, C.M., Q.S. and D.Z.; visualization, K.S.; supervision, C.M., Q.S. and D.Z.; project administration, Q.S.; funding acquisition, Q.S.; All authors have read and agreed to the published version of the manuscript.

Funding: This research was funded by the National Natural Science Foundation of China under grant number 42176183.

Data Availability Statement: Dataset available upon request from the authors.

Acknowledgments: We acknowledge the data support from the National Satellite Ocean Application Service (http://www.nsoas.org.cn/) and NASA GSFC (https://oceancolor.gsfc.nasa.gov/). We thank the reviewers for their valuable feedback on the paper.

Conflicts of Interest: The authors declare no conflicts of interest.

References

1. Zhao, D.; Feng, L.; He, X. Global gridded aerosol models established for atmospheric correction over inland and nearshore coastal waters. *J. Geophys. Res. Atmos* **2023**, *128*, e2023JD038815. [CrossRef]
2. Wei, J.; Lee, Z.; Shang, S. A system to measure the data quality of spectral remote-sensing reflectance of aquatic environments. *J. Geophys. Res. Oceans* **2016**, *121*, 8189–8207. [CrossRef]
3. Mukai, S.; Sano, I.; Toigo, A. Removal of scattered light in the Earth atmosphere. *Earth Planets Space* **1998**, *50*, 595–601. [CrossRef]
4. Gordon, H.R. Atmospheric correction of ocean color imagery in the Earth Observing System era. *J. Geophys. Res. Atmos* **1997**, *102*, 17081–17106. [CrossRef]
5. Mobley, C.D.; Werdell, J.; Franz, B.; Ahmad, Z.; Bailey, S. *Atmospheric Correction for Satellite Ocean Color Radiometry*; Goddard Space Flight Center: Washington, DC, USA, 2016.
6. Shi, C.; Nakajima, T. Simultaneous determination of aerosol optical thickness and water-leaving radiance from multispectral measurements in coastal waters. *Atmos. Chem. Phys.* **2018**, *18*, 3865–3884. [CrossRef]
7. Kompalli, S.K.; Suresh Babu, S.; Krishna Moorthy, K.; Gogoi, M.M.; Nair, V.S.; Chaubey, J.P. The formation and growth of ultrafine particles in two contrasting environments: A case study. *Ann. Geophys.* **2014**, *32*, 817–830. [CrossRef]
8. Sayer, A.; Hsu, N.; Bettenhausen, C.; Ahmad, Z.; Holben, B.; Smirnov, A.; Thomas, G.; Zhang, J. SeaWiFS Ocean Aerosol Retrieval (SOAR): Algorithm, validation, and comparison with other data sets. *J. Geophys. Res. Atmos* **2012**, *117*, D3. [CrossRef]
9. Bassani, C.; Manzo, C.; Braga, F.; Bresciani, M.; Giardino, C.; Alberotanza, L. The impact of the microphysical properties of aerosol on the atmospheric correction of hyperspectral data in coastal waters. *Atmos. Meas. Tech.* **2015**, *8*, 1593–1604. [CrossRef]
10. Gordon, H.R.; Du, T.; Zhang, T. Remote sensing of ocean color and aerosol properties: Resolving the issue of aerosol absorption. *Appl. Opt.* **1997**, *36*, 8670–8684. [CrossRef]
11. Omar, A.H.; Won, J.G.; Winker, D.M.; Yoon, S.C.; Dubovik, O.; McCormick, M.P. Development of global aerosol models using cluster analysis of Aerosol Robotic Network (AERONET) measurements. *J. Geophys. Res. Atmos* **2005**, *110*, D10S14. [CrossRef]
12. Shettle, E.P.; Fenn, R.W. *Models for the Aerosols of The Lower Atmosphere and the Effects of Humidity Variations on Their Optical Properties*; Optical Physics Division, Air Force Geophysics Laboratory: Birmingham, AL, USA, 1979.
13. Tanré, D.; Kaufman, Y.; Herman, M.; Mattoo, S. Remote sensing of aerosol properties over oceans using the MODIS/EOS spectral radiances. *J. Geophys. Res. Atmos* **1997**, *102*, 16971–16988. [CrossRef]
14. Logan, T.; Xi, B.; Dong, X.; Li, Z.; Cribb, M. Classification and investigation of Asian aerosol absorptive properties. *Atmos. Chem. Phys.* **2013**, *13*, 2253–2265. [CrossRef]
15. von Bismarck-Osten, C.; Weber, S. A uniform classification of aerosol signature size distributions based on regression-guided and observational cluster analysis. *Atmos. Environ.* **2014**, *89*, 346–357. [CrossRef]
16. Davies, C. Size distribution of atmospheric particles. *J. Aerosol Sci.* **1974**, *5*, 293–300. [CrossRef]
17. Chomko, R.M.; Gordon, H.R. Atmospheric correction of ocean color imagery: Use of the Junge power-law aerosol size distribution with variable refractive index to handle aerosol absorption. *Appl. Opt.* **1998**, *37*, 5560–5572. [CrossRef] [PubMed]
18. Deirmendjian, D. Scattering and polarization properties of water clouds and hazes in the visible and infrared. *Appl. Opt.* **1964**, *3*, 187–196. [CrossRef]
19. Junge, C.E. Our knowledge of the physico-chemistry of aerosols in the undisturbed marine environment. *J. Geophys. Res.* **1972**, *77*, 5183–5200. [CrossRef]
20. Yu, Q.-R.; Zhang, F.; Li, J.; Zhang, J. Analysis of sea-salt aerosol size distributions in radiative transfer. *J. Aerosol Sci.* **2019**, *129*, 71–86. [CrossRef]
21. Gordon, H.R.; Wang, M. Retrieval of water-leaving radiance and aerosol optical thickness over the oceans with SeaWiFS: A preliminary algorithm. *Appl. Opt.* **1994**, *33*, 443–452. [CrossRef]
22. Ahmad, Z.; Franz, B.A.; McClain, C.R.; Kwiatkowska, E.J.; Werdell, J.; Shettle, E.P.; Holben, B.N. New aerosol models for the retrieval of aerosol optical thickness and normalized water-leaving radiances from the SeaWiFS and MODIS sensors over coastal regions and open oceans. *Appl. Opt.* **2010**, *49*, 5545–5560. [CrossRef]
23. Frouin, R.; Deschamps, P.-Y.; Gross-Colzy, L.; Murakami, H.; Nakajima, T.Y. Retrieval of chlorophyll-a concentration via linear combination of ADEOS-II Global Imager data. *J. Oceanogr* **2006**, *62*, 331–337. [CrossRef]
24. Bru, D.; Lubac, B.; Normandin, C.; Robinet, A.; Leconte, M.; Hagolle, O.; Martiny, N.; Jamet, C. Atmospheric correction of multi-spectral littoral images using a PHOTONS/AERONET-based regional aerosol model. *Remote Sens.* **2017**, *9*, 814. [CrossRef]
25. Montes, M.; Pahlevan, N.; Giles, D.M.; Roger, J.-C.; Zhai, P.-w.; Smith, B.; Levy, R.; Werdell, P.J.; Smirnov, A. Augmenting heritage ocean-color aerosol models for enhanced remote sensing of inland and nearshore coastal waters. *Front. Remote Sens.* **2022**, *3*, 860816. [CrossRef]
26. Molero, F.; Pujadas, M.; Artíñano, B.J.R.S. Study of the Effect of Aerosol Vertical Profile on Microphysical Properties Using GRASP Code with Sun/Sky Photometer and Multiwavelength Lidar Measurements. *Remote Sens.* **2020**, *12*, 4072. [CrossRef]
27. Mobley, C.D.; Zhang, H.; Voss, K.J. Effects of optically shallow bottoms on upwelling radiances: Bidirectional reflectance distribution function effects. *Limnol. Oceanogr.* **2003**, *48*, 337–345. [CrossRef]
28. Dutton, E.G.; Reddy, P.; Ryan, S.; DeLuisi, J.J. Features and effects of aerosol optical depth observed at Mauna Loa, Hawaii: 1982–1992. *J. Geophys. Res. Atmos* **1994**, *99*, 8295–8306. [CrossRef]

29. Holben, B.N.; Eck, T.; Slutsker, I.; Smirnov, A.; Sinyuk, A.; Schafer, J.; Giles, D.; Dubovik, O. AERONET's version 2.0 quality assurance criteria. In Proceedings of the Remote Sensing of the Atmosphere and Clouds, Goa, India, 13–16 November 2006; pp. 134–147.
30. Holben, B.N.; Eck, T.F.; Slutsker, I.a.; Tanré, D.; Buis, J.; Setzer, A.; Vermote, E.; Reagan, J.A.; Kaufman, Y.; Nakajima, T. AERONET—A federated instrument network and data archive for aerosol characterization. *Remote Sens. Environ.* **1998**, *66*, 1–16. [CrossRef]
31. Thuillier, G.; Hersé, M.; Labs, D.; Foujols, T.; Peetermans, W.; Gillotay, D.; Simon, P.; Mandel, H. The solar spectral irradiance from 200 to 2400 nm as measured by the SOLSPEC spectrometer from the ATLAS and EURECA missions. *Sol. Phys.* **2003**, *214*, 1–22. [CrossRef]
32. Dubovik, O.; Lapyonok, T.; Litvinov, P.; Herman, M.; Fuertes, D.; Ducos, F.; Lopatin, A.; Chaikovsky, A.; Torres, B.; Derimian, Y. GRASP: A versatile algorithm for characterizing the atmosphere. *SPIE Newsroom* **2014**, *25*, 2-1201408. [CrossRef]
33. Torres, B.; Dubovik, O.; Fuertes, D.; Schuster, G.; Cachorro, V.E.; Lapyonok, T.; Goloub, P.; Blarel, L.; Barreto, A.; Mallet, M. Advanced characterisation of aerosol size properties from measurements of spectral optical depth using the GRASP algorithm. *Atmos. Meas. Tech.* **2017**, *10*, 3743–3781. [CrossRef]
34. Dubovik, O.; Fuertes, D.; Litvinov, P.; Lopatin, A.; Lapyonok, T.; Doubovik, I.; Xu, F.; Ducos, F.; Chen, C.; Torres, B. A comprehensive description of multi-term LSM for applying multiple a priori constraints in problems of atmospheric remote sensing: GRASP algorithm, concept, and applications. *Front. Remote Sens.* **2021**, *2*, 23. [CrossRef]
35. Moula, M.; Verdebout, J.; Eva, H. Aerosol optical thickness retrieval over the Atlantic Ocean using GOES imager data. *Phys. Chem. Earth. Parts A/B/C* **2002**, *27*, 1525–1531. [CrossRef]
36. King, M.D.; Dubovik, O. Determination of aerosol optical properties from inverse methods. In *Aerosol Remote Sensing*; Springer: Berlin/Heidelberg, Germany, 2013; pp. 101–136.
37. Dubovik, O.; King, M.D. A flexible inversion algorithm for retrieval of aerosol optical properties from Sun and sky radiance measurements. *J. Geophys. Res. Atmos* **2000**, *105*, 20673–20696. [CrossRef]
38. Eck, T.F.; Holben, B.; Sinyuk, A.; Pinker, R.; Goloub, P.; Chen, H.; Chatenet, B.; Li, Z.; Singh, R.P.; Tripathi, S.N. Climatological aspects of the optical properties of fine/coarse mode aerosol mixtures. *J. Geophys. Res. Atmos* **2010**, *115*, D19. [CrossRef]
39. Kahn, R.A.; Gaitley, B.J.; Garay, M.J.; Diner, D.J.; Eck, T.F.; Smirnov, A.; Holben, B.N. Multiangle Imaging SpectroRadiometer global aerosol product assessment by comparison with the Aerosol Robotic Network. *J. Geophys. Res. Atmos* **2010**, *115*, D23. [CrossRef]
40. Giles, D.M.; Holben, B.N.; Eck, T.F.; Sinyuk, A.; Smirnov, A.; Slutsker, I.; Dickerson, R.; Thompson, A.; Schafer, J. An analysis of AERONET aerosol absorption properties and classifications representative of aerosol source regions. *J. Geophys. Res. Atmos* **2012**, *117*, D17. [CrossRef]
41. Carrico, C.M.; Rood, M.J.; Ogren, J.A. Aerosol light scattering properties at Cape Grim, Tasmania, during the first Aerosol Characterization Experiment (ACE 1). *J. Geophys. Res. Atmos* **1998**, *103*, 16565–16574. [CrossRef]
42. Zieger, P.; Fierz-Schmidhauser, R.; Weingartner, E.; Baltensperger, U. Effects of relative humidity on aerosol light scattering: Results from different European sites. *Atmos. Chem. Phys.* **2013**, *13*, 10609–10631. [CrossRef]
43. Kinne, S.; Lohmann, U.; Feichter, J.; Schulz, M.; Timmreck, C.; Ghan, S.; Easter, R.; Chin, M.; Ginoux, P.; Takemura, T. Monthly averages of aerosol properties: A global comparison among models, satellite data, and AERONET ground data. *J. Geophys. Res. Atmos* **2003**, *108*, D20. [CrossRef]
44. Kim, S.-W.; Yoon, S.-C.; Kim, J.; Kim, S.-Y. Seasonal and monthly variations of columnar aerosol optical properties over east Asia determined from multi-year MODIS, LIDAR, and AERONET Sun/sky radiometer measurements. *Atmos. Environ.* **2007**, *41*, 1634–1651. [CrossRef]
45. Lenoble, J.; Remer, L.; Tanré, D. *Aerosol Remote Sensing*; Springer Science & Business Media: Berlin/Heidelberg, Germany, 2013.
46. Zarzana, K.J.; Cappa, C.D.; Tolbert, M.A. Sensitivity of aerosol refractive index retrievals using optical spectroscopy. *Aerosol Sci. Technol.* **2014**, *48*, 1133–1144. [CrossRef]
47. Zhang, M.; Cui, Z.; Han, S.; Cai, Z.; Yao, Q. Inversion and extinction contribution analysis of atmospheric aerosol complex refractive index in Tianjin urban area. *Res. Environ. Sci* **2019**, *32*, 1483–1491. [CrossRef]
48. Vermote, E.; Tanré, D.; Deuzé, J.; Herman, M.; Morcrette, J.; Kotchenova, S. Second simulation of a satellite signal in the solar spectrum-vector (6SV). *6s User Guide Version* **2006**, *3*, 1–55.
49. Gordon, H.R. Radiative transfer in the atmosphere for correction of ocean color remote sensors. In *Ocean Colour: Theory and Applications in a Decade of CZCS Experience*; Springer: Berlin/Heidelberg, Germany, 1993; pp. 33–77.
50. Wang, M.; Son, S.; Shi, W. Evaluation of MODIS SWIR and NIR-SWIR atmospheric correction algorithms using SeaBASS data. *Remote Sens. Environ.* **2009**, *113*, 635–644. [CrossRef]
51. Hu, C.; Carder, K.L.; Muller-Karger, F.E. Atmospheric correction of SeaWiFS imagery over turbid coastal waters: A practical method. *Remote Sens. Environ.* **2000**, *74*, 195–206. [CrossRef]
52. Fan, Y.; Li, W.; Gatebe, C.K.; Jamet, C.; Zibordi, G.; Schroeder, T.; Stamnes, K. Atmospheric correction over coastal waters using multilayer neural networks. *Remote Sens. Environ.* **2017**, *199*, 218–240. [CrossRef]
53. Mao, Z.; Chen, J.; Hao, Z.; Pan, D.; Tao, B.; Zhu, Q. A new approach to estimate the aerosol scattering ratios for the atmospheric correction of satellite remote sensing data in coastal regions. *Remote Sens. Environ.* **2013**, *132*, 186–194. [CrossRef]

Ahn, J.-H.; Park, Y.-J.; Kim, W.; Lee, B. Simple aerosol correction technique based on the spectral relationships of the aerosol multiple-scattering reflectances for atmospheric correction over the oceans. *Opt. Express* **2016**, *24*, 29659–29669. [CrossRef]

5. Singh, R.K.; Shanmugam, P. A novel method for estimation of aerosol radiance and its extrapolation in the atmospheric correction of satellite data over optically complex oceanic waters. *Remote Sens. Environ.* **2014**, *142*, 188–206. [CrossRef]

5. Chen, S.; Zhang, T.; Hu, L. Evaluation of the NIR-SWIR atmospheric correction algorithm for MODIS-Aqua over the Eastern China Seas. *Int. J. Remote Sens.* **2014**, *35*, 4239–4251. [CrossRef]

7. Pahlevan, N.; Roger, J.-C.; Ahmad, Z. Revisiting short-wave-infrared (SWIR) bands for atmospheric correction in coastal waters. *Opt. Express* **2017**, *25*, 6015–6035. [CrossRef] [PubMed]

8. Wang, M.; Shi, W. Sensor noise effects of the SWIR bands on MODIS-derived ocean color products. *IEEE Trans. Geosci. Remote Sens.* **2012**, *50*, 3280–3292. [CrossRef]

9. Wang, M.; Shi, W. The NIR-SWIR combined atmospheric correction approach for MODIS ocean color data processing. *Opt. Express* **2007**, *15*, 15722–15733. [CrossRef] [PubMed]

0. Liu, H.; Zhou, Q.; Li, Q.; Hu, S.; Shi, T.; Wu, G. Determining switching threshold for NIR-SWIR combined atmospheric correction algorithm of ocean color remote sensing. *ISPRS J. Photogramm.* **2019**, *153*, 59–73. [CrossRef]

1. Zhang, M.; Ma, R.; Li, J.; Zhang, B.; Duan, H. A validation study of an improved SWIR iterative atmospheric correction algorithm for MODIS-Aqua measurements in Lake Taihu, China. *IEEE Trans. Geosci. Remote Sens.* **2013**, *52*, 4686–4695. [CrossRef]

2. Wang, M.; Shi, W. Cloud masking for ocean color data processing in the coastal regions. *IEEE Trans. Geosci. Remote Sens.* **2006**, *44*, 3105–3196. [CrossRef]

MDPI AG
Grosspeteranlage 5
4052 Basel
Switzerland
Tel.: +41 61 683 77 34

Remote Sensing Editorial Office
E-mail: remotesensing@mdpi.com
www.mdpi.com/journal/remotesensing

www.ingramcontent.com/pod-product-compliance
Lightning Source LLC
Chambersburg PA
CBHW041352210926
51850CB00013B/3061